Vegetable Plants and their Fibres as Building Materials

Other RILEM Proceedings available from Chapman and Hall

1 Adhesion between Polymers and Concrete. ISAP 86
Aix-en-Provence, France, 1986
Edited by H.R. Sasse

2 From Materials Science to Construction Materials Engineering
Proceedings of the First International RILEM Congress
Versailles, France, 1987
Edited by J.C. Maso

3 Durability of Geotextiles
St Rémy-lès-Chevreuse, France, 1986

4 Demolition and Reuse of Concrete and Masonry
Tokyo, Japan, 1988
Edited by Y. Kasai

5 Admixtures for Concrete
Improvement of Properties
Barcelona, Spain, 1990
Edited by E. Vázquez

6 Analysis of Concrete Structures by Fracture Mechanics
Abisko, Sweden, 1989
Edited by L. Elfgren and S.P. Shah

7 Vegetable Plants and their Fibres as Building Materials
Salvador, Bahia, Brazil, 1990
Edited by H.S. Sobral

Publisher's Note
This book has been produced from camera ready copy provided by the individual contributors. This method of production has allowed us to supply finished copies to the delegates at the Symposium.

Vegetable Plants and their Fibres as Building Materials

Proceedings of the Second International Symposium sponsored by RILEM (The International Union of Testing and Research Laboratories for Materials and Structures) and co-sponsored by CIB (International Council for Building Research Studies and Documentation) and organized by UFBa (Universidade Federal da Bahia, Brazil).

Salvador, Bahia, Brazil
September 17–21, 1990

EDITED BY
H.S. Sobral

LONDON AND NEW YORK

Published 1990 by Taylor & Francis
2 Park Square, Milton Park, Abingdon, Oxon, OX14 4RN
52 Vanderbilt Avenue, New York, NY 10017, USA

First issued in paperback 2020

Taylor & Francis is an imprint of the Taylor & Francis Group, an informa business

First edition 1990

ISBN 13: 978-0-367-58011-7 (pbk)
ISBN 13: 978-0-412-39250-4 (hbk)

British Library Cataloguing in Publication Data

Vegetable plants and their fibres as
building materials.
1. Construction. Use of vegetable
fibres.
I. Sobral, H.S. (Hernani S).
II. RILEM. III. International Council
for Building Research, Studies and
Documentation.
624.1897

ISBN 0-412-39250-X

Library of Congress Cataloging-in-Publication Data
Available

ERRATUM

pp. 288-289
The material shown on page 288 should be on page 289, and vice versa.

Vegetable Plants and their Fibres as Building Materials
Edited by H.S. Sobral
Published by Chapman and Hall, 2-6 Boundary Row, London SE1 8HN

Publisher's Note
The publisher has gone to great lengths to ensure the quality of this reprint but points out that some imperfections in the original may be apparent

Scientific Committee

V. Agopyan	***Universidade de São Paulo*** *IPT – Instituto de Pesquisas Tecnológicas, Setor de Edificações, Cidade Universitária Armando Sales de Oliveira, São Paulo, Brazil*
M. Fickelson	***RILEM*** *Cachan, France*
E. Giangreco	***Universita di Napoli*** *Instituto di Teccnica della Construzioni, Napoli, Italy*
C. Gesteira	***Universidade Federal da Bahia*** *Salvador, Bahia, Brazil*
S. Guimarães	***Centro de Pesquisas e Desenvolvimento da Bahia*** *Camaçari, Bahia, Brazil*
J. Hellmeister	***Universidade de São Paulo*** *Laboratório de Estruturas de Madeira, São Carlos, São Paulo, Brazil*
A.C.Q. Mascarenhas	***Universidade Federal da Bahia*** *Departamento de Construção e Estruturas, Salvador, Bahia, Brazil*
M.A. Moragues	***Instituto de la Construcción y del Cemento Eduardo Torroja*** *Madrid, Spain*
C. Pauw	***Centre Scientific et Technique de la Construction*** *Brussels, Belgium*
M.A. Samarai	***University of Baghdad*** *Civil Engineering Department, Engineering College, Jadiriyah, Baghdad, Iraq*
Gy. Sebestien	***CIB*** *Rotterdam, Netherlands*

H. S. Sobral **Chairman of Scientific Committee**	***Universidade Federal da Bahia*** *Núcleo de Serviços Tecnológicos, Salvador, Bahia, Brazil*
M. Sosa	***Universidad de Caracas*** *Calle Chacao Parque Res. el Bosque, Maracuay, Caracas, Venezuela*
R.N. Swamy	***University of Sheffield*** *Department of Mechanical and Process Engineering, Sheffield, United Kingdom*
Y. Tezuka	***Associação Brasileira de Cimento Portland*** *São Paulo, Brazil*

Organizing Committee

Chairman	
J.R. da C. Vargens	*Rector of Universidade Federal da Bahia, Brazil*
N.V. Viana	*Vice-Rector of Universidade Federal da Bahia, Brazil*
M. Fickelson	*RILEM, Paris, France*
A.M. Barreto	*Associação Brasileira de Cimento Portland, São Paulo, Brazil*
F.L.L. Carneiro	*Universidade Federal do Rio de Janeiro, Brazil*
C. Lapostol M.	*Universidad de Concepción, Chile*
E. Pousada P.	*Universidade Federal da Bahia, Brazil*
R. Rivera V.	*Universidad Autonoma de Nuevo Leon, Monterrey, Mexico*
H.S. Sobral	*Universidade Federal da Bahia, Brazil*
C.E. de M. Strauch	*Universidade Federal da Bahia, Brazil*

Contents

Preface xv

Introduction xvii

PART ONE INTRODUCTION AND BACKGROUND 1

1 **Vegetable fibre reinforced cement composites – a false dream or a potential reality?** 3
R.N. SWAMY
Department of Mechanical and Process Engineering, University of Sheffield, England

2 **First International Symposium on the Use of Vegetable Plants and Their Fibres as Building Materials. Baghdad, Iraq, 1986** 9
M.A. SAMARAI
Consulting Engineering Bureau, Engineering College, Baghdad University, Iraq

PART TWO SURVEY OF PRESENT SITUATION IN RELATION TO FIBRES 19

3 **Vegetable fibres in craftwork techniques for building care** 21
M. FOTI and A. GILIBERT
Turin Polytechnic, Italy

4 **Experimental methods for the preparation of palm fruit and other natural fibres for use in reinforced cement composites** 29
J.G. CABRERA and S.O. NWAUBANI
Department of Civil Engineering, University of Leeds, England

PART THREE PROPERTIES OF VEGETABLE FIBRE COMPOSITE MATERIALS 37

5 **Effects of moisture content on mechanical properties of wood fiber reinforced cement** 39
P. SOROUSHIAN and S. MARIKUNTE
Department of Civil and Environmental Engineering, Michigan State University, East Lansing, USA

6 **Étude des possibilites d'utilisation du bambou dans les bétons de fibres** 50
(Study of the possible use of bamboo in fibre concretes)
R. CABRILLAC, F. BUYLE-BODIN, R. DUVAL and W. LUHOWIAK
Paris University X, LEEE, IUT Génie Civil, Cergy, France

7 **Fire resistant materials made with vegetable plants and fibres and inorganic particles** 60
T. SUZUKI
Institute of Technology, Shumizu Corporation, Tokyo, Japan
T. YAMAMOTO
Tohoku Electricity Technology Research Development Center, Tokyo, Japan

8 **Composite materials made from vegetable fibres as agglomerated irregular micro-reinforcement and Portland cement used in pieces for low-cost housing** 69
A.A. HESS and M.L. BUTTICE
Faculty of Engineering, UNNE, Chaco, Argentina

9 **Prospects for coconut-fibre-reinforced thin cement sheets in the Malaysian construction industry** 77
M.W. HUSSIN and F. ZAKARIA
Faculty of Civil Engineering, Technological University of Malaysia, Johor, Malaysia

10 **Durability of blast furnace-slag-based cement mortar reinforced with coir fibres** 87
V.M. JOHN and V. AGOPYAN
Institute of Technological Research and University of São Paulo, Brazil
A. DEROLLE
Institute of Technological Research, São Paulo, Brazil

11 **Vegetable fiber-cement composites** 98
S. Da S. GUIMARÃES
Research and Development Centre (CEPED), Camaçari, Bahia, Brazil

12 **Limit state of crack widths in concrete structural elements reinforced with vegetable fibres** 108
A. La TEGOLA and L. OMBRES
Department of Structures, University of Calabria, Cosenza, Italy

13 **Possible ways of preventing deterioration of vegetable fibres in cement mortars** 120
M.F. CANOVAS, G.M. KAWICHE and N.H. SELVA
Polytechnic University of Madrid, Spain

14 **Mortar reinforced with sisal – mechanical behavior in flexure** 130
A.C. De C. FILHO
Federal Technical School of Pernambuco, Recife, Brazil

15 **Application of sisal and coconut fibres in adobe blocks** 139
R.D.T. FILHO
Department of Agricultural Engineering, Federal University of Paraiba, Campina Grande, Brazil
N.P. BARBOSA
DTCC, Federal University of Paraiba, João Pessoa, Brazil
K. GHAVAMI
DEC, Pontifical University of Rio de Janeiro, Brazil

16 **The use of coir fibres as reinforcement to Portland cement mortars** 150
H. SAVASTANO Jr
Academia da Força Aérea, Pirassununga, Brazil

PART FOUR BUILDING COMPONENTS WITH VEGETABLE FIBRE COMPOSITE MATERIALS 159

17 **Comparison between gypsum panels reinforced with vegetable fibres: their behaviour in bending and under impact** 161
R. MATTONE
Turin Polytechnic, Italy

18 **Sisal-fibre reinforced lost formwork for floor slabs** 173
H.G. SCHÄFER
University of Dortmund, Federal Republic of Germany
G.W. BRUNSSEN
Failure Analysis Associates, Dusseldorf, Federal Republic of Germany

19 **Stabilisation d'un torchis par liant hydraulique** 182
(Stabilization of adobe with hydraulic binder)
F. BUYLE-BODIN, R. CABRILLAC, R. DUVAL
and W. LUHOWIAK
Paris University X, Laboratoire LEEE, IUT, Cergy, France

20 **Preliminary work to produce papyrus-cement composite board** 193
K.S.J. Al-MAKSSOSI and W.A. KASIR
College of Agriculture and Forestry, Hamman Al-Alil, Mosul, Iraq

21 **Fibre-concrete roofing tiles in Chile** 199
S. ACEVEDO, M. ALVAREZ, E. NAVIA and R. MUÑOZ
University of Santiago, Chile

22 **From research to dissemination of fibre concrete roofing technology** 204
T. SCHILDERMAN
Intermediate Technology Development Group, Rugby, England

23 **Effect of reed reinforcement on the behaviour of a trial embankment** 214
N.M. Al MOHAMADI
Department of Civil Engineering, College of Engineering, Baghdad University, Consultant (NCCL), Iraq

24 **Reed fibers as reinforcement for dune sand** 224
T.O. AL-REFEAI
King Saud University, Riyadh, Saudi Arabia

PART FIVE BUILDING COMPONENTS MADE WITH WOOD 237

25 **Properties and durability of rapidly curing cement-bonded particle boards, manufactured by addition of a carbonate** 239
M.H. SIMATUPANG
Institute of Wood Chemistry and Chemical Technology of Wood, Federal Research Center for Forestry and Forest Products, Hamburg, Federal Republic of Germany

26 **The longitudinal modulus of elasticity of wood** 248
E. CHAHUD
Structures Department, School of Engineering, Federal University of Minas Gerais, Brazil

27 **Multi-storey lightweight panel buildings** 255
H. GALLEGOS
Catholic University, Lima, Peru

28 **The scale influence on laminated glued timber beams strength (Glulam)** 266
E.V.M. CARRASCO
Structures Department, School of Engineering, Federal University of Minas Gerais, Brazil

29 **Durabilité des composants en bois dans habitations à loyer moderé** 277
(Durability of wood composites in low cost housing)
A.C.Q. MASCARENHAS and M.J.A. SANTANA
Polytechnic School, Federal University of Bahia, Salvador, Brazil

30 **Research on elements of a one family low-cost building system** 286
Z. MIELCZAREK
Technical University of Szczecin, Poland

PART SIX BUILDING WITH CULMS AND STICKS 293

31 **The use of forestry thinnings and bamboo for building structures** 295
P. HUYBERS
Civil Engineering Faculty, Delft University of Technology, The Netherlands

32 **The use of timber and bamboo as water conduits and storage** 305
T.N. LIPANGILE
Ministry of Water, Iringa, Tanzania

33 **Use of vegetable plants in housing construction in northern Iraq** 314
M.R.A. KADIR
University of Anbar, Ramadi, Iraq

PART SEVEN RECYCLING OF AGRICULTURAL WASTE AND RELATED TOPICS 319

34 **Study on the use of rough and unground ash from an open heaped-up burned rice husk as a partial cement substitute** 321
G. SHIMIZU and P. JORILLO Jr
College of Science and Technology, Nihon University, Tokyo, Japan

35 Optimization of rice husk ash production 334
M.A. CINCOTTO, V. AGOPYAN and V.M. JOHN
Institute of Technological Research and Polytechnic School of University of São Paulo, Brazil

36 Fique liquor as raw construction material 343
R. De GUTIERREZ
University of Valle, Cali, Colombia

37 The performance of a pulp and paper industry by-product as a water-reducer in concrete and a corrosion inhibitor to steel reinforcement 350
H.A. EL-SAYED
Building Research Institute, Dokki, Cairo, Egypt

38 Study for Brazilian rice husk ash cement 360
J.S.A. FARIAS
Federal University of Rio Grande do Sul, Porto Alegre, Brazil
F.A.P. RECENA
Foundation for Science and Technology, RGS, Porto Alegre, Brazil

Index 370

Preface

It is not yet possible to conceive of the construction of our modern cities, our industrial infrastructure and our engineering structures without total dependence on cement or steel. But for buildings which claim only to provide us with shelter and where in fact we make our home, there is available to us a whole range of less domineering materials which make no demands for massive manufacturing or fabrication facilities.

These are often traditional materials which have been adapted and improved, or substitutes such as industrial by-products or even agricultural waste, used to alleviate shortage or reduce cost.

Apart from recycled agricultural waste, the plant world offers a multitude of resources still not properly appreciated, except for the case of timber which benefits from a healthy industry developing under the impetus of research.

These resources from the plant world and the practical potential they constitute, are well worthy of interest from our research institutes, of close attention from the public authorities and investment from industry. Here lies a wealth of new materials waiting to be formulated from the gifts of nature through systematic improvement of their initial mechanical, chemical and biological characteristics using a strict scientific approach. At a time when our industrialized societies are endeavouring to comprehend the meaning and import of aid to development, we consider that our approach must focus on full use of local resources and proper adaptation of the products they offer, fostering their insertion into modern practice with the concourse of science and advanced technology.

There comes a time when this approach must become a policy and this is precisely when and where this Symposium of Salvador de Bahia becomes a landmark on our path towards this goal.

M. Fickelson

Introduction

This book contains the papers presented at the Second International Symposium on Vegetable Plants and their Fibres as Building Materials, held in Salvador, Bahia, Brazil, from 17–21 September, 1990, organized by the Federal University of Bahia, Brazil, under the sponsorship of the International Union of Testing and Research Laboratories for Materials and Structures (RILEM) and of the International Council for Building Research Studies and Documentation (CIB).

Following the First Symposium held in Baghdad, in 1986, this Symposium is intended to be a forum to discuss the role of vegetable plants and their fibres as building materials for low cost construction, using the available modern technology of materials.

The technical sessions of the Symposium examine the state-of-art in the field of vegetable plants and their fibres as building materials, emphasizing their use, properties, fabrication, new procedures and future developments. Procedures for making vegetable fibres and wood usual building materials associated with traditional and synthetic materials are also analysed.

The papers in this book are arranged in the following sessions:

- introduction and background;
- survey of present in relation to fibres;
- properties of vegetable fibre composite materials;
- building components made with vegetable fibre composite materials;
- building components made with wood;
- building with culms and sticks;
- recycling of agricultural waste and related topics.

The research results, as presented in these papers, show the technical feasibility of using many of these alternative building materials and technologies in important construction projects of social interest, mainly for developing countries.

As chairman of the Scientific Committee of the Symposium, I have to express my deepest gratitude to all my fellow members of the Scientific Committee whose contributions and participations were essential for the analysis and selection of the papers published in this book.

I also thank M. Fickelson, from RILEM, and Dr M.A. Samarai, from the University of Baghdad, for their special attention and support to my work concerning the Symposium planning.

Hernani S. Sobral
Salvador, Bahia, Brazil
May 1990

PART ONE

INTRODUCTION AND BACKGROUND

1 VEGETABLE FIBRE REINFORCED CEMENT COMPOSITES – A FALSE DREAM OR A POTENTIAL REALITY?

R.N. SWAMY
Department of Mechanical and Process Engineering,
University of Sheffield, England

Abstract
The use of natural fibres in concrete matrices poses a special challenge to science and technology. Their use can save energy, conserve scarce resources and protect environment whilst alleviating the housing problem and enhancing a country's infrastructure. The inherent weaknesses of the fibres such as low modulus, lack of adequate interfacial bond and long term stability can be overcome by microstructural studies and micromechanics, but ultimately, it is their long term engineering performance that needs to be ensured. Natual fibre cement composites can contribute significantly to enhance the quality of life and of living.
Keywords: Natural Fibres, Cement Composites, Durability, Cement Matrix, Low Modulus, Microstructure, Engineering Performance, Housing.

1 Introduction

The development and understanding of building materials has generally received much less attention in the last few decades compared to sophisticated analysis and design procedures. The latter is much more exciting, and often considered intellectual, whereas the former is labelled as mundane and experimental. Probably one area where the building materials technology has been much neglected is in the realm of housing. The human habitat has become an almost intractable world problem. It is only when one realises that housing is as much a problem for the developed societies as for the developing countries of the Third World, and that some six hundred million houses need to be built before the end of the century if every family in the world is to have a roof over its head, that the enormity of the problem, and of the challenge to engineers and architects, dawns on all of us.

Natural organic fibres have a very important and unique role in the contribution they can make to alleviate the housing problem. They not only occur in luxurious abundance in many parts of the world, but they also can directly lead to energy savings, conservation of the world's more scarce resources and protect the environment. The fact that one of the most easily and readily replenishable earth's resources can be used to solve, at least in part, one of the most acute forms of human misery, is just as challenging not only to the basic human instinct of fellow feeling but also to the science and skills of developed technologies. Bamboo, for example, is one of the fastest-growing, highest-yielding and easily renewable of natural resources, and yet our ability to use them in durable construction is far from reality. There is also the other point about these materials - Nature will continue to be bountiful to mankind so long as we do not wantonly consume, misuse or destroy the vast and rich resources that it so generously provides. Natural and vegetable plants and fibres have thus a unique irreplaceable role in the ecological cycle, and their natural abundance, plentiful supply, relative cheapness and swift replenishability are the strongest arguments to utilise them in the construction industry.

Cement and concrete matrices reinforced with short, discrete or long single/bundles of fibres present exciting and challenging new construction materials. The major role of the fibres is in delaying and controlling the tensile cracking of the matrix. This controlled multiple cracking reduces deformations at all stress levels, and imparts a well-defined post-cracking and post-yield behaviour. The fracture toughness, ductility and energy absorption capacity of the composite are then substantially improved. These technical benefits can be utilised both in semi-structural elements such as thin sheets and cladding panels as well as in load bearing members.

Cement composites reinforced with steel, glass and polypropylene fibres have seen extensive development and a wide range of practical applications, Swamy et al (1986). There is currently an enormous body of data available on these composites, and yet, natural fibre reinforced concrete composites have not enjoyed the same sort of development and applications they deserve, in spite of the fact that there is also sufficient research and practical experience in the use of natural fibres in concrete, Gram (1983), Gram et al (1984), Gram et al (1986), IPT (1987), Mwamila (1984), RILEM (1986), Swamy (1984a, 1988b). By comparison with many natural fibres, and vegetable plants, composites incorporating wood fibre and cellulose fibres are well advanced in the manufacture of both autoclaved and air-cured products and flat sheets, corrugated roofing, moulded products and non-pressure pipes are produced by industrialized production processes, Akers et al (1989), Coutts (1988), Fordos (1986). Their use at village-level small scale industries, on the other hand, is much less developed. This paper is primarily addressed to the latter: its aim is to make a critical evaluation of the present status of the use of natural fibres in cement and concrete, mainly for housing units, the problems to be tackled, and point the way forward to make natural fibre cement composites economic and durable construction materials.

2 The problems

Although natural fibres exist in abundance and are readily available at low cost, they have many inherent weaknesses such as low elastic modulus, high water absorption, susceptibility to fungal and insect attack, lack of durability in an alkaline environment and variability of properties amongst fibres of the same type. It is not surprising therefore that natural fibres have not always been the ready or automatic choice as a reinforcing medium in cement matrices despite widespread interest, numerous research efforts and many trial applications.

A major factor contributing to this slow development is the lack of precise scientific information on the structure and properties of natural fibres, their compatibility with the various matrices and the properties of the composites themselves. In published literature, a wide range of values exist for any given property both within a given class of fibre and across classes of fibres. The reasons for this are not difficult to find. Fibres can be derived from leaf, stem or wood, and both fibre diameter and length are dependent on age of growth. Abacas, for example, occurs in as many as 200 varieties, with reported fibre lengths varying from 5.0-9.0mm, and fibre widths of 12.4-21.5mm. Bamboos, on the other hand, can be subdivided into four families and an estimated fifty genera: further, over 1250 individual species of bamboo have been identified so far. Another factor making studies on composites difficult and complex is that natural fibres are hollow; the central lumen changes in size with age, and may sometimes collapse, altering the cross-sectional area of the fibre within a given fibre type.

There has also been a mistaken and misguided belief that chopped material fibre composites have similar properties of strength and durability as asbestos cement sheets. The destructive effects of the alkaline pore solution on organic fibres is well established. The net effect of these chemical degradation processes is that with time the fibres will cease to exist and provide no reinforcing property. In the short term the fibres provide cohesion and workability to the cement matrix, and help to arrest plastic shrinkage and initial drying shrinkage cracks. In the long term, prolonged reinforcing action betwen the matrix and the fibre can only be ensured by carefully designed procedures calculated to enhance long term stability of the fibre in the cement matrix.

Many plant fibres contain hemicellulose, starch, sugar, tannins, certain phenols and lignins, all of which are known to inhibit the normal setting and strength development properties of the cement matrix. Apart from increasing the handling time, the water soluble extractives also prevent the composite from attaining its full strength and durability characteristics. The decomposition of the fibre is much faster in an alkaline medium than in water, and the rate and degree of decomposition will depend upon the nature of the fibre. The degree of polymerisation, and the location or accessibility to the hemicellulose and lignin will both affect adversely setting time and strength development, and these need to be controlled. In practice, both setting time and strength loss can be controlled by low water-binder ratios, and a compatible accelerator can also make good further the loss of strength.

2.1 Engineering aspects

The study of the microstructure of the fibre cement matrix and their interaction and their interface has a very important role in the development of these fibre composites. However, microstructural investigations cannot and should not be isolated from engineering performance. There are several lessons to be learnt from previous developments, where emphasis and confidence in one at the expense of the other has led to many practical difficulties and economic implications unforeseen by microstructural or micromechanics studies. The effects of thermal moisture gradients, of fixings and external stresses, and of the production processes on anisotropic properties, in addition to the effects of realistic exposure conditions are sufficiently well established to warn producers and users alike of the need to link changes in microstructural characteristics to behaviour at engineering level, Swamy (1986).

2.2 Size effects

In many laboratory studies, size effects are ignored. Size effects on strength are well known, strength increasing as the size of the specimen is reduced. This is particularly relevant when composites are judged from flexural strength tests. For example increases in strength of up to 50% or more can be observed when the size of specimen is reduced from 1000x300x10mm to 500x100x10mm, and strength increases of 100% or more may be observed when the size is reduced to 150x50x10mm, the usual size adopted in many studies. The size of specimen will also affect its mode of failure, shear failures and delamination occurring instead of the usual flexural failure.

The differences in strength and mode of failure will generally be accentuated if the fibres fail to provide any reinforcing effect. A larger product like a sheet is thus more likely to give trouble, crack and fail suddenly than a geometrically smaller product like a tile. This partly explains why a high incidence of cracking and considerable performance deficiencies have been reported by users of natural fibre reinforced sheets and tiles, limiting useful life of these products to no more than 2 to 4 years. Extensive surveys have shown clearly that it is not the technology or the product that failed but the inherent destructive effect on the fibres of the alkalinity of the cement, particularly in tropical environments, and the lack of adequate standards of roof construction and installation that were responsible for these failures. In addition, there has been a general lack of appreciation of the role of the natural fibre in the cement matrix, and the need to relate the geometrical shape of the product, and how it is fixed, to the inherent decomposition of the fibre and lack of quality control which inevitably occurs in less developed technologies.

3 Cement matrix - the key to composite stability

The main concern with natural fibres is one of long term strength and stability. Marked embrittlement of sisal fibres, for example, embedded in thin roofing sheets exposed to a tropical climate has been observed within a period of months. The embrittlement is known to be caused directly by the alkaline environment of the cement system. There is also evidence, on the other hand, that natural fibres remain practically intact in carbonated concrete. Several methods are currently available for waterproofing and for prophylactic treatment against deterioration. Protective impregnation of the fibres with various chemical treatments have, however, been unable to prevent the chemical decomposition of the fibre components completely, and have resulted in poorer composite properties due to impairment of fibre tensile strength and the effects on fibre-matrix interfacial bond properties. While further research can probably develop other protection methods that are simple, effective and economic, it is very unlikely that such methods result in viable technologies at an acceptable cost to developing countries.

Since fibre embrittlement is caused primarily by the high alkalinity of the cement system, the solution to the deterioration of the natural fibre lies in the matrix itself. There are several simple techniques readily available to reduce the alkalinity of the cement matrix. Cements other than portland cement such as aluminous cements and cements with negligible alkali contents can now be manufactured, but these methods may not offer a ready solution to countries where such cements are not available.

Cement replacement materials such as fly ash, ground granulated blastfurnace slag and microsilica offer one of the best means of reducing cement alkalinity to sufficiently low levels to enable the fibres to contribute to composite strength for a long time. Microsilica is a highly reactive pozzolan compared to fly ash and slag, and can be very effective, although it can be relatively more expensive. Fly ash and slag on the other hand, are more readily available at affordable prices, and there is no reason why 70 to 80% of the cement should not be replaced by these mineral admixtures. High replacement levels have, however, their own problems mainly due to the fact that the cement-ash and cement-slag hydration processes are in fact two stage processes which are highly susceptible to poor or inadequate curing. Mineral admixtures can produce stable and durable natural fibre cement composites provided four basic principles are observed, namely: (i) low water-binder ratio, (ii) use of a superplasticizer, (iii) early and longer water curing, and (iv) early strength development. Even if the initial material costs are slightly increased by this methodology, this will be more than compensated for by longer service life.

Inert fillers like crushed fines can also be used to reduce cost and the alkalinity of cement. Such composite cements may contain limestone powder, graded sand or selected laterite fines. These composite cements can also substantially reduce plastic and drying shrinkage.

4 The bamboo

The use of bamboo as reinforcement for concrete materials deserves special mention and special study, Swamy (1984a, 1988b). It is the fastest-growing and highest-yielding renewable natural source, and can be used for a variety of concrete structures such as housing, slabs on grade, septic tanks, etc. Lack of adequate bond and elastic stiffness as well as lack of long term stability in concrete are the main factors inhibiting its use. Although these aspects have been studied, further research is urgently required as bamboo can effectively and economically provide stable reinforcement for concrete.

5 Conclusions

Natural fibre reinforced cement and concrete have special relevance to developing countries because of their low cost, ready and plentiful availability, and savings in energy. They can significantly contribute to the rapid development of a country's infrastructure. The need to develop locally available building materials, improve the quality of products and understand better how building materials behave in real environments cannot be overemphasized. Research cooperation and the use of advanced scientific methods can help to evaluate existing materials and adapt them for durable construction, but ultimately it is engineering performance that really needs to be understood. Fibre cement sheeting, corrugated roofing and precast building components represent a global activity. They need planned resource adaptation, but above all, they cater to a much needed human activity, the quality of life and of living. Science and technology can contribute significantly to a betterment of human life.

6 References

Akers, S.A.S. et al (1989) Long term durability of PVA reinforcing fibres in a cement matrix, Int. J. Cement Composites and Lightweight Concrete, 11, 79-91.

Coutts, R.S.P. (1989) Wood fibre reinforced cement composites, Concrete Technology and Design, Vol.5, 1-62.

Fordos, Z. (1989) Natural or modified cellulose fibres as reinforcement in cement composites, Concrete Technology and Design, Vol.5, 173-207.

Gram, H.E. (1983) Durability of natural fibres in concrete, Swedish Cement and Concrete Research Institute, Sweden.

Gram, H.E. et al (1984) Natural fibre concrete, Swedish Agency for Research Cooperation with Developing Countries, Sweden.

Gram, H.E. (1986) Fibre concrete roofing, Swiss Centre for Appropriate Technology, Switzerland.

IPT (1987) Fibre-agro industrial byproducts bearing walls, Instituto de Pesquisas Technolgicas do Estado de Sao Paulo, Brazil.

Mwamila, B.L.M. (1984) Low modulus reinforcement of concrete - special reference to twines - Swedish Council for Building Research, Sweden.

RILEM (1986) Use of vegetable plants and their fibres as building materials, Joint Symposium, Baghdad, Iraq.

Swamy, R.N. (1984a) Editor, New Reinforced Concretes, Concrete Technology and Design, Vol.2, Blackie and Sons, Glasgow, U.K.

Swamy, R.N. (1988b) Editor, Natural Fibre Reinforced Cement and Concrete, Concrete Technology and Design, Vol.5, Blackie and Sons, Glasgow, U.K.

Swamy, R.N. et al (1986) Developments in Fibre Reinforced Cement and Concrete, Vols. 1 and 2.

Swamy, R.N. and Hussin, M.W. (1986) Effect of curing conditions on the tensile behaviour of fibre cement composites, RILEM Symposium, Sheffield, U.K.

2 FIRST INTERNATIONAL SYMPOSIUM ON THE USE OF VEGETABLE PLANTS AND THEIR FIBERS AS BUILDING MATERIALS. BAGHDAD, IRAQ, 1986

M.A. SAMARAI
Director, Consulting Engineering Bureau, Engineering College, Baghdad University, Iraq

Abstract
The first International symposium on the use of vegetable plants and their fibers as building materials was held in Baghdad during 7-10 Oct. 1986. The main aim was to discuss the role of vegetable plants such as : bamboo reed papyrus, rice, paddy or husk, palm leaves and sugar cane bagasse as building materials, and also to evaluate the use of these vegetable plants and their products to improve low-coat housing in different countries.
These materials are cheap; very popular, easily available and renewable, in many countries especially in developing countries. Such properties, and others of constructional engineering point of view would enhance their use as building materials. However, more knowledge and improvements are needed to fulfill such task. For this purpose it is required to make a conversion towards the technical use of these vegetable plants from their locally widely dispersed applications into scientifically-based engineering ones. Through clear principles and with improved technical applications it would be possible to make use of these materials economically and with great benefit. The main topics were divided as follows :

1 Mechanical properties and testing.
2 Durability and production methods.
3 Thermal properties of vegetable fibers
4 Performance of elements, components and systems.

Many recommendation were given which were concerned with the following subjects :

1 Durability problem.
2 Enhancing the application.
3 Economical, sociological and cultural aspects.
4 Educational aspects of this science.

More than 300 experts representing 33 different countries attended the symposium.

1 Introduction

The first symposium on the "Use of Vegetable Plants and their Fibers as Building Materials" was sponsored by the International Union of Testing and Research Laboratories for Materials and Structures (RILEM), the Iraq National Centre for Construction Laboratories (NCCL), the International Council for Building Research Studies and Documentation (CIB), the Iraqi Engineering Society and Tourism Authority. The expected shortage of the well-known construction materials is leading to a large demand for new low-cost building materials in developing countries. For this purpose the first symposium on the use of vegetable plants as construction materials was held. Many papers were presented dealing with the efficient use of such plants in constructional engineering field, and aiming at reducing the cost of their elements for economical housing accessible to more occupants.

This could be done by developing new materials, and technical methods, making maximum possible use of agricultural products and by-products which could be substituted for more conventional materials such as, steel, cement, brick, etc. The presentation discussed were be divided into three sections as follows :

1 Work done to improve traditional materials and techniques.
2 Enhancing the use of traditional materials by employing innovatory techniques.
3 Research aiming towards producing new materials from plants and agricultural or industrial residues available locally.

2 Main Topics

Due to the large number of papers presented they were divided into five themes.

2.1 Theme A

Survey on the present practice and techniques. In this theme the use of plants as building materials without any binding matrix or manufacture process was dealt with.

A-1 Yahya[1] presented the importance of grass thatching as roofing material for low-cost housing in Kenya. In his work he discussed the various types of roofing used there; the paper included a study on the service-life of those materials. Such service-life did not last to the effect of the climatic, cultural and economic conditions of the country; but with careful protection from rain and insects and with good maintenance the grass roof could easily last for 10 years.

A-2 The next paper was presented by Laakkonen[2] who declared that 78% of houses are in rural area in Zambia and used grass or thatch as roofing materials; the life span is between 2-30 years, such variation is attributed to several factors such as; the technique used; grass species, design, the use of the building, and the treatment of the material.

A-3 The paper by Hasem[3]; stated that about 90 million people in Bangladesh lived in rural areas using materials like, bamboo, Jute stalks, straw, palm-leaves, thatches, etc. These materials are very popular because they are cheap, easily available and renewable. But, due to their organic nature, they are attacked by, insects, rot-fungi and fire. They are characterized by their short service life (2-3 years) therefore by providing some improvements to these materials more durable houses can be erected lasting for longer service life ((4 to 5 times than the untreated materials)). A soaking method for the treatment of the plants by using a solution of composition : copper sulphate, sodium diehromate, acetie acid and water was suggested. Before performing the treatment the plants are to be dried out.

A-4 The using of reed as construction material was presented by Samarai[4], referring to the techniques used for traditional houses with reeds which include using reed as soil reinforcement of the floating island on which the house will be constructed and also using reed in all the housing units. The paper provides a good solution to improve this technique by using steel skeleton and panels of reed treated against, insect, rot and fire. The suggested method was semi-industrialized for the production of more durable low-cost housing.

2.2 Theme B "Internationalization of traditional materials"

A lot of research was done on vegetable plants and their wastes to be used with a suitable binding material to produce building element such as building blocks and panels for partition; insulation and ceiling sheets, etc. The presented papers dealt with the following headings :-

B-1 Panels prefabrication, from Peru, Gallegos[5] presented his paper on the traditional formal and informal construction of houses using untreated culms of grasslike woody plants in walls domes and cupolas, here cane or reed grass plastered with mud and/or gypsum or cement mortar. Using wooden or bamboo skeletons would introduce a system known as "quincha"; manufactured fibrous cement panels using mineralized vegetable fibers were also manufactured. From Poland, Mielczarek[6] has described the state of the resources of fiber crops and wood waste and their utilization such as; flax and hemp harle which were used in the production of flaxboards; and reed which could be used in the production of insulating mats, in addition to using it in concrete construction blocks and few ceiling-roofing structure. In his paper he presented many boards made of wood waste, and discussed the research carried on the application of prefabricated wall plates made from small size timber and wood waste.

Suzuki[7] presented test data on (S-T panel) using reeds and rice husks. He dealt with the design and cost of using that panels in housing construction. Plant fiber were used, as core and spraying of plastering mortar on the outside of space truss given S-T panel greater strength.

The panels were tested for their heat insulation; and the effect of atmospheric carbonization processes. The results estimated that it would take more than 50 years to corrode the mortar [with governing thickness of 20 mm] which was sprayed on the panel surface.

The last paper on panel production was presented by Mariotti[8], who carried out his work on the "Ivory Cost" grasses and included bibliographical survey and experimental study on the physical and mechanical testes on grasses, prefabricated panels and roof units. His work included test on protection against biological action.

B-2 Clay reinforced by vegetable fibers

Berhane[9], from Ethiopia discussed the familiar building material "chika" which was a mixture of clay, chid and water. Chid is millet straw used as a reinforcement of clay to minimize the loss of strength of clay when it comes in contact with water, and it reduces the excessive amount of drying shrinkage that leads to cracking occasionally. Some precautions to improve the service -life of the material where given; such as protecting it from the influence of the moisture ((rainfall)); and termite and insect attack. These precautions extend the service life from [5-10 years before protection] to about 50 years.

Martine and Rubaud[10], studied the influence of vegetable fibers reinforced laterite. He used coir (coco fiber) due to its availability, good durability and mechanical properties.

The tests performed were the proctor and CBR tests, crushing and shearing tests. The influence of water and fiber content on the composite material were discussed and solutions suggested.

2.3 Theme C Vegetable plant fibers in composite materials

Using the principle of composite materials vegetable plant fibers can be utilized to improve the properties of the matrix. Many papers dealt with this aspect, among them are the following:-

C-1 Shah[11]; from U.S.A. presented a theoretical analysis of fiber reinforced concrete (FRC) with models. The mechanical behavior of FRC was studied through the pull-out fracture, uniaxial tensile, and crack resistace mechanism. The load-deflection response of such composite material was predicted and measured experimentally.

C-2 Many papers on wood fibers were presented.
The first paper was prepared by AGRWAL[12] who discussed the use of cement as a matrix with wood chips and saw dust to introduce chip boards and panels in India. Such elements are characterized by having good density and bending strength, low moisture content, very low thickness and swelling and can be used in exposed condition. Moreover, they have good fire resistance.
Al-Makssosi[13]; from Iraq presented the use of wood-cement composite board, the influence of wood particle size and wood/cement ratio was examined. The effect of the wood fines was also dealt with. The mechanical and physical properties of these boards were calculated. The main conclusions were that with increasing the size of wood particle better properties (except the bond) could be obtained, and that the presence of wood fines would have an adverse affect on most of the board properties.
Hachmi[14]; discussed the scientific means of evaluating the compatibility of wood and cement from the setting and hardening point-of-view.
The last paper on this subject was presented by SARJA[15]; from Finland, who studied the effect of using wood fiber on concrete properties evaluating the compressive and tensile strength, ductility, workability and thermal conductivity. Furthermore, some of the composite were tested for weather resistance, termite attack and non-flammability. It was found that such composite could be used in; bearing and non-bearing walls, foundations and slabs. By using modern production techniques the wood fiber concrete could be utilized in low-cost housing.

C-3 Ribas[16]; from Brasil studied the effect of sisal fibers on the properties of cement paste and mortar, and the effect of fiber length was mentioned. It was found that the compressive strength increases to a limit when the fiber length increases. It was concluded the alignment of fibers had a great influence on the overall properties. Sisal fibers reinforced slab was found to crack and rupture gradually and slowly.

2.4 Theme D - Vegetable plants as a reinforcement in concrete and as pozzolanic material :-

Due to the shortages in steel reinforcement, many vegetable plants were used for the same function as steel. Some papers reflected this matter, while others dealt with the use of some plants as a pozzolanic material.
D-1 Plants as reinforcement :-
Berwanger; from Canada presented two papers on using bamboo as reinforcement. The first one[17] dealt with, bamboo sticks,

splints or strips used in the form of woven meshes describing the tests on the physical and mechanical properties of cement mortar or concrete panels and slabs.

In his second[18] paper the flexural strength of bamboo reinforced concrete inverted T-beams and the behavior of such beams on being overreinforced and underreinforced were evaluated. The latter case was found to be unsatisfactory due to higher deflection and wider cracking accompanying it.

Krishnamurthy[19]; from Jordan presented the use of bamboo as reinforcement, solving the problem of shrinkage-bond loss that is associated with that technique. Three simple inexpensive techniques, regarded as a solution to this problem are mentioned, depending on the use of bittumen treatment to the bamboo so that improved bond and structural behavior are achieved simultaneously.

Raouf[20]; from Iraq presented his paper on the use of reeds as a reinforcement for concrete. It included an experimental and theoretical analysis for tensile strength, pullout tests, dimensional stability due to rain and wetting. The flexural behavior of slabs and joists, with the design calculation of reed reinforced concrete were provided.

Raauf presented another paper[21] on using reeds in the form of mats for ferrocement suspension roofs as a replacement of steel wire meshes. The design calculation; durability and the advantage of such ferrocement roof were covered.

D-2 Plants as pozzolanic materials :-

Smith[22]; from England discussed the use of rice husks as cementitious material, by the incineration of husks removing lignin and other organic components leaving a silica-rice ash, then introducing rice husk ash cement and studying all properties of such cement. An operation of incinirator for the production of rice husk ash (RHA), which can be considered as cheap and simple was suggested.

Natalini[23], from Argentina discussed in his paper the production of "cellular concrete mixes" which is composed of rice husk, wood saw dust, and silica particles with air-entrained bubbles. Such concrete had super fluid characteristics and would need minimum compaction energy, with acceptable properties.

Raouf[24]; described an experimental investigation of the use of rice husks to improve the quality of local bricks ground and underground rice husks were added in different volume percentages, then the compressive strength; water absorption; efflorescence and the percentage of soluble salts were investigated. The results confirmed that better surface finished higher density and compressive strength bricks made of unground rice husks could be produced.

2.5 Theme E Durability and general properties of vegetable plants used as building materials and other related topics.

E-1 Durability research :
Groot[25], from Netherlands, presented a study on thatch roof durability. The results discussed were those of the decay and weathering influences; the decay is mainly a result of fungal attack and ultra-violet radiation. A treatment method was suggested to improve the durability and enhance the use of local vegetable materials in developing countries. The improvement covered the physical behavior; quality control and evaluation of new technology.

Singh[26]; from India, presented a study of the main problem faced when vegetable fiber is used in concrete, namely the alkali resistance. In his work different plant fibers and a solution of sodium hydroxide and saturated dispersion of lime in water are considered. Then a study of the change in their tensile properties due to the processes of placing the fiber in this solution, using sixty cycles of wetting and drying was performed.

The effect of weathering on fiber concrete roofing sheet, and the use of water repellents, ultrafines, silica fume and pozzolanic cement (to reduce alkalinity of fiber-cement composite and the embrittlement of the fiber) were all discussed.

E-2 The properties of plants used as building materials :-
Samarai[27]; presented in his work a survey of reeds available in the southern part of Iraq. The chemical data of reeds varied owing to geographical location, cultivation, harvesting methods and spaces. The kind and physical properties of the reeds were described to state their suitability for use in industry. The study included a treatment to improve the resistance of reed to fire, inscts and fungi.

Tawfiq[28], presented in his paper the mechanical properties of the reeds such as, compressive and flexural strengths measured by three and four-point tests.

E-3 The use of bamboo as water supply pipes.
Lipangile[29]; from Tanzania discussed the use of bamboo as water supply pipes. Such a practice became common during the past 10 years in Tanzania. A wide research was done on this aspect. The flow, system design, construction of the network system, treatment of the bamboo to have a long service life were all studied, in addition to the precautions against health hazards. He adds that now they have a good staff to construct and maintain such systems. About 200km of pipe networks which was acceptable by public and the government, have been successfully constructed.

3 Recommendations[30]
During the discussion that followed the presentations of the papers, participants made the following recommendations :

1 There is a wide field for the utilization of the vegetable plants and their fiber such as reeds, bamboo, sisal, papyrus etc. in the construction industry.
2 More research with experimental work on the suitable application in construction is still needed.
3 The durability aspects of the vegetable materials need to be verified with accelerated tests designed to take into account the effects of UV, heat, and humidity. The durability of these materials against the insects, animal attacks, and fungious needs further research. A need to understood the mechanisms of those attack is still present.
4 It is recommended to provide a standard code of a practice for the vegetable plants and their fibers dealing with their application as building materials.
5 Since these materials are mainly used on low cost housing and other buildings there is a need for many examples of socioeconomic studies showing and proving a good adequation of materials and building processes cost to the financial availability of low income of populations.
6 There was a need for solution of environment problems.
7 For the large development of construction methods using vegetable plants and their fibers, there is an essntial need for training of engineers and architectures with the aim of establishment of a new branch of engineering dealing with these materials in construction.
8 The last recommendation was to contact the NN, UNIDO and UNESCO to establish a research institute in Africa to conduct an immediate and long term research in that subject. Such institute should develop a program for the training of unskilled labor.

4 References

4.1 S. Yahya; "Those who live in grass houses" pp. A-51.
4.2 N.O. Laakkonen; "The use and technique of thatched roofs in Zambia"; pp. D-31.
4.3 A. Hashem;"Use of vegetable plants in rural housing in Bangaladesh and their improvement"; pp. C-139.
4.4 M. Samarai;J. Al-Taei and R. Sharma; "Use and technique of reed for low-cost housing in the marshes of Iraq"; pp. D-47.
4.5 H. Gallegos;"Use of vegetable fibers as building materials in Peru"; pp A-25.
4.6 Z. Mielczarek, "Application of fiber crops and wood waste in polish building industry"; pp. D-41.
4.7 T.Suzuki,T. Furuta and M. Obata;"Development of multipurpose panel (S-T panel)"; pp. D-63.
4.8 M. Mariotti; "Grass roofing in tropical regions"; pp. D-89.

4.9 Z. Berhane;"The traditional use of millet ("teff") stalk asreinforcement for clayey building materials"; pp. C-37.
4.10 M. Martin and M. Rubaud; "Reinforcing literate with vegetables fibers"; pp. C-79.
4.11 S. P. Shah; "Theortical models for predicting performance of fiber reinforced concrete"; pp. B-1.
4.12 L. K. Agarwal;"Wood fiber reinforced cement boards"; pp.C-25.
4.13 K. Al-Makssosi, and C.E. Shuler; "Properties of wood-cement board with variations in particle size and wood/cement ratio"; pp. C-29.
4.14 M. Hachmi; A. A. Moslemi, A. G. Campbell; "An improved classification method for ranking wood species based on wood-cement compatibility"; pp. C-61.
4.15 A.Sarja; "Structural concrete with wooden or other vegetable fibers"; pp. C-117.
4.16 M. Riba Silva and A.M. Compolina; "The use of sisal fibre in comentitation pastes and mortars physical and mechnical properties"; pp. C-97.
4.17 C. Berwanger; "Bamboo mesh as reinforcement for concrete"; pp. C-45.
4.18 C. Berwanger, A. Alaeddine, I. G. Indra, and E. Purwanto; "The flexural strength of bamboo reinforced concrete inverted T-beams"; pp. C-51.
4.19 D. Krishnanurthy; "Use of bamboo as a substitute for steel in conventional reinforced concrete"; pp. C-71.
4.20 Z. Raouf; "Structural qualities of reed reinforced concrete" pp. C-89.
4.21 Z. Raouf; "Examples of building construction using reeds"; pp. A-35.
4.22 R. G. Smith, C. A. Kamwanja; "The use of rice husks for making a cementitious material"; pp. E-85.
4.23 M. B. Natalina, M. Sabesinsky, O. Gauto, R. A. Mayer and G.M. Gomez; "Portland cement cellular composite material with organic and inorganic partcles";pp.E-79.
4.24 Z. Raouf, J. Kachachi; "The use of rice husks for the improvement of the quality of local brick in Iraq"; pp. E-61.
4.25 C. J. Groot; "The progress of decay and weathering in thahched roofs"; pp. D-23.
4.26 S. M. Singh; "Studies on the durability of plant fibers reinforced concrete products"; pp. C-127.
4.27 M. Samarai; M. Al-Taey and A. Kassir; " Some chemical data and operational tests for Iraqi reed and reed products"; pp. C-107.
4.28 H. Tawfiq, and H. Bakir; "Mechanical properties of Iraqi reeds"; pp. C-131.
4.29 T. Lipangile; " The use of bamboo pipes in water supply systems"; pp. D-37.
4.30 The final report on Baghdad symposium. "The use of vegetable plants and their fibres as building materials".

PART TWO

SURVEY OF PRESENT SITUATION IN RELATION TO FIBRES

3 VEGETABLE FIBRES IN CRAFTWORK TECHNIQUES FOR BUILDING CARE

M. FOTI and A. GILIBERT
Turin Polytechnic, Italy

Abstract
Vegetable fibres were widely used in the past for reinforcement as stabilizing agents in many types of mortar and concrete. The adoption of modern techniques with cement and polymer mortars has led in many countries to traditional techniques with natural materials being abandoned. The growing production of artificial and synthetic fibres has resulted in the development of new industrial processing techniques and the creation of composite materials with extremely high performance. This paper examines a number of examples of the use of these artificial fibres and attempts to identify possible similar applications of natural non-woven fibres. Several such examples are described, involving, for instance, the use of non-skilled labour and natural materials for the restoration of damaged surface in old buildings and the stabilization of slopes.
Keywords: Composite materials, Vegetable fibres, Sisal, Gypsum, Developing countries, Building care, Craftwork techniques, External walls.

1 Introduction

Until now, in developing countries, the prevailing attitude towards less significant buildings has been to demolish and then replace: at least this has been the intention because in reality we know how difficult it is to be commited on a large scale as the situation would require.

However, in cities where new buildings have gone up, these have very often taken the place of what was there before, contributing to upsetting the pre-existing urban context; every district, however spontaneous and lacking in infrastructure, achieves a certain homogeneity in appearance as time passes, takes, so to speak, some form or other, even if it has no functional organization and, in some way, due either to natural ageing or to the modification brought about little by little, by community life, succeeds in assuming the form of a precarious but recognizable environmental balance with the particular caracteristics of an urban district.

At times, it has to do with a building of little value, at times with image, but each case would justify taking into consideration the possibility of preserving and improving what nevertheless constitutes an existing property notwithstanding the burden imposed by the variety of problems to be faced and the task of adapting possibly inappropriate materials and techniques.

The problem is interesting from many points of view, even if it is approached from a technological aspect in this paper. Existing architectural structures are a continuity factor. They form the image of a city, or village, for which they often represent the dwelling and building traditions, that a community tends to take with it even if it moves from its place of origin (as often happens when the outskirts of an urban centre are formed by farm workers who emigrate from the country).In short,it is a question of something, potentially valuable, even from an economic point of view, owned by a community, that cannot be neglected lightly.

While numerous techniques for the oldest forms of architecture have been perfected, safeguarded only because they are exceptional, up to now, very little has been done for more or less recently built common dwellings that make up the connective tissue. The individual physiognomy of single regions is determined by the overall stratification that as a whole is impressed on the collective memory.

While, for buildings that for some reason or other do not follow the usual pattern, diverse interests capable of bearing the burdens of maintenance or restoration, even when the building itself does not represent a directly productive investment, may converge. However, in the case of more or less deteriorated anonymous buildings, a well-defined political willingness is needed to guide and support cautious coordinated preservation and renovation action.

Experience of many years, in countries of very different types, has taught that only mixed management of public administration and private property can control anarchy and the lack of individual initiative (Gilibert 1990). The simpler the techniques proposed and the cheaper the materials to be used, the easier it will be to guarantee the social and economic effectiveness of an operation of this kind that should be both ductile and pressing.

Here the intention is to examine some possibilities that vegetable fibres, used with gypsum, can offer both in prefabrication and working on site (Foti et al. 1987a).

2 The use of fibres in recuperating buildings

2.1 The use of mineral and artificial fibres

The numerous possible applications of artificial and mineral fibres in building today are by now well-known; when speaking about natural fibres, the fields of application already proposed appear limited and in a certain sense reduced with respect to their possibilities. Whoever handles the latter is reluctant to move towards varied and at times spectacular applications, already experimented with the former, for fear of a comparison that could in some way prove to be unequal. For this reason the use is repetitive and limited with respect to the potentiality that composite materials with vegetable fibres have, or, at least, according to past evidence, should have.

There is no doubt that vegetable fibres normally present fewer characteristics of resistance and durability, but these deficiencies can be corrected. In considering the use of artificial fibres, it is already known that gypsum, as a matrix for fibreglass reinforced composites offers some advantages over cement and plastic materials (Brokes 1986). Glass Reinforced Gypsum, that can be produced in extremely thin sheets and in highly resistent complex shapes, offers consistent cost saving in production, because gypsum does not require special alkali-resistent fibreglass, as in Glass Reinforced Concrete and also because it avoids using expensive resins as in Glass Reinforced Plastics.

Now, it has been shown that natural fibre reinforced gypsum has good workability and resistance, confirmed, at least as regards gypsum and sisal composites, by numerous lab tests carried out in the Politecnico di Torino (Mattone 1987). Of course it is not possible to obtain very reduced thicknesses as in the case of GRC, nor wide spans with limited weights as in the case of GRP but, bearing in mind the possibility of performance obtainable from the single gypsum-sisal mixtures, it is possible to plan a very wide range of finished objects with fairly predictable behaviour and durability characteristics (Foti 1988).

2.2 The use of vegetable fibres in recuperating buildings

In terms of maintenance or improvement, existing buildings can be treated by using gypsum-sisal components diversified to cope with the vast number of very varied problems that come up in buildings constructed with traditional techniques or even more so in self-constructed ones. Using gypsum-sisal it is possible to produce, for instance, sheets for floors, external panels, covering and even structural load-bearing elements (D'Alfonso et al. 1986a). Fibres to be used directly on site, can be prepared in different shapes, woven to form sheets, knotted to make up nets, tied together to make bundles, arranged in tape form for the use of long fibres and cut to be used at random.

The preparation of fibres adopted depends on the defects to be corrected or on the improvements to be made and depends on the position of the particular part to be treated. The fibres are held in the desidered arrangement, possibly by hooks or nails or various binders and then the necessary quantity of mixed gypsum is modelled onto it.

Integrating these techniques with other devices it is even possible to obtain both flat and curved false ceilings and also internal partitions. In general, it can be said that the gypsum-sisal composite is fairly widely adaptable. If the working techniques of it are known, useful ideas could evolve from examining a specific case. Naturally, no mention of these techniques, widely described elsewhere, will be made.

To delimit the subject, only the problem of load-bearing walls of buildings will be examined, a problem in itself very wide and differentiated, but in each case a central point in recuperating work.

We know the pathology of what can affect load-bearing walls in a bulding but above all which and how much building variety these can present. It is possible to find, for instance, in the case of earth walls in rather precarious conditions both from a static and from a surface deterioration aspect; on the other hand, a priority objective in recuperating work seems to be the preservation of the specific building expression. It is a field of investigation still open in which research on the use of natural fibres can be a useful contribution, following the example of bio-geo technology that in few decades has radically innovated the range of proposed solutions to stabilize slopes subject to landslides (Gilibert 1977). At times, heavy structural intervention techniques (for example caging in reinforced concrete or piling for anchorage to solid supports no matter how deep) have been substitued by light interventions adopting surface impregnation using resins in biodegradable solutions sprinkled onto the surfaces to be treated, previously sprayed with loose fibres.

Natural earth walls, particularly typical of settlements on sloping ground (from certain outskirts of large Latin American cities to Kurd villages) and, above all, surviving troglodyte dwellings, deteriorate progressively, due to lack of surface protection. There is no lack of specific case studies but they are little known and, because of this, remain extremely limited with respect to the diffusion of what is still standing: the Sassi of Matera in the south of Italy, Matmata in Tunisia together with the most famous Peruvian monuments are examples to be kept in mind to safeguard the cave dwellings of China, Malaysia, Caucasus and the north of Africa from Libya to the Red Sea (A. Alva et al. 1985a).

3 Improving external walls

3.1 External walls: the reasons for their conservation

Generally speaking, the major part of walls should be preserved in recuperating work because they are characteristic of building typology and measurement of volume, that is they are the determining factor in recognizing living space. These walls need, however, to undergo interventions of consolidation, adaptation to new distributive arrangements, special finish, protection, adjustment to the desired levels of heat insulation and deadening.

It is important to bear in mind that the above refers to various types of architecture existing in developing countries: from spontaneous buildings in districts on the outskirts to those already recognized as being of historic or environmental interest. According to the cases, it is a question of intervening in self-construction or with more or less skilled workers, of producing gypsum-sisal components in the building yard or in more or less equipped labs, of preparing them with several types of surface finish, of using coatings to make them suitable for current use. Generally speaking, there seem to be cases where using fibre reinforced gypsum is particularly suitable:

- surface finish or coating of walls to provide more effective protection of them, externally and internally;
- adjustment of the level of heat insulation proper to the wall itself.

3.2 Coating and treating walls

In coating walls, small prefabricated gypsum-sisal elements, for instance, could be useful: tiles, baseboards, cornices, door or windowsills, elements to adapt or finish door or window openings. When these elements are destined to be exposed to atmospheric agents, they should be treated with a water-proofing protective covering.

Innumerable research has been carried out at the Politecnico di Torino on this problem. Various industrial products (with acrylic, polyurethane or epoxy bases) and some materials of vegetable origin (with latex rubber and linseed-oil bases), easily found in several parts of the world, were tested for use as coatings. Both types of products gave interesting results which are applicable even when the material to be protected has been produced in self-building. After the first test experiments of water-proofing treatments, which were briefly mentioned in the report of the research group, that we form part of, presented to the Joint Symposium of Baghdad in 1986, further encouraging reports were obtained by some researchers of the same group (Comoglio et al. 1986a). Doorsills which need to be resistant to wear and tear as well as waterproof can be treated with the same products with the addition of appropriate loads and the application of more layers (Musadeh et al. 1988a).

Working techniques that use vacuum process systems have been followed up by other researchers with the aim of increasing performance in mechanical resistance and surface hardness, to check the suitability of the composite in its resistance to wear and tear and impact strain as is necessary in the case of floor tiles, doorsills, steps, external wainscoting (Mattone 1990). The possibility of intervening on site was experimented using loose fibres, fixed with gypsum-based paints suitable for restoring coatings that have minute widespread cracks. Here, particular reference is made to false ceilings in lathing or matting frequent in past eras in Europe both in rural and urban building and still in use today in developing countries (Gilibert 1978).

4 Heat insulation of walls

The gypsum-sisal composite can improve the performance of exisisting walls in several ways. The good properties of gypsum-sisal as a heat insulator can be exploited by applying, on existing walls, sheets or panels that, besides having the function of coating, help to achieve the desired values of comfort inside the rooms. Even better results can be obtained by inserting insulating material between the walls and the panels.

If desired, industrial products can be applied; for instance, if panels in gypsum-sisal forming a C-shape in section are used, it is possible to place a layer of heat insulator between the wings of the "C". The latter can be found with a proper shape which thus maintains stability after being placed. But it is important to consider, first the possibility of using insulators easily preparable with the materials available where working even if the materials are incoherent (weakened wood, cork or other types of bark, dry leaves, rice chaff, nutshells, pomice granules, etc.). In this case, the insulating materials can be poured to fill the airspace that forms between the gypsum-sisal panels and the walls.

A technique to obtain gypsum-sisal panels with differentiated density that offer in themselves better heat insulation values, has also been experimented. In practice, it is the preparation of finished components whose external faces are made up of compact gypsum-sisal layers, while the core is made up of a layer of sisal fibre sized with gypsum. In the experiments carried out, these elements were prepared first by casting one of the external compact gypsum-sisal layers and then bundles of fibres, well impregnated with gypsum, were placed in a sinusoidal arrangement onto the still fresh layer. Then the bundles were worked manually and lifted to make them light and spongy creating a ventilated framework serving to form a space between the most compact, resistant skins.

5 References

Gilibert Volterrani A. (1977) Prima rassegna di sistemi a scomparsa per intervento di controllo dei pendii, in **Tutela dell'ambiente: programmi e attività delle regioni**, Torino, Regione Piemonte, pp. 1-10.

Gilibert Volterrani A. (1978) Materiali speciali per il recupero di soffitti composti da incannucciati o stuoie o tele intonacate, in **I Borghi rurali Friulani** (eds. A. Bragutti and A. Nicoletti), Regione Autonoma Friuli-Venezia Giulia, Udine, pp. 81-84.

A. Alva Baltarrama and J.M. Teutonico (1985) Notes on the manufacture of adobe blocks for the restoration of earthern architecture, in **Adobe** (ed. J. Truel), UNDP/ UNESCO/ICCROM, Lima, pp. 41-54.

Brokes A. (1986) Glass reinforced gypsum. **The Architect's Journal**, 22, 69-74.

Comoglio Maritano D., Foti M. and Gilibert Volterrani A. (1986a) Matériaux composites avec fibres de sisal: étude de faisabilité, in **Use of Vegetable Plants and Their Fibres as Building Materials** (ed. M. Samarai), RILEM/NCCL/CIB, Baghdad, Session A, pp. 11-24.

D'Alfonso M., Mattone R. and Pasero G. (1986) **Self-help manual proof for a gypsum-sisal conoid,** C.L.U.T., Torino.

Foti M. and Gilibert Volterrani A. (1987) Matériaux traditionnels ou avancés: un choix pour les pays en voie de développement. **Matériaux et Constructions**, 20, 255-259.

Mattone R. (1987) Operational possibilities of sisal fibre reinforced gypsum in the production of low-cost housing building, in **Building Materials for Low-income Housing.** Asian and Pacific Region, (ESCAP/RILEM/CIB) E. & F.N. Spon, London, pp. 47-56.

Foti M. (1988) **La progettazione dei componenti in "Gesso-sisal"**, C.L.U.T., Torino.

Masaedeh H. and Omar M. (1988) **Analisi di sistemi costruttivi per paesi tropicali, con attività di laboratorio** (Tutorship of G. Ceragioli, G. Canavesio, D. Comoglio Maritano), Degree Thesis, Faculty of Architecture, Politecnico di Torino.

Gilibert Volterrani A. (1990) From defferred maintenance to planned maintenance, in **International Symposium on Property Maintenance Management & Modernisation** (ed.

L.K. Quah), Longman, Singapore, vol. 1, pp. 298-304.
Mattone R. (1990) Comparison between Gypsum panels reinforced with vegetable fibres: their behaviour in bending and under impact, in this 2nd Symposium on **Vegetable Plants and Their Fibres as Building Materials.**

4 EXPERIMENTAL METHODS FOR THE PREPARATION OF PALM FRUIT AND OTHER NATURAL FIBRES FOR USE IN REINFORCED CEMENT COMPOSITES

J.G. CABRERA and S.O. NWAUBANI
Department of Civil Engineering, University of Leeds, England

Abstract
A review of methods of extraction of oil, total and alkali soluble carbohydrates and in general, the preparation of palm fruit fibres, grass, cane sugar and maize fibres is presented.

Preliminary results of extraction results are included and a tentative set of requirements for acceptance of these natural fibres for use in cement composites are proposed.

Available building materials using organic fibre composites are reviewed and requirements for the assessment of their performance and durability highlighted.
Keywords: Vegetable Fibres, Cellulose, Chemical Composition, Acid Extraction, Alkali Extraction, Water Extraction, Hemicellulose, Lignin, Alkali Resistance, Durability.

1 Introduction

The concept of incorporating fibres of Cellulose origin in building composites is very old, it dates back to the old Egyptian civilization.

Current interest in the application of the technology to cement composites stems mainly from the acute shortage of housing and high cost of building materials in the less technologically advanced countries. The use of indigenous knowhow and resources in manufacturing building products with low energy demand is the most adequate alternative.

Cellulose and vegetable fibres are available in relatively large quantities in many developing countries. They have been extensively investigated by various research groups in many countries to ascertain their suitability as substitutes for conventional synthetic fibres such as glass and asbestos in the production of roof slabs and thin-sectioned wall and ceiling panels [Gram, H.E., et al, 1984; Gallegos, V.H., 1986; Subrahmanyan, B.V., 1984; Nwaubani, S.O. and Sliwniski, M.A., 1985].

Fibres from coconuts, sisal, jute, sugar cane, banana, wood, palms, flax, Akwara and Elephant grass are among the many studied by various investigators.

Aziz et al [Aziz, M.A., 1984], has reviewed the role played by the above fibre types. The indications are that although most of these may be used successfully in fibre-reinforced cement composites, their long term durability is in question. This is due mainly to the instability of the cellulose fibres in the high alkaline environment which usually exists in the cement matrix [Gram, H.E. and Nitnityongskul, P., 1987].

This paper therefore attempts to address two issues;

1) It looks at the properties of oil palm fibres (OPF), sugar cane and corn fibres and how each may be prepared for use in fibre reinforced cement composites.

2) It reviews the inherent cause of fibre embrittlement and suggest ways of improving the long term performance and

3) Presents preliminary results on chemical extractions carried out on the fibres mentioned.

2 Oil-Palm Fibres

Oil palm fibres [OPF], are derived from the palm fruits. The palm tree which produces the fruits are cultivated in plantations but most grow wild in several hectares of wild groves throughout the tropics.

The OPF consists of a series of thin strands and leathery skin which surrounds the kernel and provides reinforcement for the outer fleshy part of the fruit. The fibres are obtained as waste products after the palm oil and kernel has been extracted in the oil mills. The fibre length varies between 10 to 40 mm while the width varies from 3-70 μm.

The oil palm fibres are potentially the largest source of vegetable fibres being produced throughout the tropical region today. Several tonnes are produced each day in the oil mills situated in several villages in palm oil producing countries.

There are no statistics on the amount of oil palm fibres produced worldwide. However, a good inference on this matter can be made by observing the annual production of palm oil in some countries (see Table 1).

Table 1. Palm Oil Production - [Butty, J. (1989)]

Country		Annual Oil Palm Production [Tonnes]
Nigeria		600,000
Ghana		25,000
Côte D'Ivore		100,000
Indonesia	)	
+	)	258,000
Malaysia	)	as export to the US alone in 1986

Traditionally, the OPF are burnt to provide fuel for household cooking but with the advent of electrical and gas alternatives surplus quantity of the OPF remain as waste.

In this regard the OPF can be expected to make cheaper cement fibre composites than most other cellulose fibres like the bamboo and Bagasse (crushed sugar cane fibre), which are used in the manufacture of paper or the coconut fibres for which there is greater demand in the manufacture of footmats and ropes.

3 Sugar Cane Fibres

Sugar cane fibres are the fibrous residue from crushed sugar cane. The sugar cane plant from which they are obtained are grown in foundations and consist primarily of cellulose, hemicellulose and lignin. They have higher moisture content, about 50%, in comparison with some hard woods which exhibit approximately 10 to 30% of moisture content [Blackburn, F., 1984].

They also contain large amounts of sucrose (sugar). The sugar cane husk as well as being an important source of raw material for paper production is used as fuel in Brazil.

Sugar cane fibres have been used in the production of 1.20 m^2 panels which are either bonded with cement or resins. The panels may have a fibre content of up to 49% [Blackburn, F. (1984)].

4 Grass Fibres

Most of the grass fibres like the corn, rice, exparto grass, elephant grass, bamboo etc, are characterised by considerable straw pulp content which as is the case of sugar cane may be used in paper production. Most of the available literature concerning the application of grass fibres in cement composites refers mainly to the use of bamboo fibres.

There are several reports describing the use of the bamboo as reinforcement for concrete beams, slabs, pavements and in making fibre boards [Pama, R.P., 1976; Ali, Z. and Pama, R.P., 1978].

5 Special Problems with Vegetable Fibres in Cement Composites

5.1 Problems due to the cellulose content of the fibres

Cellulose fibres of vegetable origins include fibres derived from Bast, wood, leaf, seed, grass or fruits of plants. They have a chemical composition which is characterised by large cellulose, hemicellulose and lignin contents [Swamy, R.N., 1986; Gram, H.E. et al, 1984]. The latter may be made up of carbohydrates and other materials including oil or wax as in the case of the oil palm seeds.

The nature of the problems which may arise in using vegetable fibres in cement composites will depend on the chemical make up of the fibre. For instance:

1) There are many reports indicating that the lignin-content of the fibres may affect hydration and setting time of the composite [Swamy, R.D., 1986].
2) In fibres like the sugar cane, the presence of carbohydrates such as starch, sugars or tannins, which are the essential compounds in commercial retarding agents, may retard hydration of cement particles [Swamy, R.N., 1986].

(Additives like accelerators ($CaCl_2$, $MgCl_2$, calcium formiate, calcium acetate)) are often used in the production of boards and

blocks. These additives are used in dilute solutions of 1 to 5% for the purpose of accelerating the hydration of cement and counteracting the retarding effect of inhibitors like phenolic compounds and free carbohydrates. In general the chemical interactions are very complex and difficult to treat theorectically. This is why most of the work available is of experimental empirical nature.
3) In the form in which they are collected from the oil mills, the oil-palm fibres will contain some oil. The presence of oil in the mix will reduce the interfacial bonds between the cement and the fibres which may then result in lower ultimate strength values.

These undesirable constituents should be ideally removed using appropriate methods, but since removal treatment is not always economically feasible, fibres should be chosen according to their composition. This option requires the application of simple methods of characterization.

5.2 Problems due to alkali attack on the fibres

A problem which is very well documented in the literature concerns the long term stability of vegetable fibres in the cement matrix.

The pore water of an ordinary portland cement is saturated with alkali ions and therefore under normal conditions the pH of the pore water is above 12.4. Acid attack or penetration of gases may change the pH to values below 9 but the environment will be always alkaline.

There are many reports which confirm that various vegetable fibres become embrittled by the action of the alkali's in the pore water thus jeopardising durability of the composite [Gram, H.E., 1983; Velparri, V., et al, 1980]. Gram, [1983] reported that high alkalinity of the pore water results in chemical attack on natural fibres causing reduction or even nullification of strengths.

6 Criteria for Accepting Vegetable Fibres in Cement Composites

Typically the ratio of cement to fibre varies from 0.4 to 2.3. At these ratios it is very important to consider the physico-chemical interaction which takes place at the fibre-cement interface since ultimately the durability and structural integrity of the composite material will depend on the reactions taken place at the interface.

Selection of fibres and/or acceptance of treated fibres for use in cement composites has inevitably been based on limiting values of the organic compounds which are soluble and retard cement hydration. It is known that lignin content is beneficial for binding while cellulose and more so hemicellulose (which is very alkali soluble), affect strongly the bond strength. For example softwoods which have more than 25% lignin and lower hemicellulose content than hardwoods are known to give better results when used as fibre reinforcement in ordinary portland cement matrices.

Solubility measurements are in some instances not clearly defined, but the general principle is that soluble compounds are directly related to the sugar content of the fibre and therefore they should be limited. An example of suggested limiting values is given in Table 2.

Table 2. Limits of glucose and starch concentration to avoid retardation of cement hydration

Compound	% based on cement weight	Effect
Glucose	>= 0.25	retards hydration
Starch	>= 0.125	retards hydration

It is necessary to point out that upon storage (which can cause moisture content changes), fibres will change composition regarding starch, sucrose, glucose and fructose contents. The degradation of carbohydrates is an enzimatic reaction and therefore it takes place if the moisture content is above the saturation point [Simatupang, M.A. et al, 1979], therefore tests should be carried out on samples which are representative of their composition when they are incorporated in the cement matrix.

It seems clear to the authors that there is need to develop methods which can be used rapidly and cheaply on the raw fibres so that their suitability is assessed for use in different composites.

7 Preliminary Experimental Work

Laboratory extraction methods have been used to determine the carbohydrate, glucose and starch content of four types of fibres. These were grass, sugar cane, maize and oil palm.

The methods used involved the following:

a) acid extraction for determination of the total carbohydrate. The method used consisted in refluxing the fibre with sulphuric acid for 16 hours. This is a very severe treatment and therefore the results represent the total carbohydrate content of the fibre.
b) alkali extraction for determination of the alkali soluble carbohydrate and alkali soluble glucose using saturated lime solution.
c) water extraction for determination of the soluble carbohydrate and glucose of the fibres.

The results are presented in Table 3.

Table 3. Preliminary results of acid, alkali and water soluble carbohydrates, glucose and starch

Treatment	Compound	Fibre, %			
		Sugar Cane	Grass	Maize	Oil Palm
Acid extraction	carbohydrate	31.0	26.0	26.0	2.1
Lime extraction	carbohydrate	6.0	6.0	5.0	0.3
Water extraction	carbohydrate	3.0	3.0	1.0	-
Lime extraction	glucose	1.2	0.7	0.0	0.0
Water extraction	glucose	1.4	1.1	0.45	0.42
	starch	3.0	4.2	3.6	1.5

8 Discussion of Results

From the results obtained it can be inferred that oil palm fibres could be tentatively selected as the fibres which will give the best bond strength and be most durable, however preliminary results on the methods of extraction of the oil contained in the fibres show that the oil can be extracted but that its content is high and could affect the energy of the bond to the cement matrix. Experiments in progress will produce data on the effect of oil content on the strength and durability of the bond at the fibre-matrix interphase. From the other three fibres the best of them appear to be the maize fibre. In fact this fibre has been used to fabricate woodwall cement slabs with a good degree of success.

9 Concluding Remarks

A simple laboratory method for the determination of lime soluble sugars appears to give good results in practice. The use of this test might be objected in strictly scientific terms, however from a pragmatic point of view this simple test can be implemented in any field laboratory and fibres can be tested and preliminarily selected in the field. Of course it is recognized that more work is required regarding the quantitative relation between alkali soluble sugars and durability of fibre-cement composites. This is being investigated.

Oil palm fibres are very abundant in many tropical regions. Their use will have economic and environmental advantages providing that economic oil extraction methods can be developed.

10 References

Ali, Z. and Pama, R.P. (1978) Mechanical properties of Bamboo as reinforcement for concrete pavements. Proc. Int. Conf., pp. 49-66.

Aziz, M.A. Paramasivam, P. and Lee, S.L. (1984) Concrete reinforced with natural fibres. New Reinforced Concretes, ed. R.N. Swamy, Surrey University Press, [Blackie], Glasgow and London, pp. 106-140.

Blackburn, F. (1984) Sugar Cane. Longman Group Limited, England, 389 pp.

Butty, J. (1989) Palm oil in the lobby, West Africa Magazine, January 30-February 5, pp. 146-147.

Gallegos Vargas, H. (1986) Use of vegetable fibres as building materials in Peru. Proc. Symp. Use of Veg. Plants and Fibres as Build. Mat., Baghdad, October, A25-A34.

Gram, H.E. (1983) Durability of natural fibres in concrete. CBI Research 1. Swedish Cement and Concrete Research Institute, Stockholm, 255 pp.

Gram, H.E. Person, H. and Skarendahl, A. (1984) Natural Fibre Concrete, SAREC Report R2, Swedish Agency for Research Cooperation with developing countries, pp. 1-139.

Gram, H.E. and Nitnityongskul, P. Durability of natural fibres in cement-based roofing shets. Proc. Symp. Building Materials for Low-income Housing, Bangkok, January, 1987, pp. 328-334.

Nwaubani, S.O. and Sliwniski, M.A. (March, 1985) The use of locally produced fibres as reinforced in cement-based matrix, - A preamble, 2nd National Conference on Materials Testing, Research and Control, Lagos, pp. 1-10.

Pama, R.P. Durrani, A.J. and Lee, S.L. (1976) A Study of Bamboo as reinforcement for pavements. Proc. 1st Conf. of the Road Engineering Association of Asia and Australia, Bangkok, pp.45-96.

Simatupang, M.H. Schwarz, G.H. and Broker, F.W. (1979) Small scale plants for the manufacture of mineral-bonded wood composites. World Forestry Congress Special paper FID-II/21-3. Jakarta.

Subrahmanyan, B.V. (1984) Bamboo reinforcement for cement matrices. New Reinforcement Concretes, Ed. R.N. Swamy, Surrey University Press, [Blackie], Glasgow and London, pp. 141-194.

Swamy, R.N. (1986) Natural Fibre Reinforced Cement and Concrete. Blackie and Son Ltd, London.

Velparri, V. Ramachandram, B.E. Bhaskaram, T.A. Pai, B.C. and Balasubramani, N. (1980) Alkali resistance of fibres in cement. J. Mater. Sci., 15, pp. 1579-1584.

PART THREE

PROPERTIES OF VEGETABLE FIBRE COMPOSITE MATERIALS

5 EFFECTS OF MOISTURE CONTENT ON MECHANICAL PROPERTIES OF WOOD FIBER REINFORCED CEMENT

P. SOROUSHIAN and S. MARIKUNTE
Department of Civil and Environmental Engineering,
Michigan State University, East Lansing, USA

Abstract
Wood fiber reinforced cement can provide the highest performance-to-cost ratio among fibrous cement composites considered for the replacement of asbestos cement. Such composites can find applications in the production of thin flat and corrugated cement sheets, non-pressure pipes and many other thin-sheet cement products. There are, however, concerns regarding the moisture-sensitivity of wood fiber-cement composites. Considerable differences in flexural strength and fracture toughness values are observed when the specimens are tested at different moisture contents. This paper presents the results of a comprehensive experimental study concerned with the effects of moisture content on flexural performance characteristics of wood fiber reinforced cement.

The cement composites considered in this investigation incorporated 0%, 1% and 2% mass fractions of mechanical pulps. The moisture conditions investigated included oven-dried, air-dried and saturated. Comprehensive sets of replicated flexural test data were generated and were analyzed statistically. The analysis of variance techniques were employed in order to derive reliable conclusions regarding the moisture-sensitivity of the flexural strength and toughness characteristics of wood fiber reinforced cement composites.

The results generated in this study were indicative of significant effects of moisture content on flexural performance of wood fiber reinforced cement. There is a tendency in flexural strength to decrease, and in flexural toughness to increase with increasing moisture content of the composite material. Microstructural studies indicated that high moisture contents tend to damage the fiber-to-matrix bond strength, leading to changes in failure mechanism which can describe the trends observed in moisture effects on the flexural performance of wood-cement composites.
Keywords: Cement, Compaction, Composite Materials, Deflection, Flexure, Pulp, Reinforcement, Strength, Toughness, Wood Fibers.

1 Introduction

Wood fibers have been used for many years as an additive in the conventional asbestos cement industry; some of the asbestos cement replacement products also utilize wood fibers. In these cases, wood fibers contributes mainly processing benefits rather than reinforcement. Recent studies have shown that the reinforcement ability of wood fibers is quite good relative to other potential asbestos replacements (Coutts 1988; Vinson et al.1989; CSIRO 1981). There are, however, concerns regarding the moisture resistance of wood fiber-cement composites. Considerable differences in flexural strength and fracture toughness values are observed when the specimens are tested at different moisture contents (Coutts 1984). High moisture contents tend to damage the fiber-to-matrix bond strength, leading to changes in failure mechanism which can describe the trends observed in moisture effects on flexural performance of wood-cement composites (Morissey et al. 1985).

2 Background

Rapid drying of cement-based materials may induce tensile cracks due to non uniform drying (and hence differences in drying shrinkage) of the specimen. The cracks do not have much effect on compressive strength but will lower the flexural and tensile strengths (Neville 1963; Mindess et al. 1981). If drying takes place very slowly, so that internal stresses can be redistributed and alleviated by creep, an increase in strength may result from drying.

Wetting of concrete may lead to losses in compressive strength as a result of the dilation of cement gel by adsorbed water, which leads to reduction of cohesion of the solid particles. Conversely, when upon drying the wedge-action of water ceases, an apparent increase in strength of the specimen is recorded. Resoaking of oven-dried specimens in water reduces their strength to the value of continuously wet-cured specimens, provided that they have hydrated to the same degree. The variation in strength due to drying is thus a reversible phenomenon.

Cellulose fiber-reinforced composites are highly sensitive to moisture variations. Considerable differences in flexural strength and fracture toughness values are observed when the specimens are tested at different moisture contents. There is a general tendency in flexural strength to decrease and in flexural toughness to increase with increasing moisture content in wood fiber reinforced cement composites (Coutts 1984). This has been attributed to moisture effects on fiber-matrix bond strength and also on the properties of wood fibers.

3 Experimental Program

The wood fiber used in this investigation was Mechanical pulp (American Fillers). Some key properties of this wood fiber is presented in Table1.

Table 1. Properties of Wood Fibers

Manufacturer	Brand Name	Type	Species	Avg. Length
American Fillers & Abrasives	1000L	Mech.	Softwood	8.0 mm

The fiber mass fractions and matrix mix proportions used in this phase of the study are introduced in Table 2. The water and superplasticizer contents were varied in order to achieve reasonable fresh mix workability characteristics represented by a flow (ASTM C-230) of 60±5% at 1 minute after mixing.

Table 2. Mix Proportions, Fresh Mix and Hardened Material Properties

Mix	water/ Cement	Flow (%)	Setting Time (min.)		Air Content (%)		Specific Gravity		Water Absorption (%)
			Initial	Final	Fresh	Hardened	Fresh	Hardened	
Plain	0.3	58±1	186±3	210±3	21.07±0.25	22.75±0.30	2.01±0.014	2.00±0.001	11.07±0.12
1% Mech.	0.3	64±1	272±4	354±3	20.86±0.27	22.10±0.43	1.85±0.026	1.86±0.064	11.45±0.02
2% Mech.	0.3	59±1	296±4	387±3	23.94±0.41	25.90±0.68	1.71±0.017	1.65±0.055	12.48±0.18

Wood fiber reinforced cement composites were manufactured in this study using a regular mortar mixer. The mixing procedure adopted is as follows: (1) Add cement, sand and 70% of water and mix at low speed (140 RPM) for about 1 minute or until a uniform mixture is achieved; (2) turn the mixer to medium speed and gradually add the fibers and the remainder of water and superplasticizer into the mixture as the mixer is running at medium speed (285 RPM) over a period of 2-5 minutes depending on the fiber content, taking care that no fiber balls are formed; (3) Add set accelerator and mix for 1 minute at medium speed; and (4) Stop the mixer and wait for 1 minute, and then finalize the process by mixing at high speed (450 RPM) for 2 minutes.

The fresh fibrous cement mixtures were tested for: (1) flow (ASTM C-230) at 1 min, 5 min. and 10 min. after the mixing process; (2) air content of hydraulic cement mortar (ASTM C-185); and (3) setting time by penetration resistance (ASTM C-403).

From each mix, molded specimens were manufactured for flexure (ASTM C-1018) tests in the hardened state. The flexural specimens were prisms with 38.1 mm (1.5 in.)

square cross sections and total length of 152.4 mm (6 in.) tested by 4-point loading on a span of 114.3 mm (4.5 in.). The void content, specific gravity and water absorption of the hardened materials were also assessed using the broken flexural specimens (ASTM C-642). An average of 10 specimens with a specific mix design were manufactured for test in each moisture condition. The specimens were selected equally from two different batches in order to generate more representative variations in results. Five specimens for test in each moisture condition were thus obtained from each batch; this means that a batch represents a block in this experimental design in which two blocks were used. This allows to remove the variations between batches in the analysis of variance on moisture effects.

All the specimens were compacted through external vibration, and were kept inside their molds underneath a wet burlap covered with plastic sheet for 24 hours. They were then demolded and moist cured for 5 days before being air cured in a regular laboratory environment (about 50±10% RH and 22±3 deg. C, 72±5 deg. F temperature)

In order to investigate the effects of moisture content a series of standard conditions for testing were established. Samples were conditioned in the following environments: (a) in lab with relative humidity of 50±10% and 22±3 deg. C (72±5 deg. F); (b) in an oven at 116 deg C for 24 hours and then cooled in the laboratory environment; and (c) in water for 48 hours with excess water removed with a cloth prior to testing.

The flexural tests were performed according to the Japanese code JCI-SF (Japanese Concrete Institute 1984). The Japanese method of measuring flexural deflections (Figure 1), is particularly effective in reducing errors associated with rigid body movements of the specimen and local deformations at the supports and loading points.

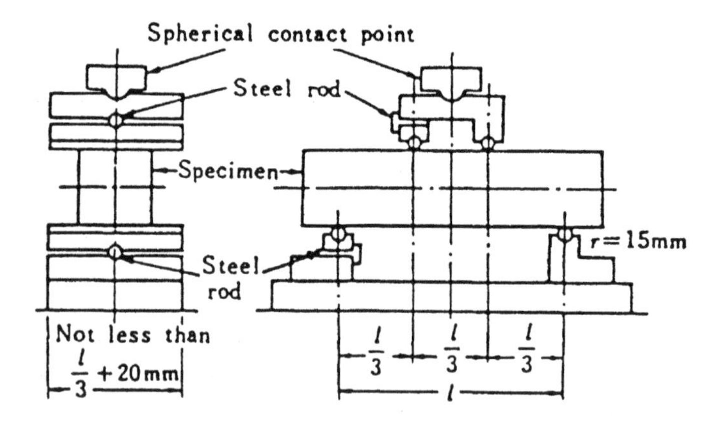

Figure 1. Japanese Standard Flexural Test Set Up.

An important consideration in flexural tests performed on fiber reinforced cement composites is the measurement of energy absorption capacity, defined as the area underneath the load-deflection curve. The japanese fixtures which monitor flexural deflections during the test give accurate results for energy absorption calculations. According to the Japanese code JCI-SF, Flexural toughness is defined as the area underneath the flexural load-deflection of 0.76 mm (0.03 in.), which is the span length divided by 150.

4 Test Results and Discussions

This section presents the test data in two parts. First the results regarding the effects of wood fiber on the water requirement, setting time, unit weight, air content, permeable void content, specific gravity and water absorption capacity of cement-based materials are briefly reviewed. Then the experimentally observed effects of moisture condition on the flexural behavior of composites with different wood fiber mass fractions are discussed and the results of statistical analysis of moisture-sensitivity test data are presented.

Table 2 presents the test data on water requirements (for achieving comparable levels of workability), air content, water absorption and setting time of composites with different fiber mass fractions. The results presented are average of two values obtained from two different batches for a specific mix, with information given on the range covered by the two replicated test points. The variations between batches are observed to be relatively small. Water-cement ratios were varied in order to obtain comparable workability levels represented by a flow (ASTM C-230) of $60 \pm 5\%$.

Setting time of mixtures containing fibers increased as fiber mass content was increased (see Table 2). The initial setting time increased by as much as 25% for the mix containing 2% mass fraction of mechanical fibers. This increase was even more for the final setting time (84% for mechanical fibers). It should be noted that fibrous mixtures had a set accelerating agent included in their composition.

Mechanical pulps retard setting time possibly because they are porous and hollow and thus absorb the water available in fresh mix, making it difficult for cement to access the water for hydration purposes. Mechanical pulps also contain lignin, which dissolves in water and interferes with the process of cement hydration, thereby further retarding the setting time of the fibrous cement composites.

Wood fibers are observed in Table 2 to reduce the unit weight of fresh mixtures. The reduction in unit weight was 15% when 2% mass fractions of mechanical pulps was added. This is due to the fact that the air content in cementitious matrices tends to increase in the presence of fibers, as discussed below.

The fresh mix air content is observed in Table 2 to increase when wood fibers are added to the mix. This increase was 14% for mechanical pulps at 2% mass fraction. The increase in air content could be attributed to the difficulty of compacting cement composites incorporating higher fiber contents, which leads to increased entrapped air content.

The hardened material void content also increased in the presence of wood fibers. The increase over plain cement-based materials was 14% for mechanical pulp at 2% mass content. Similar trends were observed in the fresh mix and hardened material air contents.

The water absorption capacity of cement based materials is observed in Table 2 to increase with increasing fiber mass fraction. The increase in water absorption over that of plain cementitious matrix was 13% for mechanical pulps at 2% mass fraction.

The specific gravity of hardened cement-based materials is observed in Table 2 to decrease with increasing wood fiber content. At 2% fiber mass content, mechanical pulps showed a decrease of 17%; similar trends were observed in the fresh mix unit weight.

In order to study the effect of fiber mass content on fresh mix and hardened material properties, a factorial analysis of variance was performed. Table3 presents the results of this factorial analysis. A "*" in this table represents moderately significant effects, and "**" represents major effects at 5% level of significance

From the results it can be concluded that the effect of fiber mass content was quite significant for setting time, fresh mix unit weight, and hardened material water absorption and specific gravity, while it was moderate for flow, air content and void content.

Table 3. Factorial Analysis of Fresh Mix and Hardened Material Properties

Property	Fiber Cont.
Flow	*
Initial Setting	**
Final Setting	**
Unit Weight	**
Air Content	*
Void Content	*
Water Absorption	**
Specific Gravity	**

In order to see if there is any relationship between the fresh mix air content, and the hardened material void content, specific gravity and water absorption correlation coefficients were calculated analyzed statistically. The results showed that the correlation is significant at 5% level of significance for all the two combinations of the properties (except for void content Vs. specific gravity in the case of mechanical pulp, which was found to be significant at 10% level of significance). The correlation analysis results indicate that the wood fiber effects on water absorption and specific gravity at least partly result from the corresponding effects of fibers on the void content of fresh and hardened materials.

5 Flexural Performance

The effects of wood fiber reinforcement on the flexural load-deflection behavior of cement-based materials are shown in Figures 2 (a), (b) and (c) for different fiber mass contents and moisture conditions.

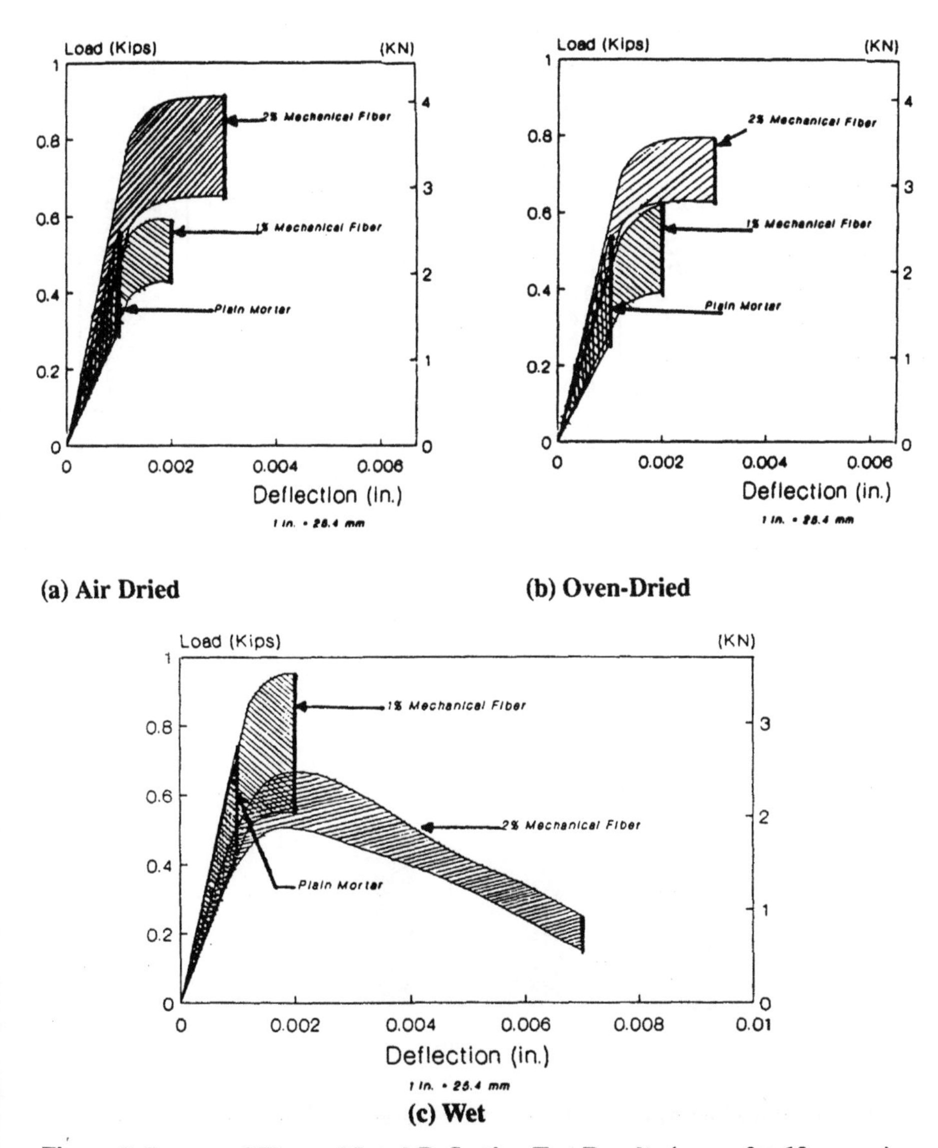

Figure 2. Ranges of Flexural Load-Deflection Test Results (range for 10 curves).

Table 4. Flexural Strength Test Results (Ksi)

Fiber Mass Content	Moisture Condition					
	Air-Dried		Oven-Dried		Wet	
	Batch 1	Batch 2	Batch 1	Batch 2	Batch 1	Batch 2
0%	0.5498	0.5996	0.5978	0.8182	0.7264	0.9607
	0.8372	0.3805	0.3098	0.6155	0.8800	0.8680
	0.7195	0.6539	0.6017	0.6781	1.0133	0.8469
	0.5697	0.5616	0.6635	0.6017	0.8197	0.8100
	0.4390	0.3835	0.6174	0.6519	0.6015	0.7520
	Mean	**0.5694**	**Mean**	**0.6156**	**Mean**	**0.8279**
	Std. Dev	**0.1453**	**Std. Dev.**	**0.1260**	**Std. Dev**	**0.1178**
1%	0.6293	0.7523	0.8180	0.8244	0.3980	1.1347
	0.7128	0.7052	0.5809	0.8494	0.8345	1.3966
	0.7881	0.6707	0.4353	0.6890	0.6387	1.2440
	0.7569	0.6281	0.7213	0.5515	0.7614	0.8347
	0.5740	0.6550	0.7737	0.6973	0.6384	0.6562
	Mean	**0.6872**	**Mean**	**0.6939**	**Mean**	**0.8573**
	Std. Dev	**0.0677**	**Std. Dev.**	**0.1347**	**Std. Dev.**	**0.3117**
2%	0.9144	1.0104	1.0687	0.8090	0.6903	0.6093
	1.2084	0.8462	1.0857	0.8827	0.5684	0.5485
	1.1179	0.9920	1.0708	1.0652	0.5439	0.6812
	1.1198	0.9365	1.1824	0.9568	0.6425	0.3174
	1.1104	0.9255	1.1961	0.8073	0.6719	0.6139
	Mean	**1.0181**	**Mean**	**1.0125**	**Mean**	**0.5887**
	Std. Dev.	**0.0135**	**Std. Dev.**	**0.0201**	**Std. Dev.**	**0.0119**

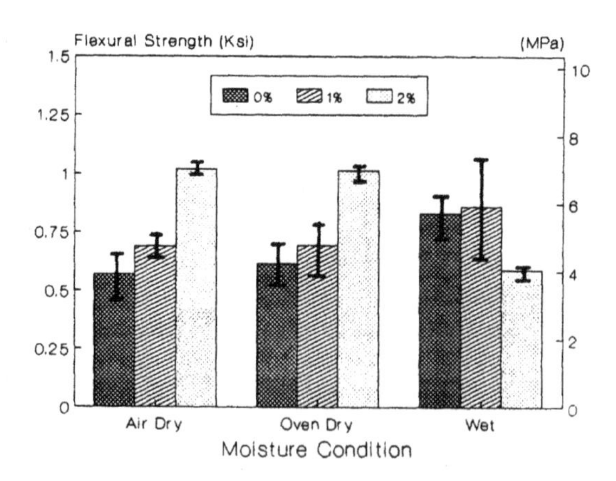

Figure 3. Effects of Fiber Mass Content and Moisture Condition on Flexural Strength.

Table 5. Fracture Toughness Test Results (lb-in.)

Fiber Mass Content	Moisture Condition					
	Air-Dried		Oven-Dried		Wet	
	Batch 1	Batch 2	Batch 1	Batch 2	Batch 1	Batch 2
0%	0.4600	0.3000	0.3170	0.6370	0.1950	0.5300
	0.3200	0.3700	0.2010	0.3450	0.3020	0.2600
	0.3100	0.3620	0.2390	0.4830	0.2100	0.3200
	0.3700	0.4100	0.4760	0.2930	0.3200	0.3010
	0.3500	0.2900	0.2630	0.3030	0.3240	0.3090
	Mean	**0.3542**	**Mean**	**0.3557**	**Mean**	**0.3071**
	Std. Dev.	**0.0528**	**Std. Dev.**	**0.1351**	**Std. Dev.**	**0.0910**
1%	0.5400	0.8400	0.4900	0.5030	0.5810	0.8360
	0.6300	0.6200	0.3670	0.4970	0.7100	0.8660
	0.7400	0.6730	0.6420	0.3470	0.7710	0.8660
	0.7300	0.5430	0.4390	0.2150	0.6020	0.7100
	0.4600	0.6230	0.5270	0.3860	0.7560	0.7710
	Mean	**0.6399**	**Mean**	**0.4413**	**Mean**	**0.7469**
	Std. Dev	**0.0124**	**Std. Dev.**	**0.0140**	**Std. Dev.**	**0.0099**
2%	0.8000	0.7100	0.6500	0.7900	2.7500	1.0000
	0.8700	0.6600	0.4300	0.6400	3.2100	3.1000
	0.8100	0.7000	0.8100	0.4300	1.9100	3.2800
	0.7700	0.6900	1.1500	0.8360	2.2100	2.5000
	0.8400	0.7200	0.5800	0.8100	3.3000	2.7500
	Mean	**0.7570**	**Mean**	**0.7126**	**Mean**	**2.6010**
	Std. Dev	**0.0050**	**Std. Dev.**	**0.0464**	**Std. Dev.**	**0.5353**

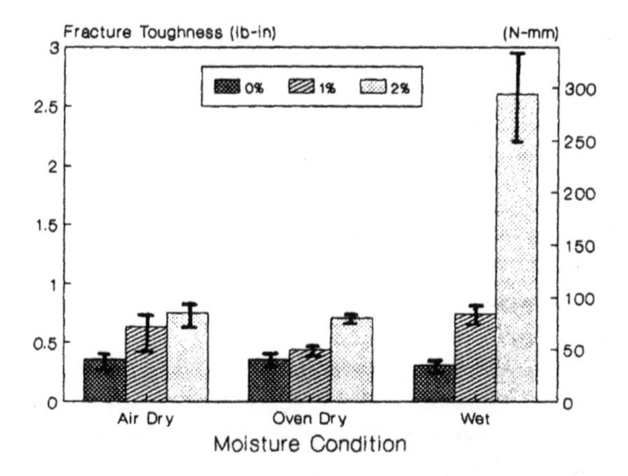

Figure 4. Effects of Fiber Mass Content and Moisture Condition on Fracture Toughness.

The following conclusions could be derived from the test data presented in Figures 3, 4 and Tables 4,5 regarding the effects of fiber mass fraction and the composite moisture condition on flexural strength:

(1) For mechanical fibers, flexural strength and toughness increase with increasing fiber mass content. The composite with 2% mechanical fibers had 79% increase in flexural strength and 114% increase in flexural toughness over the plain matrix.

(2) Oven drying of the composite caused only a slight increase in flexural strength in wood fiber composites, while plain mortar specimens showed approximately 8% increase in flexural strength over the air dried ones. Wetting, on the average, reduced the flexural strength by 42% for composites incorporating mechanical pulp at 2% mass content when compared with air dried ones.

(3) Oven drying of specimens reduced flexural toughness of composites by about 6% for mechanical pulp at 2% fiber mass content, while wetting of specimens produced a considerable increase in fracture toughness (by as much as 244% over air-dried specimens at 2% fiber mass content). Plain mortar specimens, however, did not show any significant variations caused by moisture effects.

6 Statistical Study of Moisture Effects

In order to study the effects of moisture condition a comprehensive statistical analysis was performed to analyze the effects of various factors involved, with due consideration given to distinguishing between the effects of different factors and the statistical variations in test results. The results of two-way factorial analysis of variance performed on flexural strength and toughness test results are presented in Table 6. The results suggest that the effects of fiber mass content were quite significant on flexural toughness, while moderately significant on flexural strength for mechanical fibers.

Table 6. Factorial Analysis of Flexural Strength and Fracture Toughness Results

Property	Moist Condn. (A)	Fiber Cont. (B)	Interaction (AB)
Flexural Strength	*	*	*
Fracture Toughness	**	**	**

The effects of moisture condition are moderately significant on flexural strength, and quite significant on fracture toughness. The interaction of moisture condition and mass content on flexural strength and fracture toughness is moderately significant, an indication that the presence of wood fibers plays an important role in deciding the trends in moisture effects on the flexural behavior of the cement-based materials considered in this investigation.

The decrease in flexural strength of wet specimens reinforced with wood fibers could be due to the reduction in the strength of wood fibers in wet condition, and also due to reduced fiber-to-matrix bond strength which results in the dominance of fiber pull-out over fiber fracture in the failure of composite in wet condition.

The increased flexural toughness values for the wet wood fiber reinforced cement composites could be mainly due to the dominance of fiber pull-out over fiber fracture, which increases the frictional energy dissipation associated with fiber pull-out.

7 References

American Fillers and Abrasives, Inc., Technical Sheet on Wood Fibers, Bangor, MI.

Coutts, R.S.P. (1988) Sticks and Stones. **Forest Products Newsletter**, (CSIRO Division of Chemical and Wood Technology (Australia), 2(1), 1-4.

Coutts, R. S. P. (1984) Autoclaved Beaten Wood Fiber Reinforced Cement Composites. **Composites,** 15(2), 139-143.

CSIRO, (1981) New-A Wood Fiber Cement Building Board. **CSIRO Industrial Research News (Australia)**, 146.

Japanese Concrete Institute (1984) **JCI Standards for Test Methods of Fiber Reinforced Concrete**. Report No. JCI-SF, 68.

Mindess, S. and Francis, J. Y. (1981) **Concrete,** Prentice-Hall, 422-425.

Morissey, F. E., Coutts, R. S. P. and Grossman, P. U. A. (1985) Bond between Cellulose Fibers and Cement. **International Journal of Cement Composites and Light Weight Concrete,** 7, 73-80.

Neville, A. M. (1963) **Properties of Concrete**. John Wiley & Sons, Inc., 409-415.

Vinson, D. K. and Daniel, J. I. (1989) Advances in the Development of Speciality Cellulose Fibers Specifically Designed for the Reinforcement of Cement Matrices: Early Strength of Composites Based Upon Ordinary Portland Cement. Presented at the **1989 Annual Convention, ACI,** Atlanta, Georgia.

8 Conversion Table

1 Ksi = 6.895 MPa
1 lb-in = 112.989 N-mm

6 ÉTUDE DES POSSIBILITES D'UTILISATION DU BAMBOU DANS LES BÉTONS DE FIBRES

(Study of the Possible Use of Bamboo in Fibre Concretes)

R. CABRILLAC, F. BUYLE-BODIN, R. DUVAL and W. LUHOWIAK
Paris University X, LEEE, IUT Génie Civil, Cergy, France

Résumé
Cette étude a pour objectif d'évaluer les possibilités d'utilisation du bambou qui présente des qualités remarquables pour un végétal dans la technique des bétons de fibres.
Notre travail a donc consisté en l'étude comparative des caractéristiques mécaniques et physiques des bétons de fibres de bambou et de verre en se référant à celles des mortiers qui en constituent la matrice.
Après quelques généralités sur la technique des bétons de fibres, présentation des caractéristiques du bambou et description du procédé d'obtention des fibres de bambou, nous présentons l'étude expérimentale.
Les résultats des essais réalisés, résistance en compression et en traction, masse volumique apparente et retrait de fabrication , tendent à montrer qu'il existe des teneurs en fibres et des quantités d'eau optimales pour les bétons de fibres de bambou qui permettent d'obtenir aux jeunes âges des caractéristiques similaires, voire supérieures, à celles obtenues grâce à l'adjonction de fibres de verre.
Mots clés : béton de fibres, bambou, fibres de bambou, fibres de verre, caractéristiques mécaniques et physiques.

1 Introduction

La technique du béton de fibres consiste à incorporer au mortier ou à la pâte de liant des fibres qui se répartissent omnidirectionnellement dans la matrice et s'opposent aux contraintes de traction et à la propagation des fissures. Malheureusement, les fibres employées jusqu'à présent et qui donnent des résultats satisfaisants sur le plan technique, fibres de verre, d'acier, de carbone ou synthétiques, confèrent au matériau un prix de revient élevé qui limite considérablement le développement et l'utilisation des bétons de fibres dans la construction. Par ailleurs, les tentatives réalisées à partir de fibres végétales se sont soldées par des échecs à cause de leurs caractéristiques mécaniques médiocres et de leur mauvaise tenue au vieillissement.

Cependant, le bambou, compte tenu de ses caractéristiques exceptionnelles pour un végétal, constitue peut être une solution économique au problème des bétons de fibres et les travaux que nous avons menés s'inscrivent dans le cadre de cette démarche.

Nous avons envisagé différentes compositions de matrices à base de ciment, quantité d'eau de gâchage variable, présence ou non de sable , et nous avons étudié l'influence de l'adjonction de fibres de verre ou de bambou, selon différentes teneurs, sur les caractéristiques mécaniques et physiques des matériaux réalisés.

Notre étude qui tend à accréditer la possibilité d'utiliser la fibre de bambou dans la technique des bétons de fibres, se limite cependant à l'évaluation des caractéristiques aux jeunes âges, et les résultats obtenus doivent être confirmés par des études de comportement au vieillissement.

2 Les bétons de fibres

L'ajout d'une certaine quantité de fibres à la matrice d'un béton ou d'un mortier n'introduit pas de nouvelles conditions sur la nature des composants de celle-ci. Par contre, l'optimisation et l'efficacité des fibres nécessitent une étude minutieuse de la composition du béton de fibres [1] [2].

Parmi les différents types de fibres utilisés [2] [3] [4], c'est la fibre de verre qui donne les résistances mécaniques intrinsèques les plus probantes, mais son coût élevé limite son utilisation à des éléments très particuliers ; par ailleurs un des problèmes essentiels réside en l'isolement de la fibre de verre vis à vis des réactions alcalines du liant hydraulique.

Les fibres métalliques quant à elles ont la particularité d'offrir la meilleure adhérence avec la matrice, le béton et l'acier ayant des coefficients de dilatation voisins. Leur coût est plus faible que celui des fibres de verre mais reste néanmoins élevé.

Les fibres de carbone ou synthétiques présentent des coût encore plus élevés que la fibre de verre et leur utilisation se limite quasiment au renforcement de matériaux constitués de liants d'origine synthétique peu utilisés dans la construction.

Quant aux fibres végétales [5] [6], elles présentent en général des caractéristiques mécaniques médiocres, insuffisantes pour améliorer les caractéristiques d'une matrice constituée de liant hydraulique ; elles ont surtout été essayées dans l'optique améliorer les performances thermiques des matériaux ou pour les alléger ou alors avec des matrices à base de terre. Par ailleurs, elles possèdent à l'état brut une très mauvaise tenue au vieillissement et, de plus, la technique de récupération des fibres s'avère parfois très compliquée à mettre en oeuvre. Cependant, parmi les végétaux susceptibles de fournir des fibres potentiellement utilisables dans la technique des bétons de fibres, le bambou n'a pas encore été envisagé.

3 Le bambou et les fibres de bambou

3-1 Le bambou

3-1-1 Généralités

On pourrait énumérer de multiples exemples concernant l'utilisation du bambou en tant que matériau de construction traditionnel depuis les âges les plus reculés qui reflètent bien les propriétés mécaniques et physiques exceptionnelles de ce végétal [7] [8] [9].

Mais, en dehors de ces qualités remarquables en tant que matériau [10], le bambou apparait également comme une ressource naturelle totalement renouvelable [7] [12]. En effet, il pousse plus vite que n'importe quel végétal comparable ; alors qu'une forêt croît en biomasse de 2 à 5 % par an, le développement comparé du bambou est de 10 à 30 %. De plus, la forte vitalité du bambou et son exceptionnelle capacité d'adaptation lui permettent de vivre dans presque tous les types de climats et de reliefs. Ainsi, avec plus de 10 000 espèces, le bambou est présent sur tout les continents à l'exception des pôles, ce qui en fait une des plantes les plus abondantes de la planète.

3-1-2 Structure du bambou

Le bambou possède un diamètre de quelques cm à quelques dizaines de cm (selon les variétés) ; sa structure est creuse et l'épaisseur de la couronne extérieure pleine est de l'ordre de plusieurs mm. Par ailleurs, à l'inverse des feuillus ou des résineux, le bambou a une croissance axiale uniquement, ce qui donne à la tige un diamètre constant sur la hauteur, une très bonne homogénéité radiale de la partie pleine et une très bonne coaxialité des fibres.

La partie pleine du bambou est constituée de deux types de fibres [8] : des fibres en paquets dont le diamètre peut atteindre le mm et des fibres de remplissage dont le diamètre est de l'ordre de quelques 1/100 de mm. De plus l'empilement des fibres est très dense par rapport aux végétaux classiques.

3-1-3 Caractéristiques mécaniques du bambou

Le tableau suivant récapitule les résistances à la rupture (en MPa) du bambou (variétés courantes) [8], du chêne et des résineux sous diverses sollicitations.

Tableau n°1

Résistance à la Rupture	Bambou	Chêne	Résineux
Compression axiale	62,1 à 86,3	27 à 52,3	25,6 à 49,5
Traction axiale	148,4 à 384,3	27 à 119,6	24 à 99,8
Flexion statique	76,3 à 276	30 à 53,8	24 à 55,6
Compression transversale	52,5 à 93	11,8 à 14,9	6,1 à 8,3

Comparativement aux végétaux classiques, le bambou présente donc des résistances de l'ordre de 2 fois supérieures en compression axiale, 3 à 4 fois supérieures en traction axiale, 5 fois supérieures en flexion statique et 6 à 11 fois supérieures en compression transversale. Il est bien évident que la supériorité des résistances mécaniques du bambou, du fait de sa structure, se retrouve au niveau des fibres.

3-2 Procédé d'obtention des fibres de bambou

Compte tenu de la structure fibreuse du bambou et en particulier de la coaxialité des fibres, il suffit de procéder par fendages successifs (fig 1) pour obtenir des lamelles de plus en plus petites. Les lamelles ainsi obtenues constituées de plusieurs fibres coaxiales et de mêmes caractéristiques peuvent elles-même être considérées comme des fibres au sens donné à ce terme pour les bétons de fibres.

Par les procédés manuels ou en utilisant des machines à bois classiques, il est difficile d'obtenir des lamelles de section inférieure à quelques mm2. Mais à ce stade un écrasement par martelage des lamelles permet de défibrer plus finement le végétal et d'obtenir des fibres de diamètre circonscrit compris entre 0,3 et 1mm.

Dans le cadre de cette étude nous avons procédé entièrement manuellement mais le procédé pourrait facilement se mécaniser en utilisant un appareillage de conception assez simple et peu couteux.

La résistance en traction des fibres de bambou obtenues varie de 250 à 400 MPa ; elle est de l'ordre de 5 à 8 fois supérieures à celles des fibres de bois classiques (50 MPa) et comparable à celle des fibres métalliques usuelles (370 MPa).

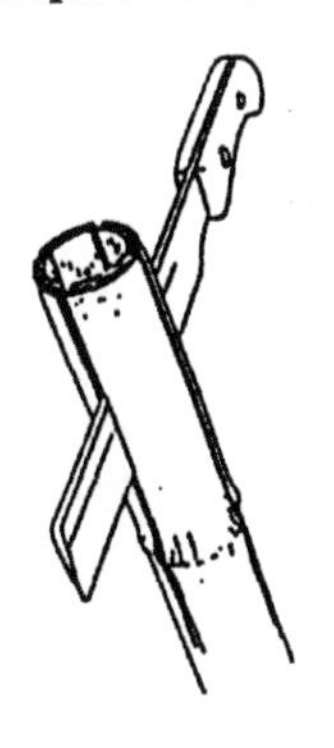
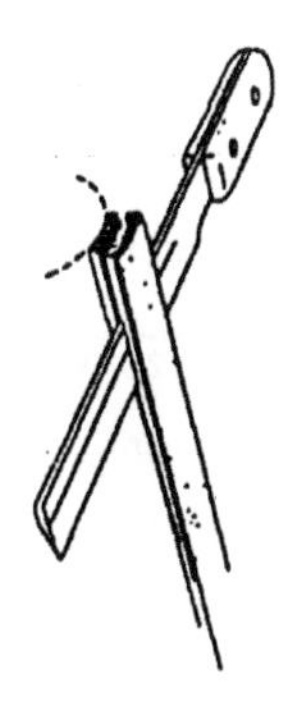

Fig.1 [8]

4 Présentation de l'étude expérimentale

Nous avons réalisé plusieurs matrices de référence dans lesquelles nous avons fait varier la quantité d'eau (E) par rapport au poids de ciment (C) de 35 à 50 % ainsi que la présence ou non de sable (S) à raison de 50 % par rapport au poids de ciment. Le ciment utilisé est du CPA 55 et le sable et un sable courant tamisé de granulométrie 0 / 1,25.

Correspondant à chacune de ces matrices de base, nous avons fabriqué des bétons de fibres en introduisant des fibres (F) de bambou (B) ou de verre (V) dont la proportion varie de 10 à 40 % par rapport au volume de ciment.

Les caractéristiques des fibres, longueur et diamètre, sont pour les fibres de verre 13 mm et 0,1 mm et pour les fibres de bambou 10 mm et 0,3 à 1 mm.

La mise en oeuvre s'est déroulé de la façon suivante : malaxage à sec : 2mn, malaxage avec eau : 3 mn, mise en place sans vibration ni lissage, conservation normale à 20 °C et 50 % d'humidité relative.

La composition de la référenciation des mélanges étudiés apparait dans le tableau 2 pour les matrices de base et dans respectivement les tableaux 3 et 4 pour les bétons de fibres de bambou et de verre.

Tableau n° 2

Mélanges	1	2	3	4	5
E/C	35 %	40 %	45 %	50 %	45 %
S/C	0 %	0 %	0 %	0 %	50 %

Tableau n° 3

B/L	10 %	20 %	30 %	40 %
1	1-10-B	1-20-B	1-30-B	1-40-B
2	2-10-B	2-20-B	2-30-B	2-40-B
3	3-10-B	3-20-B	3-30-B	3-40-B
4	4-10-B	4-20-B	4-30-B	4-40-B
5	5-10-B	5-20-B	5-30-B	5-40-B

Tableau n° 4

V/L	10 %	20 %	30 %	40 %
2	2-10-V	2-20-V	2-30-V	2-40-V
3	3-10-V	3-20-V	3-30-V	3-40-V
4	4-10-V	4-20-V	4-30-V	4-40-V

5 Résultats des essais et commentaires

5-1 Résistance en compression

Les résistances en compression à 7 et 14 jours des mélanges de base (fig 2 et 3, F/C=0%) diminuent de façon importante quand la quantité d'eau de gâchage augmente.

Pour les bétons de fibres de bambou (fig 2 et 3) on constate que pour des quantités d'eau de 35 % à 40 % l'introduction de fibres fait décroître les résistances en compression. Pour des quantités d'eau de 45 à 50 % l'introduction de 10 à 20 % de fibres augmente les résistances en compression, mais au delà de 20 % de fibres, ces dernières décroissent rapidement. Pour 45 % d'eau l'introduction de sable améliore les résistances en compression mais leur évolution en fonction de la teneur en fibres reste inchangée.

Pour les bétons de fibres de verre (fig 4 et 5) on enregistre un phénomène identique à celui des bétons de fibres bambou à 7 j, mais à 14 j, quelles que soit la quantité d'eau, l'introduction de fibres de verre fait décroître les résistances en compression.

Si l'on compare l'influence du type de fibres (fig 4 et 5) sur un mélange à 40 % d'eau, on observe une décroissance des résistances en compression aussi bien avec les fibres de verre qu'avec les fibres de bambou ; mais si cette décroissance est similaire à 7 j, à 14 j, elle est beaucoup plus importante avec les fibres de verre. Pour un mélange à 45 % d'eau, à 7 j l'introduction de fibres améliore la résistance en compression jusqu'à une teneur en fibres de 40 % pour le verre et de 30 % pour le bambou. Dans les deux cas la résistance maximale est obtenue avec 20 % de fibres. A 14 j, l'introduction de fibres de bambou jusqu'à 20 % améliore la résistance en compression alors que cette dernière décroît quelle que soit la teneur en fibres de verre. On obtient cependant des résistances analogues avec 40 % de fibres pour le bambou et le verre. L'introduction de 50 % de sable dans le mélange ne perturbe pas ces phénomènes.

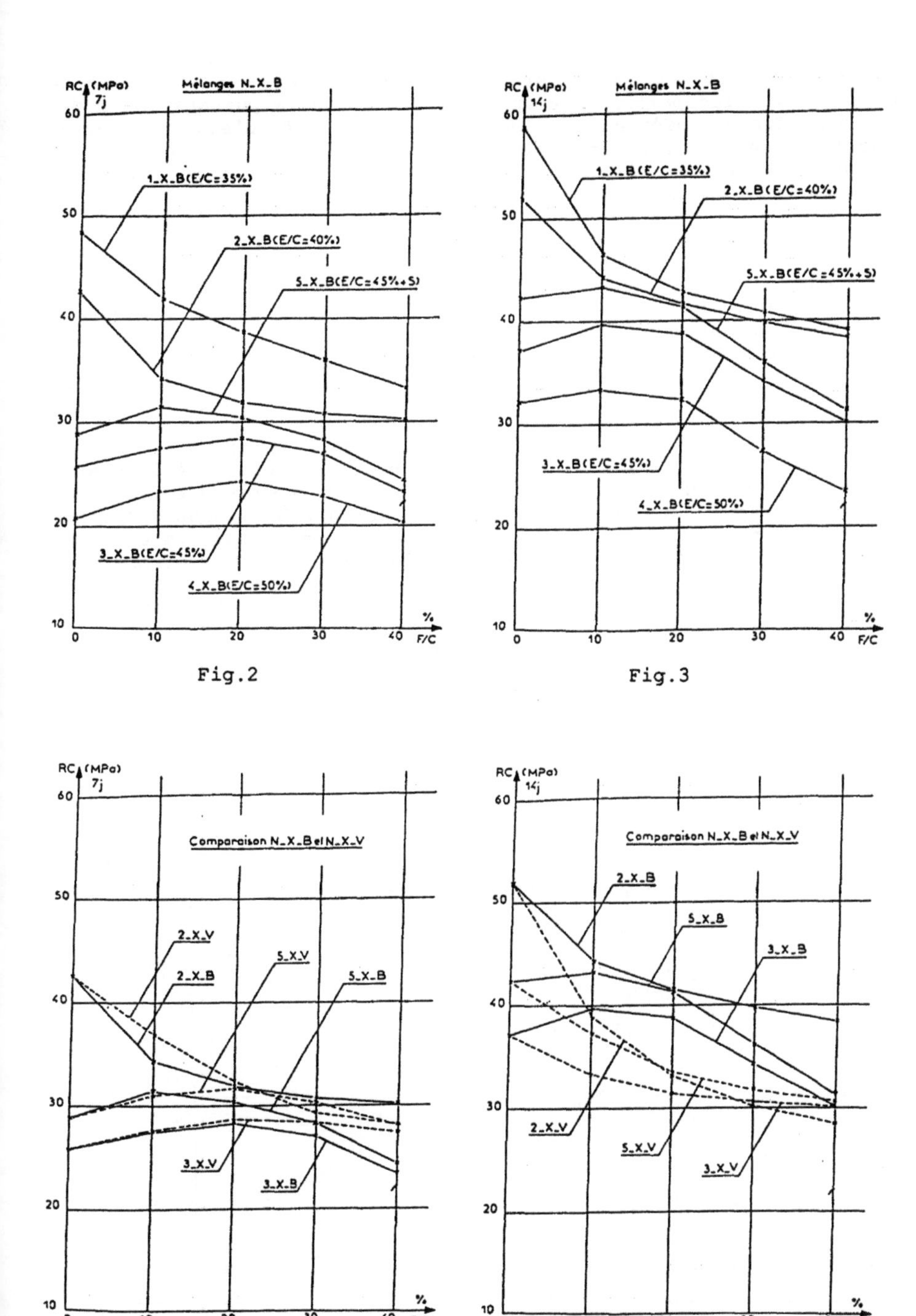

Fig.2

Fig.3

Fig.4

Fig.5

Finalement, si l'on se réfère aux résultats à 14 j, la fibre de bambou parait plus performante que la fibre de verre puisque, pour des teneurs en fibres inférieures à 20 % et des quantités d'eau supérieures ou égales à 45 %, elle améliore la résistance en compression de la matrice de base. De plus, pour des quantités d'eau inférieures la décroissance est moins importante qu'avec la fibre de verre

5-2 Résistance en traction

La résistance en traction à 14 j des mélanges de base en fonction de la quantité d'eau de gâchage (fig 6, F/C=0%) a une évolution analogue à celle des résistances en compression, à savoir une diminution importante lorsque la quantité d'eau augmente.

En ce qui concerne les bétons de fibres de bambou (fig 6), on constate avec des quantités d'eau de 35 % et 40 %, comme pour les résistances en compression, une décroissance avec l'introduction de fibres, mais qui semble se stabiliser aux alentours de 20 à 30 % de fibres. Pour 45 % d'eau, avec ou sans sable, on enregistre à l'inverse une amélioration des résistances en traction par rapport a celle de la matrice de base, quelle que soit la teneur en fibres dont l'optimum se situe aux alentours de 30 %. Pour 50 % d'eau on enregistre une très légère amélioration avec 10 % de fibres et ensuite une décroissance jusqu'à 40 %, décroissance somme toute très progressive.

Pour les bétons de fibre de verre (fig 7), il y a amélioration des résistances en traction par rapport à celles des matrices de base, quelle que soit la teneur en fibres pour 40 % d'eau et 45 % d'eau, et la teneur en fibre optimale semble se situer aux alentours de 30 %.

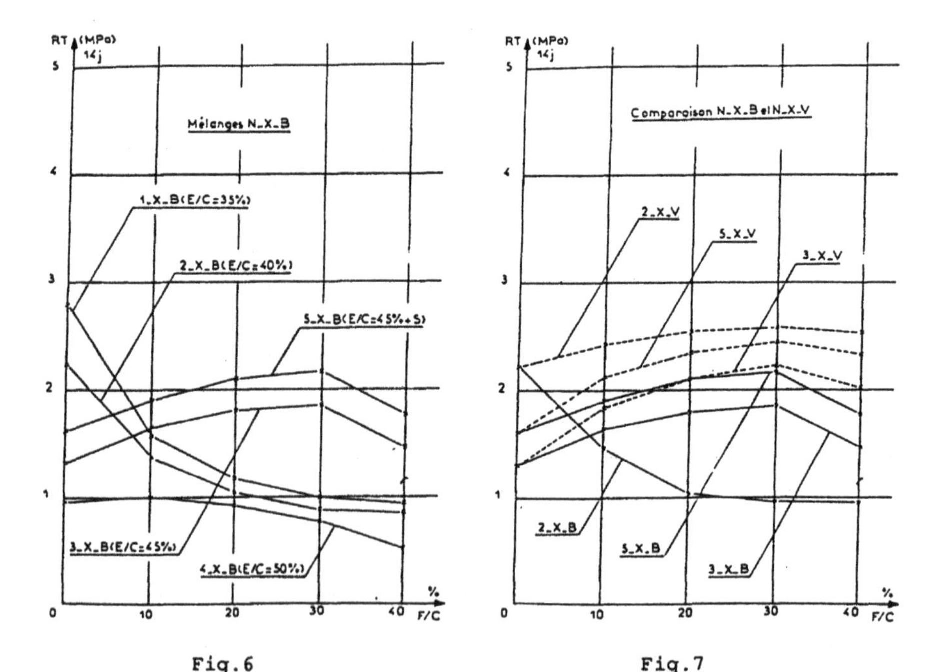

Fig.6

Fig.7

Si l'on compare l'influence du type de fibres sur la résistance en traction des mélanges (fig 7), la fibre de verre apparait plus performante que la fibre de bambou. Pour les mélanges à 40 % d'eau l'introduction de fibre de verre améliore la résistance en traction de la matrice de base alors que la fibre de bambou provoque l'effet inverse. Pour les mélanges à 45 %, avec ou sans sable, l'introduction de fibres , qu'il s'agisse de bambou ou de verre, améliore la résistance en traction de la matrice de base, mais pour des teneurs en fibres identiques les résultats obtenus sont légèrement supérieurs avec la fibre de verre.

5-3 Masse volumique apparente

L'évolution des masses volumiques apparentes à 14 j pour les mélanges de base et les bétons de fibres de bambou en fonction de la quantité d'eau de gâchage est transcrite sur la figure 8. On constate une diminution de la masse volumique apparente lorsque la quantité d'eau augmente. De plus, pour une quantité d'eau fixée, la masse volumique apparente est d'autant plus faible que la teneur en fibres de bambou augmente mais cette diminution reste quand même relativement faible.

Si l'on compare l'influence en type de fibres sur la masse volumique apparente à 14 j des différents mélanges (fig 9), on constate que, quelle que soit la teneur en fibres, les bétons de fibres de verre sont légèrement moins lourds que les bétons de fibres de bambou, ceci pour des quantités d'eau de 40 % à 45 % avec ou sans sable. On remarque par ailleurs que l'adjonction de sable pour les bétons de fibres à 45 % d'eau augmente les masses volumiques apparentes.

On obtient aussi une fourchette de masse volumiques allant de 2,1 t/m3 pour la matrice de base la plus lourde à environ 1,75 t/m3 pour le béton de fibres de verre le plus léger.

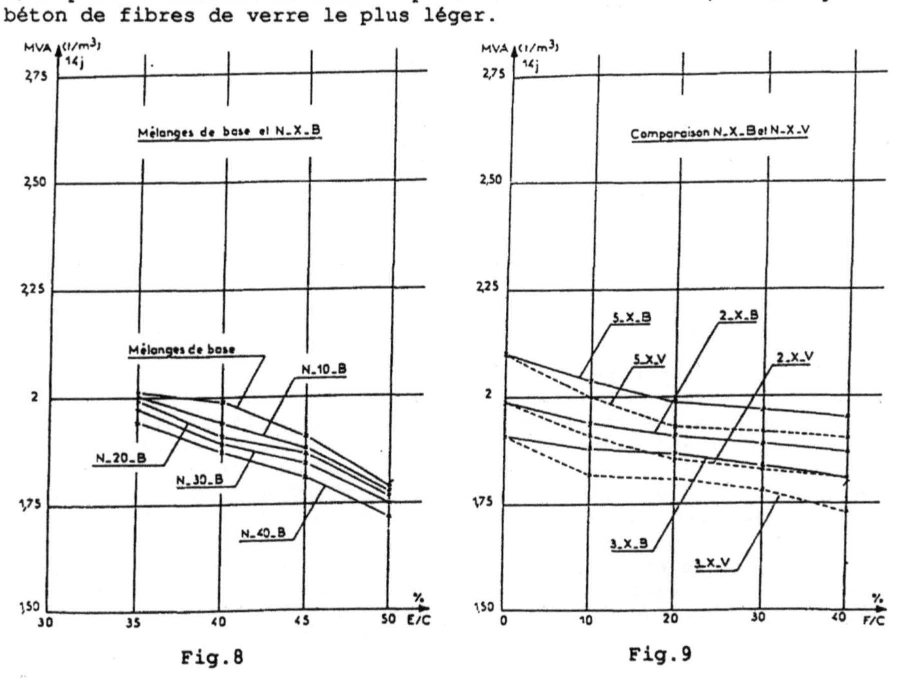

Fig.8 Fig.9

5-4 Retrait de fabrication

Si l'on étudie l'influence de la teneur en fibres de verre sur le retrait d'un mélange à 45 % d'eau (fig 10), on s'aperçoit que jusqu'aux alentours de 14 j, les retraits, quelle que soit la teneur en fibres, restent inférieurs à celui de la matrice de base. Ce phénomène s'inverse progressivement entre 14 j et 8 semaines, date à laquelle les retraits augmentent avec la teneur en fibres mais restent proches du retrait de la matrice de base.

Si l'on s'intéresse aux mêmes comparaisons pour les bétons de fibres de bambou (fig 11), les retraits à 7 j sont très faibles par rapport à celui de la matrice de base et ils restent inférieurs jusqu'à environ 5 semaines. De façon générale les retraits à 8 semaines des bétons de fibres de bambou, comme ceux des bétons de fibres de verre, sont du même ordre de grandeur que celui de la matrice de base, environ 2 mm/m.

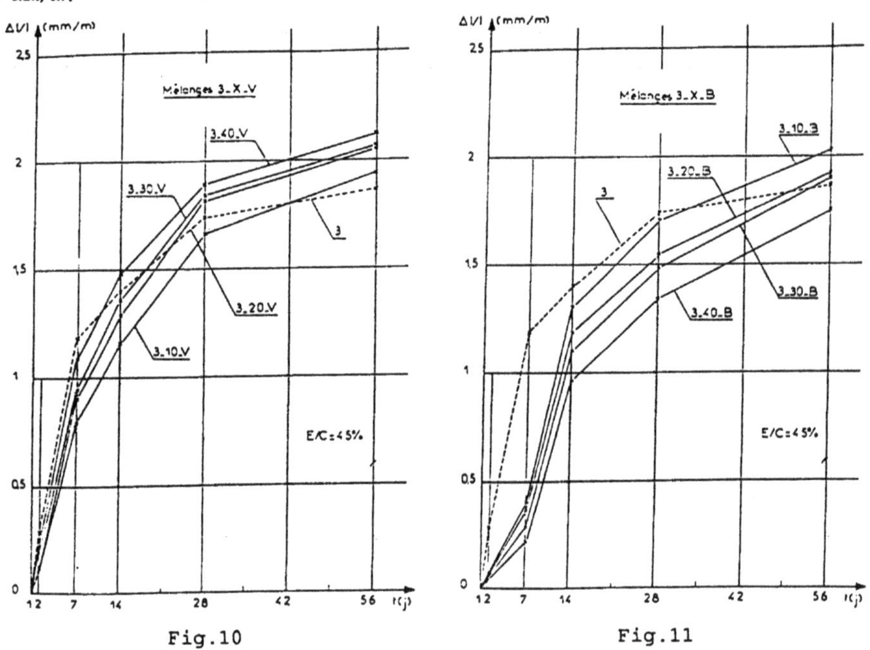

Fig.10 Fig.11

6 Conclusion et perspectives

En ce qui concerne les résistances en compression à 14 j, les fibres de bambou donnent de meilleurs résultats que les fibres de verre. Pour des quantités d'eau inférieures ou égales à 40 %, la diminution de résistance est plus faible avec les fibres de bambou qu'avec les fibres de verre ; pour des quantités d'eau supérieures à 45 %, l'introduction de fibres de bambou améliore la résistance de la matrice de base jusqu'à des teneurs en fibres de l'ordre de 20 à 30 %, ce qui n'est pas le cas avec des fibres de verre. La présence de sable améliore légèrement les résistances mais ne modifie pas ce phénomène.

Pour ce qui est des résistances en traction à 14 j, ce sont les fibres de verre qui donnent de meilleurs résultats. Pour des quantités d'eau inférieures ou égales à 40 %, seules les fibre de verre améliorent légèrement la résistance en traction de la matrice de base ; pour les mélanges à quantité d'eau supérieure ou égale à 45%, avec ou sans sable, les deux types de fibres améliorent la résistance en traction de la matrice de base mais les fibres de verre conduisent à teneur égale à des résistances légèrement supérieures.

Pour les résistances en compression à 7 j, on constate un phénomène analogue à celui des résistances en compression à 14 j pour les mélanges à faible quantité d'eau et un phénomène analogue à celui des résistances en traction à 14 j pour les mélanges à forte quantité d'eau.

Ces constatations signifient que l'influence des fibres sur les résistances mécaniques dépend de la résistance de la matrice de base, elle même très dépendante de la quantité d'eau.

A propos des masses volumiques apparentes, l'introduction de fibres quelle qu'en soit la teneur, allège sensiblement la matrice de base, les fibres de verre conduisant à des masses volumiques apparentes légèrement plus faibles que les fibres de bambou.

Enfin, pour ce qui est du retrait de fabrication, les deux types de fibres, quelle qu'en soit la teneur, conduisent à 8 semaines à des valeurs analogues à celle de la matrice de base ; par contre , aux jeunes âges, le retrait diminue avec l'adjonction de fibres mais le phénomène est plus marqué avec les fibres de bambou.

Finalement cette étude montre qu'aux jeunes âges les fibres de bambou sont à même, comparativement aux fibres de verre, de conférer aux bétons de fibres des caractéristiques mécaniques et physiques analogues sinon meilleures. Ceci laisse présager de l'utilisation possible du bambou dans la technique des bétons de fibres, mais cette possibilité doit être confirmée par des études de vieillissement et des expérimentations dans ce sens sont actuellement en cours.

7 Bibliographie

[1] Fissuration du béton : du matériau à la structure. Application de la mécanique linéaire de la rupture. P.ROSSI. **Rapport de recherche du LPC**, n°150, Juin 1988.

[2] Mortiers et bétons de fibre en France. Journée d'études du 19 Mars 1987. PARIS. **Centre Scientifique et Technique du Bâtiment. Ministère de la Recherche et de l'Enseignement Supérieur, Plan Construction.**

[3] **Les bétons légers d'aujourd'hui**. P.CORMON. Eyrolles 1973.

[4] L'intérêt du béton armé de fibres de verre. **Cahiers Techniques du Bâtiment**, n°69, Février 1985.

[5] Comportement des bétons armés de fibre métalliques. Th. VALADE. **Mémoire, ENS CACHAN**, Décembre 1987.

[6] Des fibres cellulosiques pour renforcer les ciments. M.KHENFER, P.MORLIER, J.POUTIS. **Recherche actualités**, n°179, Mai 1989.

[7] **The book of bambou**. D.FARRELY. Sierra Club Books. San Francisco, 1984.

[8] **Bamboo**. K.DUNKELBERG. Institut for leichte Flachentragwerke. Stuttgart, 1985.

[9] **Revue Bambou**. Société Européenne du Bambou. n°1 et 4 1988.

[10] Application of bamboo as a low-cost energy material in civil engineering. G.KHOSROW. **Symposium CIB.RILEM. MEXICO**. Novembre 1989.

[11] Improving the performance of low cost bambou houses. G.C.MATHUR.**Symposium CIB.RILEM. MEXICO**. Novembre 1989.

7 FIRE RESISTANT MATERIALS MADE WITH VEGETABLE PLANTS AND FIBRES AND INORGANIC PARTICLES

T. SUZUKI
Institute of Technology, Shumizu Corporation, Tokyo, Japan
T. YAMAMOTO
Tohoku Electricity Technology Research Development Center, Tokyo, Japan

Abstract
In this paper, the following points are summarized on the utilization of plants and fibres in fire resistant materials.
(1) Production methods by mixing vegetable plants, fibres, fly ash, cement etc are described. As an example, a description is given of blocks made from rice husk ash with compressive strength of 40 to 50 kgf/cm^2.
(2) Methods of making incombustible light-weight building materials from plants that are chipped, mixed with water-soluble ceramics and formed, are outlined.
(3) Methods of making incombustible light-weight materials by preparing reinforcing bamboo coated with water-soluble ceramics and pouring them into a mortar of plant and fibre mixture, are explained.
Keywords: Incombustible materials, Plant, Fly ash, Water-soluble ceramics, Rice husk, Fibre, Bamboo.

1 Introduction

Environmental destruction such as pollution of air and water has been occurring in some regions by rapid development and production of materials like iron, steel, glass, cement and aluminum that use limited mineral resources. On the other hand, plants and fibres are annually reproducible clean resources.

In very many areas, annually harvested plants and their fibres are utilized for human life. In the regions where rice husks, bamboos, and ditch reeds are produced, people utilize them as house building materials and living tools. Until several decades ago, wooden straw-thatched houses were widely constructed also in Japan, but by the desire of people to get rid of the influence of weather changes, substitute materials now in use have been developed and utilization of these plants and their fibres has been diminishing.

Presently these botanical materials are not much used, but they are being re-evaluated as clean reproducible resources, along with the trend towards more effective utilization of limited mineral resources.

People think that plants and their fibres are weak against fire and corrosion. For this reason, incombustibility has been required for house construction materials that utilize plants and their fibres.

For making incombustible materials, use of incombustible materials or coating combustible materials with incombustible substances can be thought of. For such a coating, the substance needs to be mass produced at low cost. Under the recent energy supply situation, consumption of coal and production of fly ash and its cinders were increasing. In our country, we call fly ash the coal ash captured by electric dust collectors at the midway of smoke dust. Particles meeting the Japan Industrial Standard (JIS) are used as fly ash cement mixed with fly ash and cement.

We studied the creation of incombustible materials by mixing substandard fly ash with plants and their fibres.

We also studied the production of incombustible building light-weight building materials by combined use of water-soluble ceramics. The results of these studies are reported here.

2 Experimental plan

2.1 Types of experiment

Experimental production of building materials utilizing plants and their fibres was conducted in the following three series classified by moulded products.

(1) Series 1
In a special mixer, knead and mix at a certain ratio, plants, and their fibres (rice husk, cotton refuses, cotton etc.) and fly ash cement. Collect samples from the mortar, make test pieces and carry out strength test. Also form a chamber block by block moulder, and conduct strength test after specified curing.

(2) Series 2
Knead and mix rice husk or bamboo chops with gel build up agent, cement and water, adding and mixing water-soluble ceramics afterwards, make into moulded forms. Strength tests were made on these moulded forms.

(3) Series 3
Place water-soluble ceramics coated bamboo square bars (5mm x 5mm) at intervals of 5cm. Make moulded forms by pouring in mixed mortar of plants, their fibres, ceramics and water. This is the bamboo reinforced mortar light-weight moulded product.

2.2 Materials used
The kinds and properties used in Series 1 to 3 are shown next:

(1) Plants and their fibres
Rice husk from Tohoku (North-East) region not crushed. Cotton waste before being processed to thread. Cotton-remaining threads after making articles of clothing. Bamboo-Moso-chiku (Phyllostachys pubescens - a species of thick stemmed bamboo), worked to square bar (5mm x 5mm) and chips thereof. Pulp-wood chip for paper mill.

(2) Fly ash
Those with loss on ignition exceeding 5% the upper limit specified by JIS. Analysis of the major consitituents is illustrated in Table 1.

Table 1. Chemical analysis of sample of fly ash (%)

sample	loss of ignition	SiO_2	Al_2O_3	CaO	MgO
A	3.8	56.3	27.1	4.9	1.3
B	15.5	17.0	12.4	34.3	0.4

(3) Gel build-up agent
Polyvinyl-alcohol or white clay used for earthenware.

(4) Water soluble inorganic polymer of boric acid fluoride.
Water dissolved clay mixture mainly composed of fluorine and boric acid compound.

(5) Cement
Early-strength portland cement

2.3 Mixing Method
The mixer is a forced-action pan-type mixer with rotating drum and inner blades. Throw in materials other than water and exercise idle kneading for 1 minute. After pouring water in, knead and mix for 4 minutes and then immediately pour the substance into mould and form it.

2.4 Forming method and curing
Forming of test pieces was done with a moulding frame (4cm x 4cm x 16cm) for strength test of mortar. 24 hours later, the test pieces were detached from the mould and cured in a room with temperature at $20 \pm 1^{o}c$ and humidity at 60% up to the test age.
30cm x 30cm x 1.5cm size board shape test pieces was

formed by pressing with 10kg/cm^2 force after poured into mould.

2.5 Test method
Mortar strength test pieces were subjected to bending and compressive test at specified age. Formed test board was cut out from the above test piece and subjected to test at specified age.

3 Results of experiments

3.1 Series 1
We poured into mixer and mixed various plants, their fibres, fly ash, cement and water duly measured, then collected mortar test pieces. The results of tests on these pieces at 28 days are shown in Table 2.

Table 2. Test results of mortar used vegitable plants

NO	name	unit weight	Bending (kgf strength /cm²)	compressive(kgf strength /cm²)
1	Fly-ash	1.23	37.9	185
2	cotton wastes	1.21	55.3	134
3	cotton	1.09	48.7	95
4	rice-husk	1.05	36.6	92
5	crush husk	1.14	42.5	109
6	pulp	1.16	59.7	154

Cement:plants:fly-ash=2 : 3 : 5. W/C= 60 %

Fibre based test pieces had a low specific gravity but good flexural strength. As for rice husk, the pulverized type was stronger than the non-pulverized type.

Plants and fibres become incombustible when their surface is covered by fly ash and cement. The mixture can be granulated in sand shape by extending the kneading and mixing time after adding the materials.

Results of experiments, changing the volume of fly ash and cement separately, are shown in Figures 1 to 3. Figure 1 shows specific gravity of 1.2 to 1.3 and compressive strength of 20 to 50 kgf/cm^2 when the volume of cement is varied from 300 to 600 kgf/cm^3 fixing the volume of rice husk and water.

Immediately after granulation by churning in mixer, the mixture was thrown into the chamber block former and made into a block by compression forming. At 28 days age, a specific gravity of 1.3 and compressive strength of 20 to 50 kgf/cm^2 were obtained. Figures 4 and 5 show the block forming state.

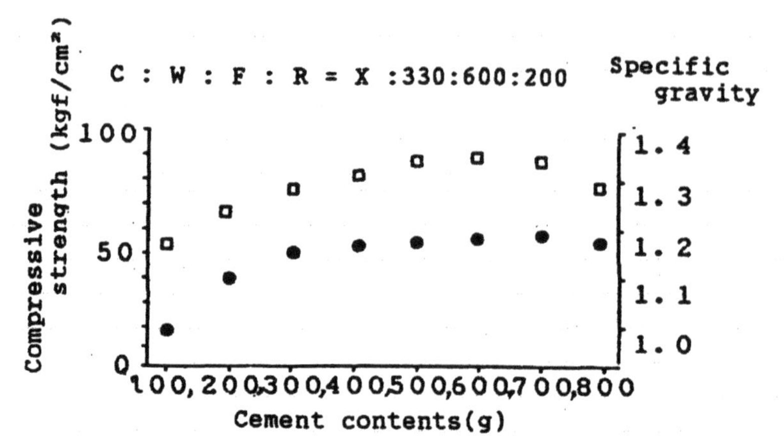

Fig.1. Efect of cement contents

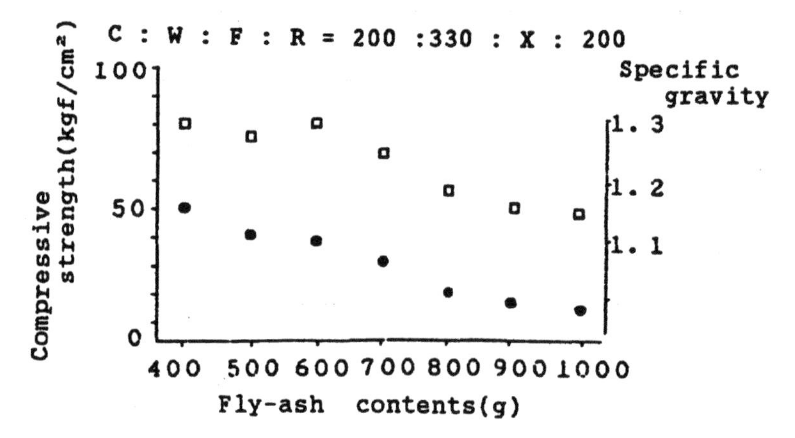

Fig.2. Efect of fly-ash contents

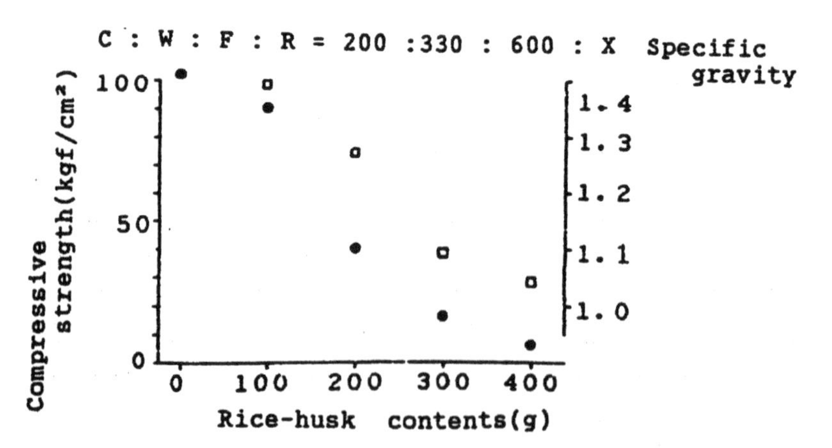

Fig.3. Efect of rice-husk contents

Measurement of thermal conductivity indicated 0.6kcal/mh°c. The results of the experiments varying mixing volume of fibres are shown in Figure 6.

Fig.4. Block used rice husk

Fig.5.Block forming state.

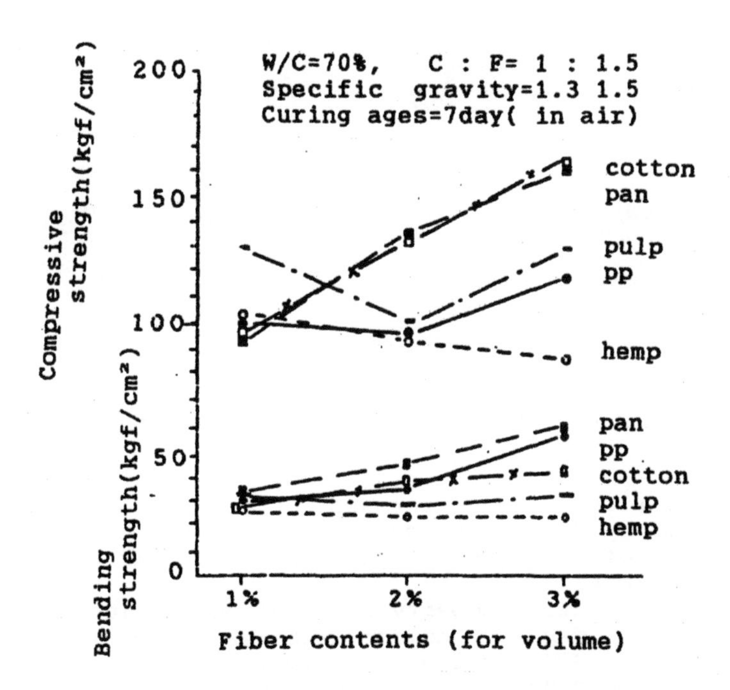

Fig.6.Test results of mortar used plants and other fibers

3.2 Series 2

The test pieces were made from rice husk, bamboo chips or pulverized chops, mixed with gel build-up agent, cement and water-soluble ceramics. Compounding of mortar and results of bending strength tests on cut-out boards are shown in Table 3.

Figures 7 and 8 show 30cm square board and its non-ignited state by nearing flame. Cement and ceramics covering the surface of the plants and their fibres contribute to incombustibility.

Table 3. Mix Proportion of Mortar (weight ratio)

NO	w/c (%)	cement	plants	PVA	PFS	unit weight	bending strength (kgf cm²)
1	35	1	0.1 *1	0.02	0.02	1.42	32
2	35	1	0.3 *2	0.02	0.02	1.57	41
3	40	1	0.5 *3	0.05	0.5	1.49	38

PVA:Increse viscosity material. *1:Risu husk
PFS:Liquids ceramics. *2:Bamboo chips
Cement:Early-strength-cement. *3:Crushed bamboo chips

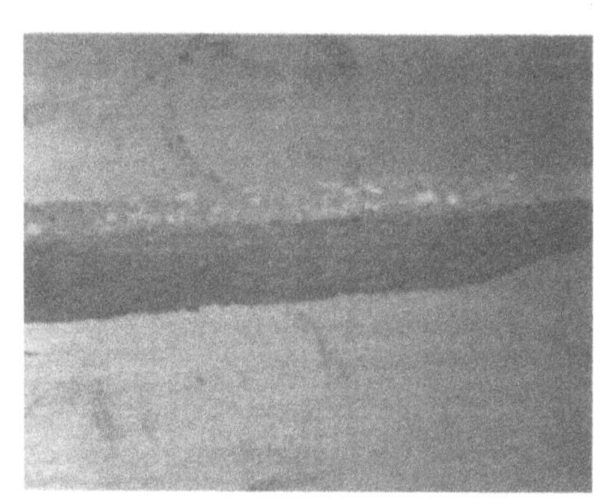

Fig.7. board style

Fig.8.Flame test

3.3 Series 3

Rice husk mixture with mortar was poured into mould frame formed by Moso-chiku square bars placed crosswise at intervals of 5cm. The proportions of mortar are shown in Table 4.

This mortar has viscosity by using gel build-up agent, and high density moulded articles can be made by press forming. 300mm x 100mm x 15mm test pieces cut out from formed board had a bending strength of 47kgf/cm^2.

It was not ignited by a close flame, proving that incombustibility was improved. Figures 9 and 10 show the states of pouring and flame experiments.

Table 4. Mix Proportion of Mortar (weight ratio)

w/c	cement	fly-ash	rice-husk	PVA	PFS
90	1	1.5	0.5	0.004	0.05

PVA:Increse-viscosity-material.
PFS:Liquids ceramics.
Cement:Early-strength-cement.

Fig.9. Board make state

Fig.10.Flame test piece

4 Summary

Results of experiments in the three series can be summarized as follows;

(1) Various plants and their fibres can be granulated in sand shape with incombustible by kneading and mixing with cement, fly ash and water at a certain compounding ratio, using a special mixer. Also, incombustibility moulded articles can be made by mixing alone instead of granulation. Though varying with the mix proportions, rice husk mortar showed a bending strength of $36kgf/cm^2$ and compressive strength of $92kgf/cm^2$.
(2) After mixing rice husk, fly ash, cement and water in a special mixer, the mixture can be thrown into a chamber block forming machine for compression moulding. After curing, the block had a specific gravity of 1.3 and compressive strength of $20kgf/cm^2$ to $50kgf/cm^2$.
Thermal conductivity was $0.6kcal/mh^{o}c$, lower than mortar, of 1.1, and equal to that of plaster.
(3) Boards formed by kneading and mixing of rice husk, bamboo chip or pulverized chip with cement, gel build-up agent and water-soluble ceramics can have a bending strength of 32 to $41kgf/cm^2$. Also, it was incombustible, not being ignited by a nearby flame.

(4) In a moulding frame, made by placing 5mm square Moso-chiku crosswise at intervals of 5cm, poured rice husk mixture mortar could be formed into panels. Bending strength upon curing was 47kgf/cm^2. Also it was incombustible, not being ignited by a near by flame.
(5) Incombustible cement paste can be made by mixing water-soluble ceramics with cement and water.

Incombustible material can be produced by spraying this paste on plants and their fibres.

5 Post-script

Incombustible granulated sand can be produced from plants and their fibres. It is inflammable when dried, mixed and churned with fly ash (the residue of coal burning) and cement.

The newly developed water-soluble ceramics is deemed to be useful to make materials incombustible.

Finally, I should like to express my sincere thanks to Mr Hiroshi Kokuda, president of Comix Co., Ltd., who collaborated with me on these new materials.

6 References

Dommalapati Krishnamurthy. (1986) Use of bamboo as a substitute for steel in conventional reinforced concrete in Use of vegetable plants and fibres as building materials, Joint symposium RILEM/CIB/NCCL Baghdad, pp. c71-c78.

Asko Sarja. (1986) Structural concrete with wooden or other vegetable fibres, in Use of vegetable plants and fibres as building materials, Joint symposium RILEM/CIB/NCCL Baghdad, pp. c117-c126. B.A.Sabrah, E.El-Didamony, M.M.El-Rabiehi. (1989) ceramic studies of the clay/rice husk/slag system and its suitability for brick making , Interbrick, vol 5, pp. 24-27.

R.J.S.Spence, D.J.Cook. Building materials in Developing Countries 1983. John Wiley & Son Ltd.

8 COMPOSITE MATERIALS FROM VEGETABLE FIBRES AS AGGLOMERATED IRREGULAR MICRO-REINFORCEMENT AND PORTLAND CEMENT USED IN PIECES FOR LOW-COST HOUSING

A.A. HESS and M.L. BUTTICE
Faculty of Engineering, UNNE, Chaco, Argentina

Abstract
In this paper, the physical-mechanical behaviour of micro-concrete containing vegetable fibres as irregular reinforcement is studied. The fibres studied were megass (the husk from sacharus-officinarum). Megass is the product from the first part of the process for obtaining sugar from this plant, this husk (megass) is usually used for making paper, for cattle feed, or as fuel.

In the first stage, available methods and mechanism for the previous treatment of fibres and removal of the sacharose and
Lignine, as well as the incorporation of inhibitor additives and the delaying effect on setting for these composites was studied.

Multiple use prism samples were made up in order to obtain the most information with the fewest specimens; these were submitted to the following physical-mechanical tests; flexure, compression, indirect tension. It is possible to develop composite materials reinforced with vegetable fibres, for use as partitions and non-structural board partitions, with appropriate allowance for traditional finishing, with the least thermal conductivity due to the presence of organic fibres. This report is part of a broader study of other vegetable fibres in North East Argentina such as "caraguate", "espartillo" and "tiger tail".
Keywords: Vegetable fibres, reinforced mortars, non-metallic reinforcement, irregular reinforcement megass.

1 Introduction

Sugar cane is a plant cultivated in most of the South American subtropical zone, and specifically in our country, in the north west and littoral zones. It is the major source of sugar. Megass is the result of the first part of the process for extracting sugar. It is generally used to manufacture paper, as cattle feed and as fuel.

2 Aim

The first objective of the present work was to evaluate experimentally the behaviour of megass fibre incorporated in micro-concrete and portland cement mortars; and to study the feasibility of its use in the construction of low cost housing, as a substitute for or alternative to so called conventional materials.

3 Background

Data on different vegetable fibres were obtained and it was found that the most suitable for our objectives were megass (of sugar cane), 'espartillo', sisal-grass, card, 'caraguata' and 'cola de tigre', all plants from this zone, some of which grow wild; others cultivated by man's hand. Bibliography about the subject was obtained, studying papers on materials made with different organic materials, such as rice husk, coconut fibre, bamboo, 'tacuara', and so on. The material employed in our work comes from Las Palmas Refinery, in a place in our Province, where the extractive process is based on milling and compression of the cane, in such a way that megass which results is a material of varied composition, because fibres from the cortex and the flesh get mixed; and a variable size, from tiny pulvurent particles to pieces with a pith length of the grindstone. These pieces are, in general, in the form

Fig. 1 Megass from sugar cane

of compact bunches (or faggots) of fibres with different thickness and reediness varying from 10 to 70. As this material is destined for paper manufacture, megass is submitted to a treatment with high purity lime, fine grinding and sand, placing it in an autoclave at 6-7 kg/cm^2 pressure for two hours, in a proportion 18% lime by weight of wet fibre. Afterwards it is filtered in a No 20 brass mesh sieve. Another procedure (used in Refinery Ledesma in Tucuman) for the elimination or residual sacharose of megass is through a microbiological process. Specific micro-organisms are maintained in a 'liquor' or culture broth by which megass is saturated, filtered and then compressed so that the digestion process can be made in a natural way. By finishing the food (sugar) the dead micro-organisms produce a hard characteristic odour. The first stage in this work used treated material given by Refinery Las Palmas.

4 Preliminary studies

The natural humidity and saturated material humidity of megass were determined as 93% and of the order of 400% respectively, both referred to the wet material. Cement and fine aggregates of rounded siliceous particles with a fineness modulus 1.8, were prepared in the following proportions:

Cement	1500 g
Air entraining additive	10 cm^3
Wet fibres	2640 g (dry 480g)
Sand	5670 g
Water	800 cm^3

At first a cement paste was prepared with initial water/cement ratio of 0.4 and the additive. Then fibre was added and at last the sand, up to a suitable workability for plate moulding. With the material prepared in this way a 255 x 255 x 83 mm plate was moulded, which was observed during its hardening process for about 25 days.

This delay should be due to the presence in the fibre of remaining sacharose which acts as a retarder in cement setting. The incorporation of additives to inhibit the setting delaying effect was studied and also fibre mineralizers for utilization in a natural state, say, with no previous treatment (see section 3).

5 Samples

This stage was made with fibres without treatment, but seasoned for a 5 year period.

In order to get most information with the least number of samples, a flexure and compression test sequence was designed using the same sample. Prismatic samples of 80 cm length were poured with a transverse section of 10 x 10 cm (Fig. 2). Cylindrical samples were also poured, having 10 cm diameter and 20 cm height.

5.1 Mix proportions

The mix proportions are shown in the following table:

Type of Mix	A	B	C	D	E	F	G
Cement	1	1	1	1	1	1	1
Sand	3	3	3	3	2	2	2
Gravel 3/4 "	-	-	-	-	-	-	3
water/Cement	0.70	0.70	0.85	0.93	0.70	0.70	0.60
Dry fibers(%)	-	10	15	15	10	10	15
Aditive (%)	-	-	-	-	-	2.5	2.5
Fresh density	-	1.87	1.81	-	1.88	1.86	-
Density	2.03	1.72	1.58	1.87	1.66	1.70	1.81

Fig 2 Moulds and sample

The 'D' dosification was previously saturated for 24 hours, it absorbs 310% of water. The amount of materials in every mix is by weight of cement.

5.2 Sample moulding

Components massing (stirring) was made in a mixer with a 120 dm^3 capacity. The materials were introduced in the following order: water, cement and sand; once the mix is homogeneized, fibre is incorporated avoiding in this way the fibre 'urchin' (brittle) process which occurs when mixing lasts more than the necessary time. This phenomenon also occurs with other kinds of fibre, such as glass, metal, named in this manner for its similarity with the sea-urchin.

5.3 Curing

Specimens were cured in chambers at constant temperature and humidity: 21 ± 2 °C and 95% relative humidity respectively.

6 Laboratory testing

6.1 Flexure

Prismatic concrete samples were tested in compound flexure with a concentrated load in the centre of the span which was of 75 cm. The remaining parts of sample were tested again in flexure, this time with a 30 cm span.

The four sections were reserved for single compression tests. The prismatic samples of mortar reinforced with fibre (with no coarse aggregates) were tested in pure flexure with a 38 cm span, which permitted at least three sections to be kept for single compression tests.

6.2 Single compression tests

Cylindrical samples and prismatic sections from the flexure tests were submitted to single compression with a load-strain curve register. The load, in this last one, was applied through metallic plates of 10 x 10 cm, at a loading speed of 2 MPa/min.

6.3 Cement mortar test

Tensile strength by a flexure test and compressive strength were determined on standard samples of 4 x 4 x 16 cm dimensions according to standard IRAM 1622.

6.4 Test machinery

The machines employed were:

Avery 100 t (10 MN)
Suzpecar 100 t with load speed control and load/displacement curve register.

Fig. 3 Fracture plane where fibers can be seen after a flexure test.

7. Test results

Water absorption of fibres	400 %
Tensile strength of mortar	3.9 MPa
Compressive strength of mortar	20.2 MPa
Fibre reediness	10-70 mm/mm
Apparent density of fibres	70 kg/m^3
Compacted density of fibres	140 kg/m^3

Type of Mix	A	B	C	D	E	F	G
Cylindr. strength	-	-	2.9	1.6	1.8	-	-
Cube strength	20.2	2.6	-	-	2.1	5.2	5.1
Bending strength	3.9	0.8	-	-	0.6	1.5	1.6

(Units are in MPa)

8 Conclusions

As can be seen in Fig. 6, the favourable effect of fibres is seen as total breakage does not occur after first cracking.

Evidently tensile strength of fibres remains among fibre-mortar adherence (see Figs. 4 and 5) because their sliding, not their breakage can be observed.

The addition of sodium silicate as an additive for the mix improves remarkably the strength of reinforced mortar, in flexure as well as in compression.

Fig. 4 Essayed sample to compression.

Fig. 5 Essayed sample, first to flexion, after to compression.

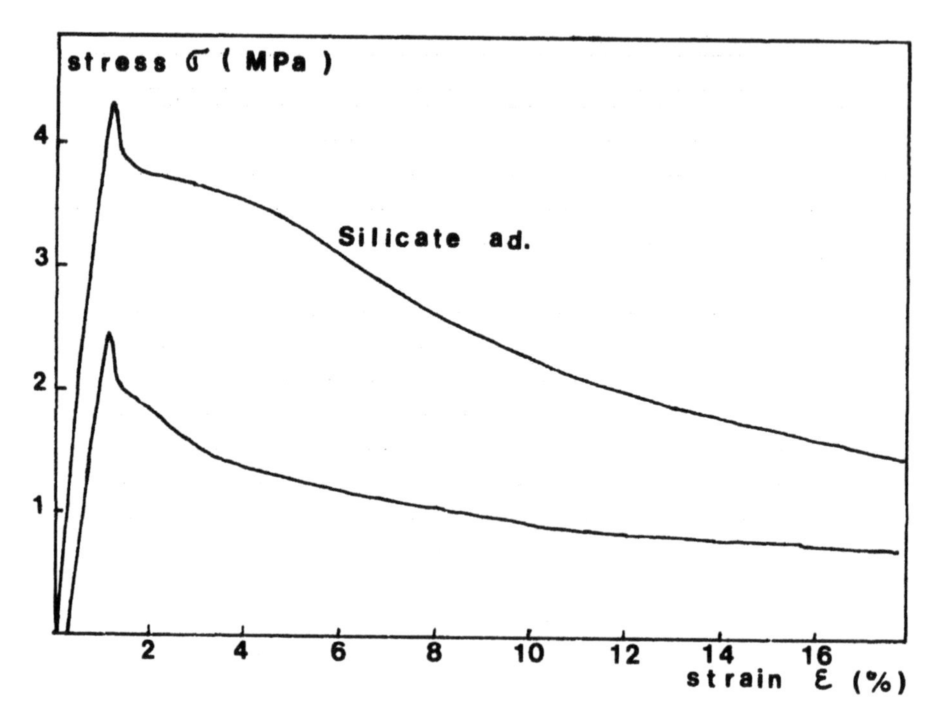

Fig. 6 Typical stress - strain curves in compression.

9 References

Klees, D., Hess, A., Natalini, M. y Gómez, G.(1987) Uso de la cascarilla de arroz como agregado granular en el mortero celular de colado superfluído para la ejecución de viviendas, en I Simpósio Internacional sobre produção e tranferência de tecnologia em habitação: Da pesquisa à prática (ed. Departamento de Divulgação do Instituto de Pesquisas Tecnologicas), São Paulo, Brasil, pp. 171-180.

Jejcic, D. et Zanghellini, F. (1977) Mortiers et ciments armés de fibres. Annales de L'Institut Technique du Batiment et des Travaux Publics, Serie: Materiaux, N° 51.

Padilla, E.(1981) Desarrollo de tecnología para la producción de elementos constructivos utilizando bagazo para su aplicación en la construcción de viviendas económicas Ciencia Interamericana , Vol. 21, N° 1, O.E.A.

Pama,R.P. and Cook, D.J.(1976) Mechanical and phisical properties of coir - fibre boards, in New Horizons in Construction Materials (ed. Hsai-Yang Fang), Envo Publishing Co. Inc. pp. 391-403.

9 PROSPECTS FOR COCONUT-FIBRE-REINFORCED THIN CEMENT SHEETS IN THE MALAYSIAN CONSTRUCTION INDUSTRY

M.W. HUSSIN
Faculty of Civil Engineering, Technological University of Malaysia, Johor, Malaysia

F. ZAKARIA
Faculty of Civil Engineering, Technological University of Malaysia, Johor, Malaysia

Abstract
This paper reports the current research and developments on the use of coconut fibres as reinforcement for thin cement sheets as roofing materials. The paper presents extensive test data on the flexural properties of cement sheets both in flat and corrugated forms at different levels of fibre concentration tested at various ages. The tests on 500 x 100 x 10 mm flexural plates and on 1220 x 630 x 10 mm corrugated sheets are reported. The load deflection curves and cracking performance are also studied based on several curing regimes. Other tests result reported are water absorption, water tightness and bulk density. It is shown that the performance characteristic of thin sheets is very much a function of fibre concentration and method of specimen fabrication. Coconut fibre reinforced thin cement sheets has a good prospect in a developing country like Malaysia especially for small scale industries.
Keywords: Coconut Fibres, Fibre Cement Composite, Flexural Strength, Thin cement Sheets, Deflection, Cracking, Water Retention, Water Absorption, Asbestos Replacement, Corrugated Sheets.

1 Introduction

Over the last decade intensive research and development work has taken place in the field of fibre reinforced concrete. Knowledge of fibre manufacturing, fibre handling and fibre concrete production has increased substantially and various aspects of applications have been studied to a great extent that now can be found in the market. Asbestos fibre reinforced cement sheets are extensively used in building construction where durability, strength and economy of asbestos cement sheets have been the main reason for their popularity. In tropical countries, natural fibres from coconut husk, bamboo, bagasse, jute, sisal, palm etc. are available abundantly and are relatively cheap. If asbestos fibres are replaced with natural fibres and reasonably good quality cement sheets or fibre boards can be produced, then it is possible to reduce the cost of building construction.

Coconut trees are widely cultivated in Malaysia with the total area of 215,000 hectares (Singh, 1978) in late seventies and is still increasing. Coconut husk is the outer covering of fibre material of coconut fruit. The husk of the mature coconut consists of a hard outer skin and numerous fibres embedded in a soft cork - like material often terms as pith and can be obtained very cheaply. The fibres are normally separated from the husk by means of mechanical extraction machine. The length of the fibres used in this investigation varies from 37 mm to 250 mm while the diameter is between 0.11 to 1.06 mm. The fibres absorb between 150 to 190% of water and the average tensile strength is about 106 N/mm^2. The studies at Universiti Teknologi Malaysia (UTM) on coconut fibre in cement - based composites mainly concentrate on the production of thin sheets either flat or corrugated in form for roofing purposes.

2 Experimental Programme

The tests reported here were carried out on 500 x 100 x 10 mm flat plates and on 1220 x 630 x 10 mm corrugated sheets tested in flexure. The 10 mm thickness was the nominal thickness. The matrix used was a cement paste with three different water - cement ratios of 0.30, 0.35 and 0.40. However, the water - cement ratio of 0.35 which produce the most encouraging results is presented here. The matrix has average density of 2.17 Kg/m^3 with compressive strength of 47.5 N/mm^2 at 28 days. All specimens for flat plates and corrugated sheets were fabricated with six different fibre contents ranging from 1% up to 6% by weight. This is to illustrate the capability of the fibre to provide strength, crack control and durability to the composite when incorporated in cement matrix. To examine the effect of curing regimes, some of the plates and sheets were exposed to three months, six months and one year of Malaysian weather after 28 days of water curing. The temperature in Malaysia varies from 24^oC to 34^oC through out the year with relative humidity around 80%. Control tests on plates were also carried out in flexure.

2.1 Preparation of Test Specimen

Prior to mixing, fibres were first boiled for 15 minutes to reduce the amount of micro organism which will destroy the fibres after prolonged period of time (Cook, 1988). The fibres which were still in saturated condition were then mixed with the calculated amount of cement and water using a mechanical mixer until a degree of uniformity was obtained. It was found that balling problem started to be serious when 5% fibre was included in the mix. After five to ten minutes of mixing, uniform mix was obtained and poured onto flat moulds and vibrated for three minutes. Vibration was needed to reduce the amount of voids present.

To fabricate the corrugated sheets, the wet specimens were left to set for about one hour before transfering to the corrugated zinc moulds. A casting pressure of 0.4 N/mm^2 was applied for 24 hours before the specimens were demoulded. The sheets were then cured in water for 28 days before testing.

2.2 Testing Procedure

Flexural tests on corrugated sheets were carried out in accordance to Malaysian Standard, MS 524:1984 (1984) under three point loading over an effective span of 1120 mm. The flat plates were tested in flexure under four point loading over an effective span of 300 mm in accordance to Rilem Recommendations (49 TFR, 1988). Measurements taken during the tests were load, deflection and strain readings. At least three specimens were tested for each variable.

Water retention tests were conducted in accordance to MS 524:1984 while other tests such as water absorption and bulk density were carried out in accordance to Rilem Recommendation.

3 Test Results and Discussion

3.1 Flexural Strength

Typical load - deflection behaviour of flat plates and corrugated sheets is shown in Figures 1 and 2. Tables 1 and 2 show the flexural strength of flat plates and corrugated sheets respectively at different age of curing. The results show that the amount of fibres used influence the strength performance of the sheets. The strength increases with the increment of fibre used but slightly reduced when the fibre content exceeds 5% where balling problem was apparent and contributed to lack of bonding between the matrix and the fibres.

The strengths were also found to be increasing with curing time especially for water cured specimens. However, the results are not consistent due to the factors such as poor distribution of fibres in the specimens and fabrication technique.

Central deflection and ductility also improved for fibre reinforced specimens as compared to control. Specimens with 6% fibre show the highest central deflection. The large deflections carried by the composites compared to control specimens also imply high surface tensile strains. Figure 3 shows the stress - strain behaviour for both flat and corrugated specimens.

Cracking behaviour of coconut fibre reinforced cement sheets also improved with the formation of a number of cracks at failure compared to a single crack for control specimen. Majority of the specimens failed by fibre rupture which indicate a good bonding between the fibre and the matrix.

3.2 Environmental Effects

The durability of coconut fibre reinforced cement sheets was studied by conducting aging tests on both flat plates and corrugated sheets. Specimens were first cured in water for 28 days and stored outdoors on a roof top for three months, six months and one year. The effects of natural weathering on load - deflection and stress - strain behaviour are shown in Figures 4 and 5 respectively. Values of the post cracking strength for the specimens aged outdoors reduced up to 30% due to the instability of the fibre itself and this can be improved by applying proper treatment to the fibres.

Past studies have shown that the natural fibres are chemically decomposed in the alkaline environment of the cement matrix resulting in brittle composite which has reduced capacity to cracking. To improve durability, it is necessary to find ways to stop or slow down the embrittlement process of natural fibre concrete. The best results were obtained with the reduction of alkalinity of the pore water in the cement matrix. This reduction in alkalinity could be achieved by replacing part of the portland cement with a highly active pozzolana such as silica fume, fly ash or rice husk ash. Sera et. al. (1989) have shown that the embrittlement of sisal fibre concrete can be completely avoided by replacing at least 40% of ordinary portland cement with silica fume.

3.3 Water Retention, Water Absorption, and Bulk Density

Table 3 shows the comparison on bulk density, water absorption and porosity between coconut fibre reinforced sheets and asbestos cement sheets. The tests were both carried out on flat plates and corrugated sheets and the results were the average of at least three number of specimens. The results show that coconut fibre reinforced specimens are less porous and absorbed less water compared to asbestos sheets. From water retention test, coconut fibre reinforced specimens can perform as good as asbestos sheets which are suitable for roofing materials.

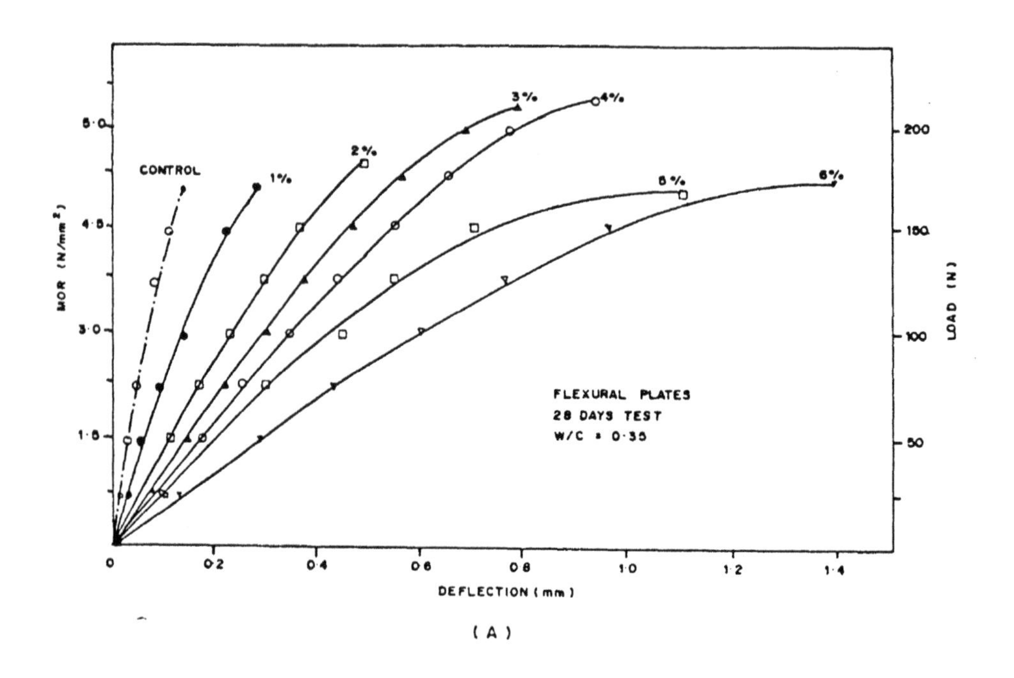

(A)

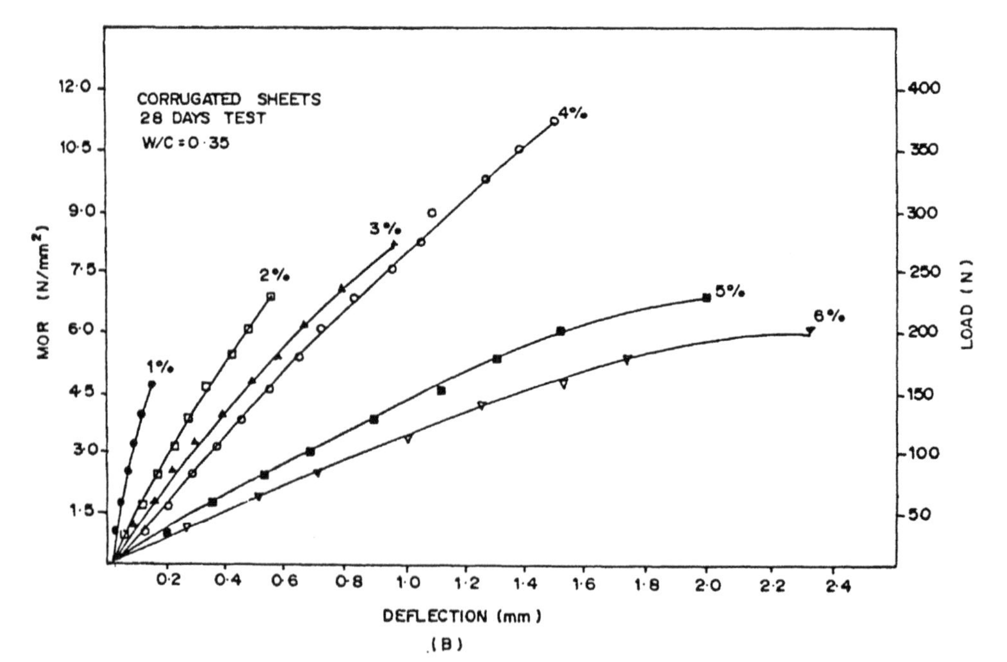

(B)

FIG. 1 LOAD DEFLECTION BEHAVIOUR OF a) FLAT PLATES
b) CORRUGATED SHEETS AT 28 DAYS.

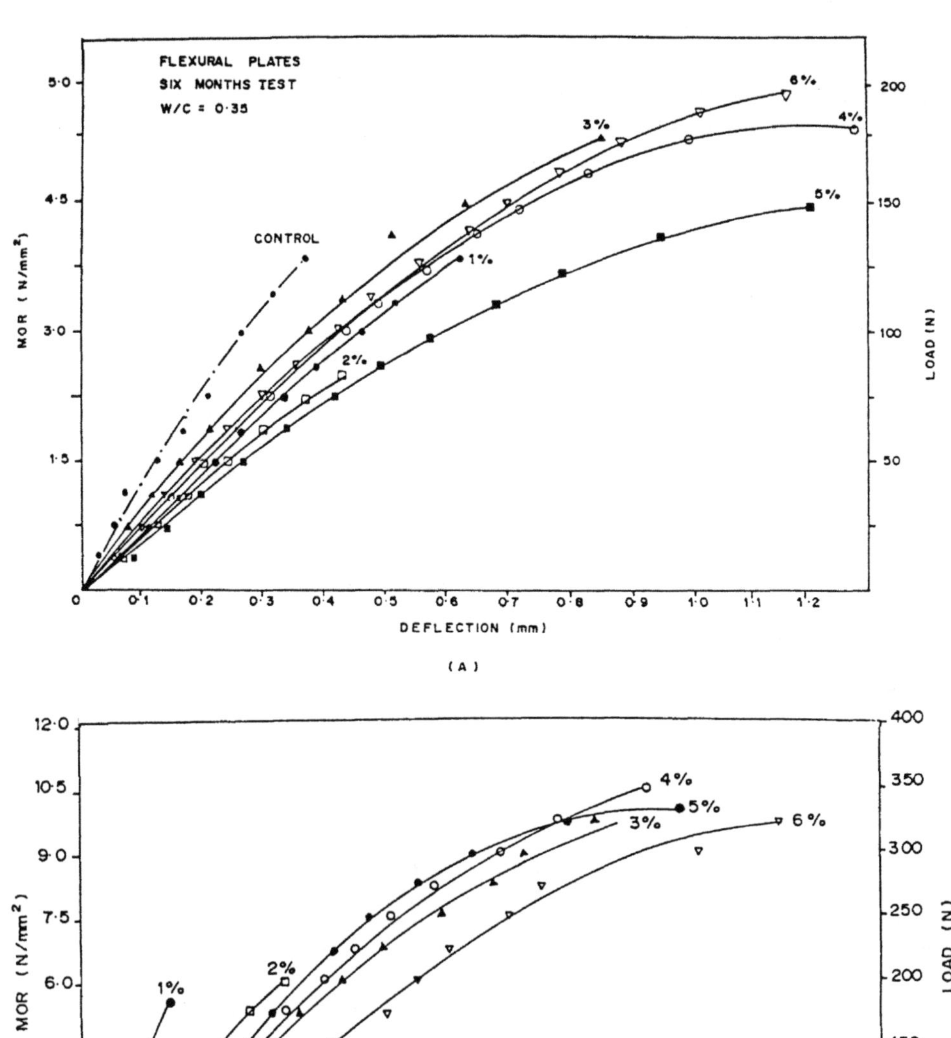

FIG. 2 LOAD DEFLECTION BEHAVIOUR OF a) FLAT PLATES b) CORRUGATED SHEETS AT 6 MONTHS.

Table 1. Flexural strength of 500 x 100 x 10 mm plates

Fibre content	Ave. self weight (Kg.)	Ave. density ($x10^3$ Kg/m^3)	Flexural strength (N/mm^2)			
			28 days	3months	6months	1 year
Fp1 - 1%	1·05	2·18	4·86	3·31	4·33	3·25
Fp2 - 2%	1·00	2·05	5·23	4·41	2·36	3·20
Fp3 - 3%	1·05	2·01	6·33	5·18	4·85	5·12
Fp4 - 4%	1·09	2·19	6·37	7·20	5·30	8·03
Fp5 - 5%	1·11	2·21	4·43	6·98	4·00	7·13
Fp6 - 6%	1·12	2·25	5·35	5·46	6·50	4·84
Control - 0%	1·08	2·17	4·33	4·08	3·32	3·63

Table 2. Flexural strength of 1220 x 630 x 10mm corrugated sheets

Fibre content	Ave. self weight (Kg.)	Ave. density ($x10^3$ Kg/m^3)	Flexural strength (N/mm^2)			
			28days	3months	6months	1year
Cs1 - 1%	16·48	1·82	4·55	5·25	5·55	6·30
Cs2 - 2%	16·86	1·77	6·05	4·50	6·00	7·62
Cs3 - 3%	17·89	1·83	7·25	7·80	9·90	8·49
Cs4 - 4%	19·31	1·78	11·80	12·75	10·50	8·70
Cs5 - 5%	17·40	1·88	6·53	11·40	10·05	7·32
Cs6 - 6%	16·41	1·85	4·87	6·98	9·37	6·63

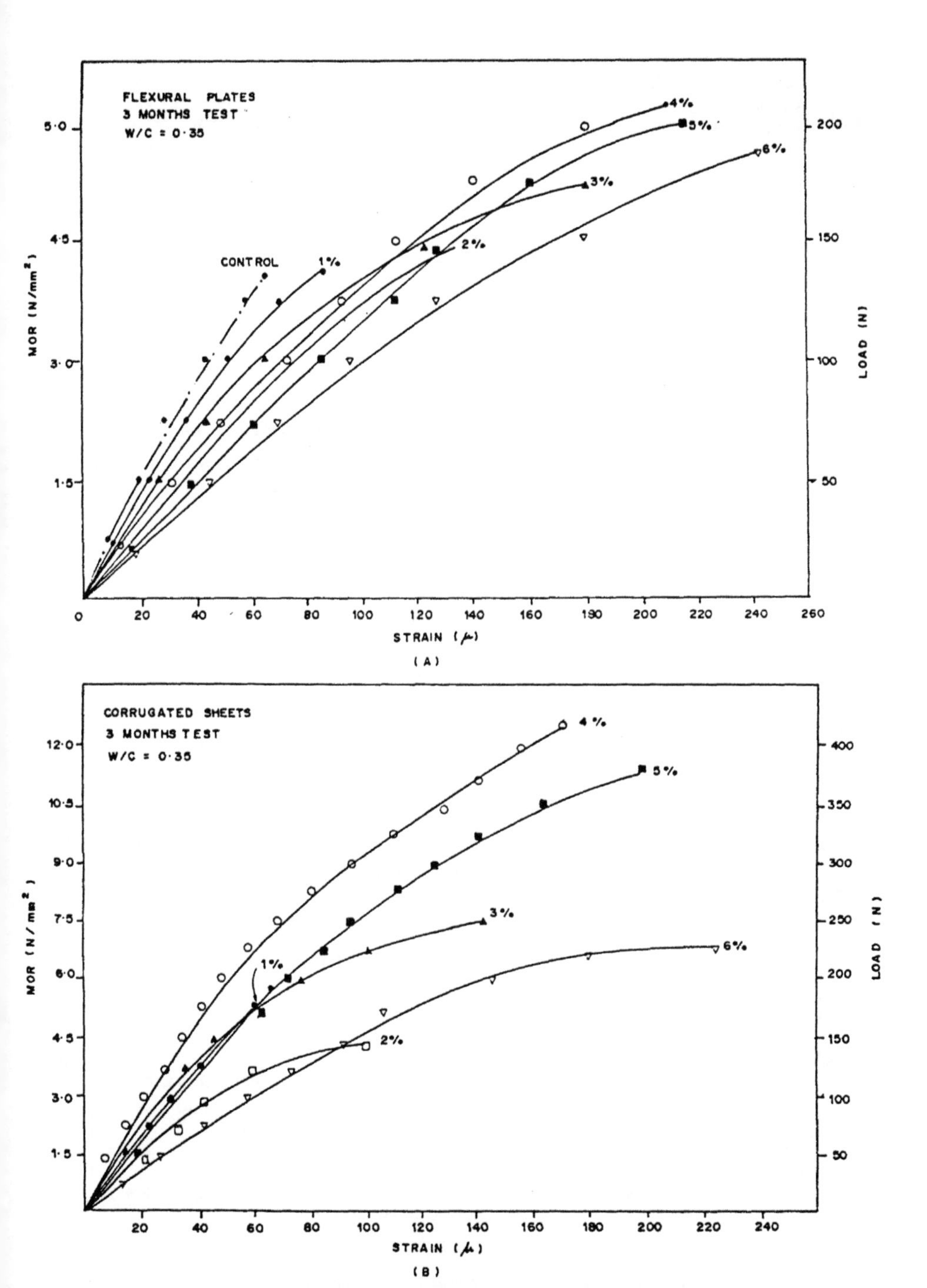

FIG. 3 STRESS STRAIN BEHAVIOUR OF a) FLAT PLATES b) CORRUGATED SHEETS AT 3 MONTHS.

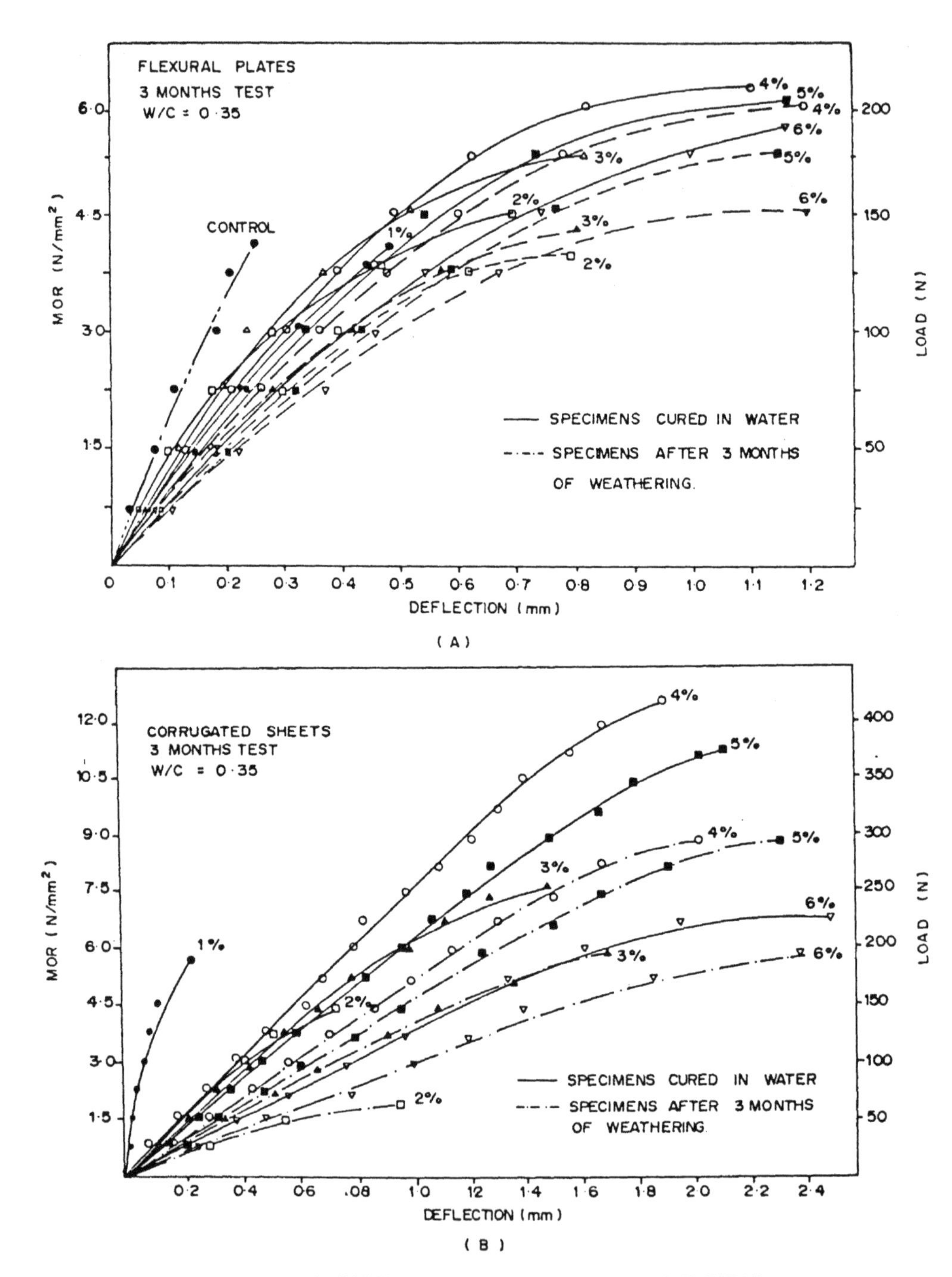

FIG. 4 EFFECT OF NATURAL WEATHERING ON LOAD DEFLECTION BEHAVIOUR IN FLEXURAL OF a) FLAT PLATES. b) CORRUGATED SHEETS.

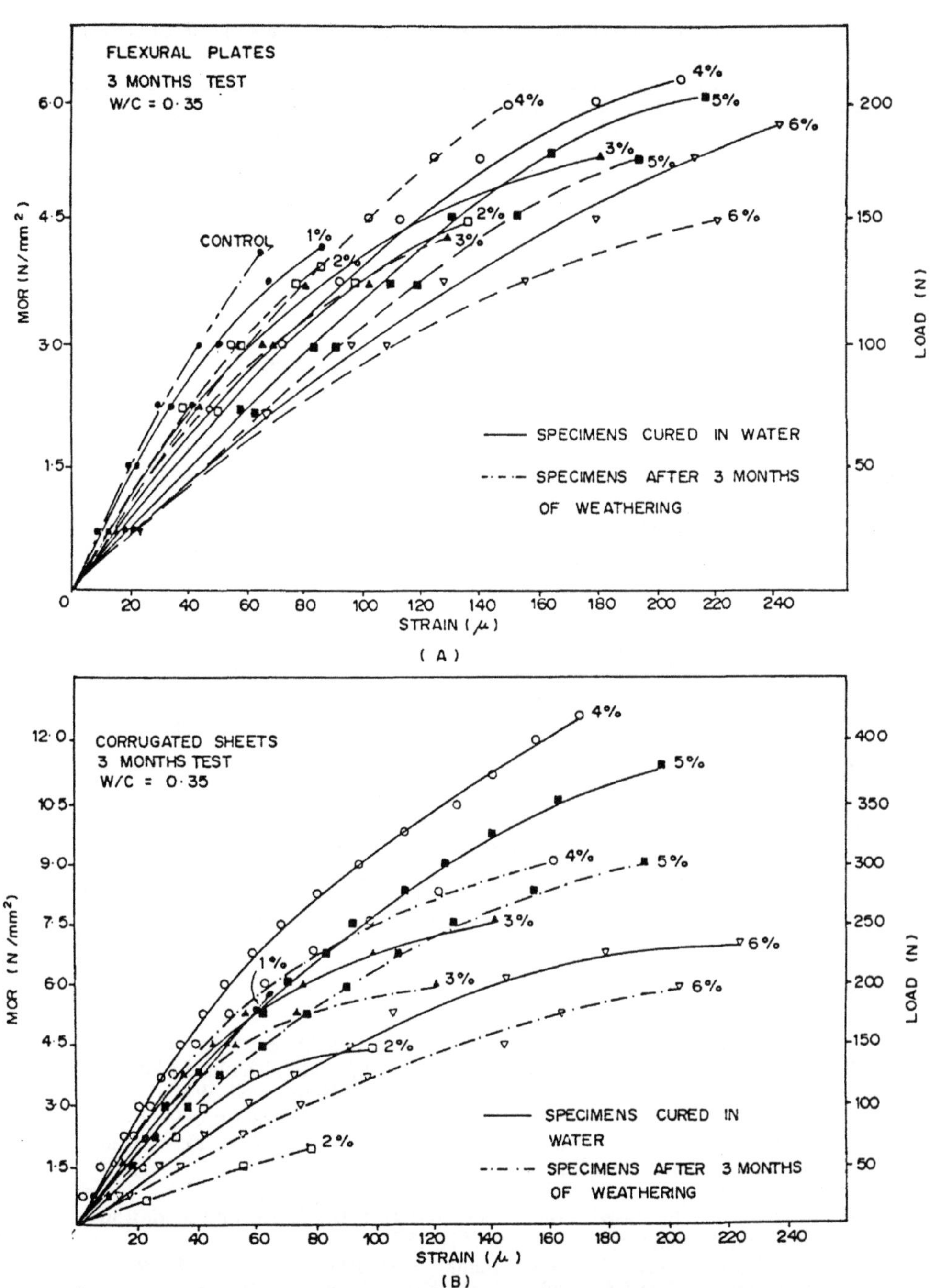

FIG. 5 EFFECT OF NATURAL WEATHERING ON STRESS STRAIN BEHAVIOUR IN FLEXURAL OF a) FLAT PLATES b) CORRUGATED SHEETS.

Table 3. Comparison of bulk density and water absorption of coconut fibre reinforced cement and asbestos.

Specimens \ Properties	Ave. dry Mass (g)	Ave. immersed mass (g)	Ave. saturated mass (g)	Ave. dry bulk density $x 10^3 Kg/m^3$	Ave. wet bulk density $x10^3 Kg/m^3$	Ave. water absorb. (%)	Ave. apparent porosity (%)
Corrugated sheets	795·00	469·80	889·50	1·89	2·12	11·86	22·50
Flat plates	197·30	132·80	224·50	2·15	2·45	13·83	29·85
Asbestos sheets	161·30	105·00	193·80	1·81	2·18	20·18	36·63

4 Conclusions

Based from the results of the studies at UTM on the use of coconut fibre as reinforcement, the following conclusions can be drawn:

a) Randomly distributed coconut fibres having different aspect ratios improved flexural strength , cracking performance and ductility of both flat plates and corrugated sheets.

b) The flexural strength of sheets increases with increasing volume of fibre included. The best result in strengh were obtained when 4% of fibre is used at 0.35 water - cement ratio.

c) Long term exposure to the Malaysian weather have affected the load deflection behaviour and strength characteristics of coconut fibre reinforced sheets but this can be reduced by replacing ordinary cement with pozzolanic materials.

d) The corrugated sheets possess inherently greater strength than flat plates and both can be used in various application in building construction.

e) At present, raw coconut fibres are available at low price in Malaysia. The cost of one square metre of corrugated coconut sheet is around US$1.20 compared to US$2.60 of the same size of asbestos corrugated sheets. In view of this, the coconut fibre reinforced corrugated sheets should be considered seriously for use in low cost housing particularly for developing countries.

5 References

Singh, S.M. (1978) Coconut Husk - A Versatile Building Material, Int. **Conf. on Building Materials For Developing Countries**, Bangkok, Thailand, August.

Cook, D.J. (1988) Concrete and Cement Composite Reinforced With Natural Fibres, **Concrete International**, Construction Press, Lancaster pp:99-114.

Malaysian Standard (1984), **Ms 524**, Specification For Asbestos Cement Symmetrically Sheets, 1st. Revision, SIRIM.

RILEM Technical Committee 49 TFR (1984) Testing Methods For Fibre Reinforced Cement Based Composites, **RILEM Draft Recommendations**, Materials and Structures, No. 102, pp:441-456.

Sera E.,Robles-Austriaco L. and Pama R.P.(1989) Natural Fibre As Reinforcement, **CONCET 89, Conf. On Concrete Engineering and Technology**, Kuala Lumpur, I1-I16

10 DURABILITY OF BLAST FURNACE-SLAG-BASED CEMENT MORTAR REINFORCED WITH COIR FIBRES

V.M. JOHN, V. AGOPYAN and A. DEROLLE
Institute of Technological Research, São Paulo, Brazil

Abstract
Durability is the main problem for vegetable fibre reinforced building materials. The IPT's approach to improve the durability of the fibres is based on the reduction of free alkaline of the matrix. A ternary binder based on blast furnace slag (BFS) activated with ground gypsum and hydrated lime is presented for this purpose. Coir fibre behaviour in alkalyne media is shown as well as the matrix composition.

Properties of specimens of coir fibre reinforced BFS based cement mortar are mentioned in comparison with those of specimens submitted to accelerated ageing (quick condensation and carbonation) and natural ageing in São Paulo (results of one year available). It is possible to increase the durability of coir fibre by changing the matrix composition.
Keywords: Durability, Composite, Natural fibre, Coir fibre, BFS cements.

Resumé
L'emploi des fibres végétales comme renforcement est limité par sa faible résistance à l'alcalinité du mortier. Dans le but d'avoir un ciment à basse teneur de chaux libre dans le produit hydraté, nous avons étudié des mélanges ternaires laitier-chaux-gypse.

Les proprietés ont été déterminés sur des mortiers tels quels et après viellissement accéléré (condensation et carbonatation) et naturel (un an d'exposition).

Les résultats nous ont permis de concluire qu'il est possible d'améliorer sensiblement la durabilité de la fibre modifiant la composition de la matrice.
Mots Clés: Durabilité, Composite, Fibres végétables, Laitier ciment.

1 Introduction

A long term research on the development of alternative building materials, using the resources available , mainly agro-industrial by-products, was carried out at IPT during the last three years. Fibres reinforced building panels for load-bearing wall were developed. The tests were carried out with low alkaline matrices such as mortars of alternative cements (OPC + rice husk ash and blast furnace slag +lime+gypsum). The fibres used were coir, sisal and desintegrated newsprint. The composite selected was blast furnace cement based mortar reinforced with chopped coir fibres. The final product was a prototype in actual size, already in use, and also a manual for the production of the panels and their assembling.

As the durability is the main problem in the study of vegetable fibres for building purposes this was studied with more emphasis. Details of the vegetable fibres degradation mainly of coir and sisal are available in Agopyan et al. (1989) and Gram (1983).

2 Fibres

2.1 Availability

The production data of vegetable fibres in Brazil are scarce and not reliable: several small producers sell to others, sometimes from different states, and there are a lot of local transaction which cannot be counted upon. Nevertheless, after visiting the production sites and having meetings with the main producers and their representatives in São Paulo, it was possible to infer that the producers are able to supply the building industry with fibres without much trouble.

Coconut is a tall cylindrical-stalked palm tree, reaching 30m in height and 67-70cm in diameter. The Brazilian annual production of coir fibres is larger than 2 millions metric tones yet only a small amount of the five hundred million coconut fruits, annually cultivated in the country, is used for fibre production. The fibres are produced mainly in the States of Bahia, Pernambuco and Sergipe, where there are two industries that produce long fibres (more than 110 mm) suitable for brushes and threads. The main production is in the form of fibre wads for upholstery.

The coir fibre production in Brazil is rudimentary. Old fashioned equipments crush the husk and separate the fibre from the mesocarp of the fruit.

The shortest fibres and the residues that are even available in the large urban centers have a low value, about US$0.15 per kg (Agopyan, 1988).

2.2 Properties
The studied fibres length varies from 10 to 20mm, and the diameter is about 0.3mm. Using the ASTM D 3822-82 test method, the average tensile strength is 107 MPa, and the elongation at break is 35%.

2.3 Durability
The results presented in the main references on this subject are quite conflicting because the durability of vegetable fibres depends on: the porosity of the fibres; the crystal structure of the cellulose and chemical composition of the fibres.

These factors may vary due to fibre growing conditions and the alkalinity of the water in contact with the fibres. For this reason Brazilian fibres were evaluated (Agopyan et al., 1989). The durability was evaluated by the reduction in the rupture load and elongation of rupture with the time for fibers submitted to three different treatments.

The Brazilian coir fibre performance is not as good as the obtained with the Asian fibres (Aziz et al. 1984), but nevertheless the durability of this kind of fibre is far better than that of the Brazilian sisal fibre.

In general the low durability of the vegetable fibres was expected as the fibres are actually strands of cellulose fibrils which disperse in aqueous matrices, moreover if the medium is alkaline. Figures 1 and 2 show the surface change of coir fibre submitted to an alkaline environment (lime solution); the surfaces of theses fibres change because of the reaction of calcium hydroxide with lignin in aqueous medium.

In spite of the lower cellulose contents of the coir fibre (50%) in comparison with sisal fibre (65%), the coir fibre is more durable probably due to the close surface structure of this fibre which is not seen in the sisal fibre surface.

3 Matrices

To improve the durability of the composites the reduction of the alkalinity of the matrix was adopted. Therefore the study of the cements with additions of by-products was pursued.

Then, matrices of mortars of alternative cements mainly those ones based on blast furnace slag (BFS) and rice husk ash (RHA) were developed.

3.1 BFS based matrix
Slags from the three main Brazilian steel industries, which represent about 50% of Brazilian steel production were analysed (Agopyan and John, 1989).

In order to accelerate the BFS hydration, lime and gypsum were added. The proportion adopted was 0.88:0.02:0.10 (BFS:lime:gypsum). This proportion was

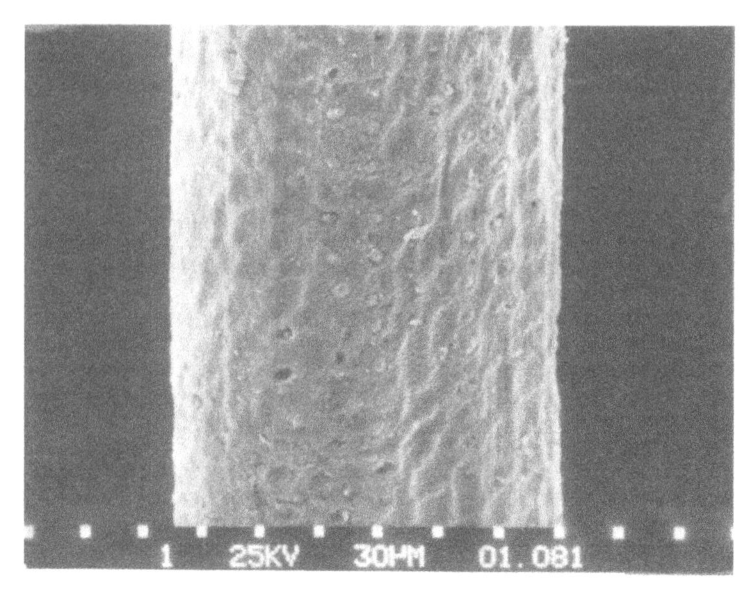

Fig. 1. Undamaged coir fibres.

Fig 2. Coir fibers before 118 days in lime solution.

suitable due to its performance: strength, setting time, exudation and the rate of free lime, still about 1.5% of the total amount of the binder at 28 days (10% for OPC).

The matrix consisting of 1:1.5;0.509 (BFS cement:sand;water) was adopted due to its performance with 2.0% of coir fibre volume (Agopyan and John, 1989).

3.2 OPC-RHA matrix

The production of the RHA was improved with the use of a fluidized bed boiler. Due to this it was possible to increase the amount of ash in the binder. At the same time with this type of boiler it was also possible to increase the amount of energy produced by the husk burning.

More details are presented elsewhere in these proceedings (Cincotto et al, 1990).

4 Composites

BFS based mortar was reinforced with coir fibres in the proportions mentioned above. Composite with fibre volume of 2% was adopted due to its performance better than those with higher fiber content: the workability was 250mm (flow table); the setting time was bellow 24 hours; the impact strength was the double of that of the unreinforced matrix; the static strengths were similar to those of the unreinforced matrix (Agopyan et al., 1989 and Agopyan and John 1989); and the air content of the fresh mixture was 6.5%.

5 Durability of the composites

After normal curing in wet chamber specimens had their durability evaluated.

Three types of tests were performed. The first one was a simple accelerated test which is known as Quick Condensation Test (Q-C-T). The other test consisted of an acceleration of the carbonation of BFS based mortar specimens. Finally, specimens were prepared for natural ageing in outdoor environment. It was planned to have specimens tested with up to 2 years of ageing.

5.1 Quick Condensation Test (Q-C-T)

As the natural degradation process is quite slow and the alterations for usual materials are only noted after long periods of time, it was decided to perform this simple accelerated ageing test. In this test one face of the specimen is submitted to cycles of wetting (100°C water moisture) and drying (40°C air). The specimens were 10x150x200mm, 28 days wet cured, and the exposed 150x200mm face is that one which was in contact with the mould. From IPT's previous studies, the duration of the test was fixed

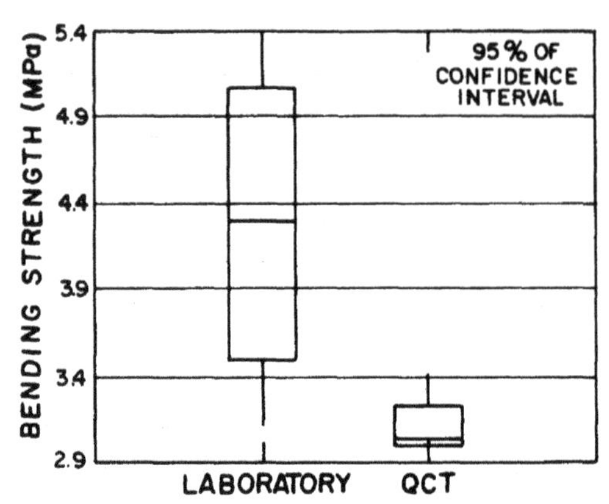

Fig. 3. Bending strengths of specimens submited to laboratory environment and 648h QCT test.

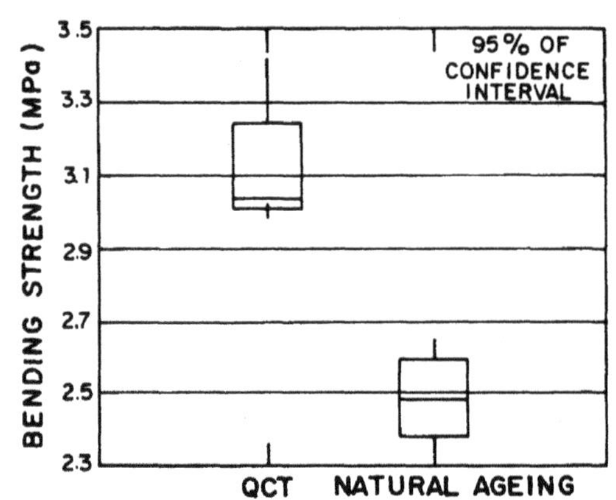

Fig. 4. Bending strengths of specimens submited to 648h QCT test and 6 months in natural ageing station.

in 300 hours with cycles of 12 hours of wetting and 12 hours of drying.

Specimens were also submitted to a longer period of exposure with a shorter cycling period. The duration of new series of tests was of 648 hours with cycles of 8 hours wetting and 8 hours drying. (Cycles of 8 hours have proved to be suitable to dry or wet these specimens.)

For the first series, the impact test was used to compare the exposed and unexposed specimens, however the test is quite crude so it was not possible to obtain conclusive results. Instead of the impact test, the RILEM bending test was performed to compare the exposed samples with unexposed ones and also with those ones left in the natural ageing station for 6, 12 and 24 months.

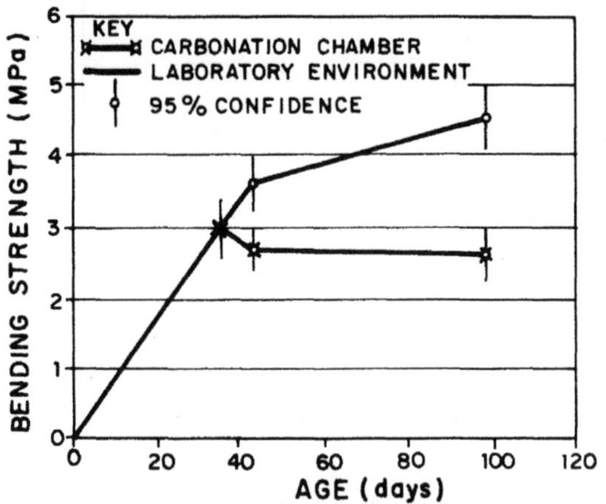

Fig. 5. Effect of accelerated carbonation in the composite.

The long term Q-C-T tests seem to be suitable for the BFS matrix as the bending strengths were similar to that of the specimens kept in the natural ageing station for 6 months with 95% confidence interval. But the specimens kept in wet chamber had a far better performance than the aged ones. The ageing also reduced the strengths of the composite specimens as can be seen in Figure 3. But this accelerated ageing did not simulate the natural ageing in Sáo Paulo's conditions (Figure 4).

5.2 Accelerated carbonation test

This research was carried out with a BFS based cement matrix which is very similar to a slag sulphate cement made with Portland cement, plaster (anhydrite) and BFS. The last one has its strength reduced due to carbonation because ettringite is decomposed in weaker products (Mann and Wesche, 1968).

The specimens were cured in humid chamber for 36 days, later they were conditioned in a carbonation chamber which was constantly 100% saturated with CO_2, at the temperature

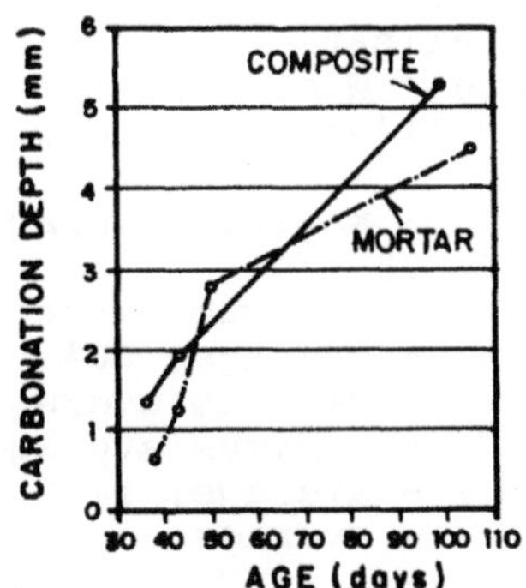

Fig. 6. Carbonation depth in laboratory environment.

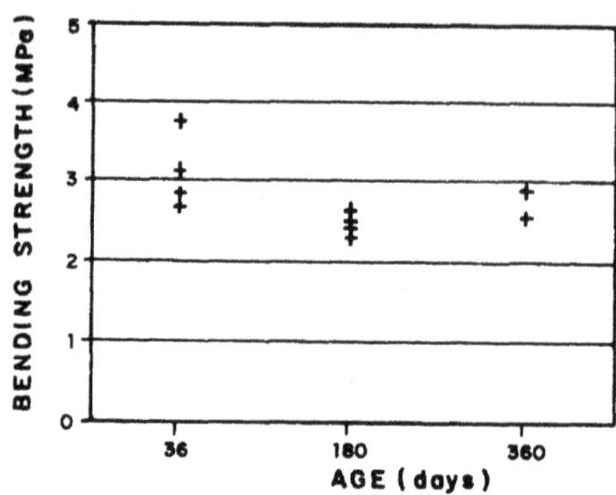

Fig. 7. Natural ageing of composite specimens.

of 35+/-1°C and a relative humidity of 75% (controlled by a saturated solution of sodium chloride).

The depth of carbonation was measured with a solution of 1% phenolphthalein in 70% ethyl alcohol.

The results are presented in Figure 5. The carbonation prevents the increase of the strength of all the specimens, even those ones reinforced with fibres. Actually in 7 days of conditioning the specimens were completely carbonated. There are not significant differences (95% of statistics confidence) between the strengths of the specimens fully carbonated (40-45 days) and those kept in the carbonation chamber up to 100 days.

In the laboratory environment the material also suffered a quick carbonation as can be seen in Figure 6. This helps the natural fibre protection.

It was also concluded that 180 days of natural ageing (outdoor) is severer for the specimens than 98 days of conditioning in the carbonation chamber. Therefore carbonation is not the major problem.

5.3 Natural ageing

The natural ageing station is located at the Campus of the IPT, in the city of São Paulo. The racks are prepared in order to allow the specimens to stay on a surface with 30° of inclination and facing North.

The geographical localization of the station is 23°34′ S of the latitude and 46o44′ W of longitude. The climatic conditions for the last 11 year period were the following:

temperature:	- maximum average for February: 28.7°C
	- minimum average for July: 11.6°C
relative humidity:	- average: 60%
	- maximum average for March: 96%
	- minimum average for July: 52%
rainfall:	- average: 1462.4 mm/year
sunshine:	- average: 5.5h/day

BFS cement specimens of the same size as those for Q-C-T tests were conditioned in natural environment at the IPT's Station. The results are presented in Figure 7. With 95% of confidence, it is possible to state that the specimens were not affected after 6 months of ageing. This statement is valid both for the mortar and for the composite specimens. Figure 8 shows the surface of a coir fibre from a BFS cement specimen conditioned in the natural ageing station for 1 year. Although the surface is covered by matrix products it looks like the undamaged fibre (Figure 1) and cannot be compared with the fibre conditioned in the lime solution for 120 days (Figure 2).
A more comprehensive analysis was made with an optical microscope. For 1 year aged composite specimens, the upper surface, with a magnification of 10 times, shows leaching and holes up to 3mm depth. The fibres (in good state) and the aggregates are exposed. With a magnification of 100 times it was possible to conclude that the coir fibres were not defibrillated however they had a lighter color than the new ones.

In the natural ageing station there are further specimens to be tested for to 2 years which will occur after the preparation of this paper.

In addition the in-use durability performance will be evaluated through the prototype (Fig. 9) (AGOPYAN et al., 1990).

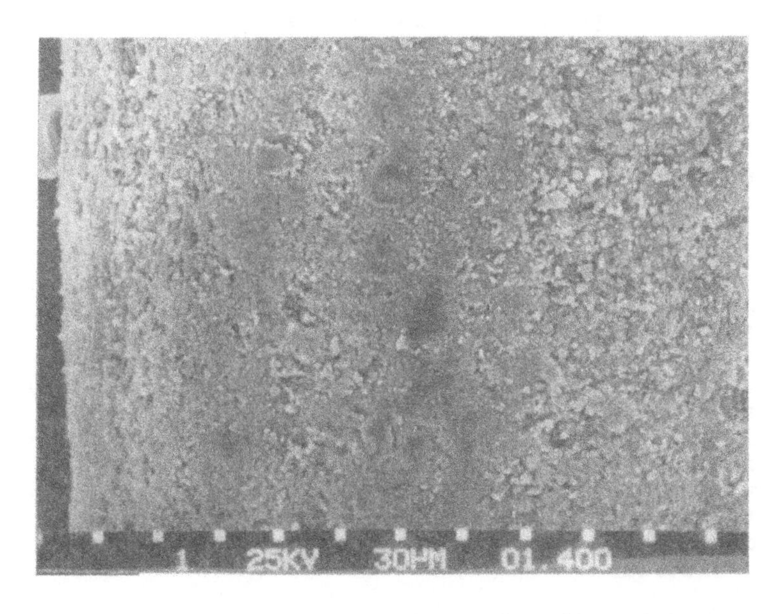

Fig. 8. SEM micrograph showing coir fibre from composite specimens submited to 360 days natural ageing.

Fig. 9. Actual size prototype.

This evaluation will take into account the following degradation factors: use, stress, physical and chemical incompatibility, biological and non regular environmental factors.

6 Comments

It is necessary to point out that the degradation of the fibre occurs while the matrix strength is reduced in the above mentioned durability tests.

The accelerated tests (QCT and carbonation) adopted were not able to simulate one year ageing in the natural environment of São Paulo, some specimens even after only six months natural ageing presented more degradation than those submitted to the accelerated tests.

The evaluation of the degradation of low modulus fibre composites like our materials is another point of concern. The impact test (falling ball) due to its crudeness was not able to measure small degradations. But also the bending test (RILEM method) was not much sensitive to this composites as the effect of the matrix in this property is very high. Perhaps the measurement of the toughness could be a better degradation indicator.

The main conclusion was to show that it is possible to increase durability of vegetable fibres only by changing the matrix composition in order to obtain low alkaline media. Moreover this research proves that it is possible to produce load-bearing panels with vegetable fibres reinforced materials.

7 References

Agopyan, V. and John, V.M. (1989) Building Panels Made with Natural Fibre Reinforced Alternative Cements, in **Fibre Reinforced Cements and Concretes** (ed. R.N. Swamy and B. Barr), Elsevier Science Publ., Essex, pp.296-305.

Agopyan, V., (1988) Vegetable Fibre Reinforced Building Materials - Developments in Brazil and other Latin American Countries, in **Natural Fibre Reinforced Cement and Concrete** (ed. R.N. Swamy), Blackie, Glasgow, pp. 208-242.

Agopyan, V. Cincotto, M.A. and Derolle, A. (1989) Durability of Vegetable Fibre Reinforced Materials, in **CIB Congress, 11, Quality for Building Users Troughout the World**, Paris, Theme II, Vol.I, pp.353-363.

Agopyan, V. John, V.M. and Derolle, A. (1990) Construindo com Fibras Vegetais (Building with Vegetable Fibres). **A Construção**, São Paulo, abril (to be published).

Aziz, M.A. Paramasivam, P. and Lee, S.L. (1984) Concrete Reinforced with Natural Fibres, in **New Reinforced Concretes** (ed. R.N. Swamy) Blackie, Glasgow, pp. 106-140.

Cincotto, M.A. Agopyan, V. and John, V. M. Optimization of Rice Husk Ash Production, in **2nd Int. Symp. on Vegetable Plants and their Fibres as Building Materials**, RILEM, Salvador (in printing).

Gram. H.E. (1983) **Durability of natural fibres in concrete.** Swedish Cement and Concrete Research Institute, Stockholm.

Mann, W. and Wesche, K. (1968) Variation in Strength of Mortars Made of Different Cements Due to Carbonation, in **The Fifth Int. Symp. on the Chemistry of Cement**, The Cement Association of Japan, Tokyo, pp.385-409.

Acknowledgements

This paper presents a part of vegetable fibre reinforcement studies done by the authors at IPT. The authors would like to thanks the Secretariat of Science, Technology and Economical Development of the State of São Paulo and the IDRC- International Development Research Centre (Canada) for the financial support.

11 VEGETABLE FIBER-CEMENT COMPOSITES

S. Da S. GUIMARÃES
Research and Development Centre (CEPED),
Camaçari, Bahia, Brazil

Abstract
A research on vegetable fiber-cement composites applied to engineering purposes has been developing since 1980. The research is summarized here and embraces:

a) determinations of physical and mechanical properties of fibers from sisal, coir, bamboo, piassava and sugar-cane bagasse,
b) the influence of parameters like fiber length,fibre volume fraction, matrix proportioning and casting processes on flexural strength, absortion and specific gravity of the composites,
c) the experiments in casting dwelling components by simplified processes for roof tiles,flumes,kitchen sinks and water-tanks.

The durability tests are not yet completed however some partial results obtained up to now are presented because they may indicate certain trends.
Keywords: Fibre Cement Composites, Vegetable Fibres, Sisal, Coir, Tensile Strength, Durability, Polyvinyl-alcohol, Polyacrylonitril.

1 Introduction

Since 1980 a research on vegetable fibre cement composites was carried out as part of an effort made by Dwelling Program - THABA- of the Research and Development Center of Bahia.

The project objective is to find cheap material for dwelling components obtained by simple processes which may be easily assimilated by the users. This implies the choice of abundant natural materials like vegetable fibres available easily on the Northeastern region of Brasil.

Natural fibres from bamboo (Bambusa vulgaris), sisal (Agave sisalana), coir (Cocos mucifera L.), piassava

(Attalea funifera), sugar-cane bagasse (Saccaharum officinarum L.) had their mechanical properties determined: tensile strength, Young modulus, elogation at break point, density, and absorption. For results and testing methods see Guimarães (1982), (84).

Coir and sisal were selected to carry out the composite research for their mechanical properties and availability.

Actually the research is directed to the long term durability of the composites and the initial approach to improve it is protecting the fibres by polymer impregnating agent.

2 Some results of the research

It´s summarized here some results of the research. Detailed results has been discussed in previous papers: Guimarães (1982),(1984),(1987),(1987a).

2.1 Fibre length

In processing techniques, where sisal fibres can be easily orientated without balling and making bunddles, better performance was obtained when the fibre aspect ratio is increased (fibre length / fibre diameter) observed the limits imposed by the volume fraction of fibre in the composite. Sisal fibres with 60 mm long showed to be very long to be used without orientation, see Table 1.

Table 1. Variation of composite flexural strength with fibre length, data from Guimarães (1982)

Volume fibre fraction (%)	fibre length (mm)	Composite flexural strength Coir (MPa)	Sisal (MPa)
5	5	5.74	5.45
5	30	5.93	6.56
5	60	5.85	5.71
4	270 (*)	4.47	12.0

(*) continuous aligned fibres
Matrix carachteristics: sand/cement = 1 (by volume); water/cement = 0.43; flexural strength = 5.62 MPa

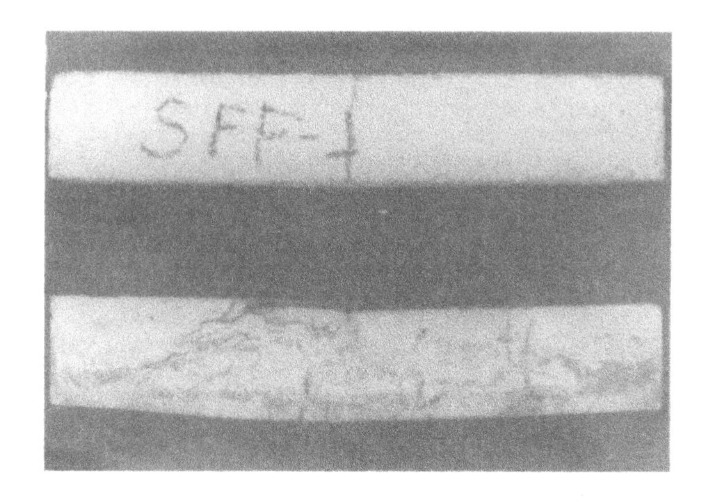

Fig.1. Sisal-cement beams specimens subjected flexural tests. Dimensions:5 x 5 x 30 cm. Above: 4% by volume continuous and aligned fibres; below 3% by volume of 3 cm long randomly dispersed fibres.

Fig.2. Coir-cement beams (5 x 5 x 30 cm) submitted to flexural tests with 4 cm long randomly dispersed fibres. Above 3% and bellow 9% fibres by volume.

Field tests carried out on flume units with coir fibres length from 1 to 4 cm indicate a increase from 46 to 86%, respectively, of its ultimate loads when compared with their plain mortar control group. The flume units were cast with sand cement ratio equal 2.6, by volume, water cement ratio equal 0.80 and a coir volume fraction of 1%, Guimarães (1987a).

From Table 1 there was no increase in flexural strength with coir fibre length, although field tests indicated an increase on ultimate load with an increase of the fibre length.

Figures 1 and 2 shows sisal fibre and coir cement beams after flexural tests.

2.2 Fibre volume content

The maximum fibres amount to be added to cement based matrices depends on the process technique used. Field experiences in casting flume units and tiles by manual mixing, shows a maximum of 1% volume coir content could be mixed to matrix in order to reach an homogeneity. When casting tiles, in a sandwich form (mortar-fibre-mortar), a maximum 2% coir, sisal or piassava (soft fibres) fibres were used leading good results, Guimarães(1987).

The fibre volume content is function of fibre type too.

The use of chopped sisal fibres best results were reached for 5% volume fibre while in coir cement composites only from 1 to 2% lead to improvement on flexural strength and post-crack ductility.

2.3 Matrix proportioning

Ordinary Portland cement (OPC) and fine sand (fineness modulus equal 1.49) were used in all experiments.

In fibre cement composites each fibre requires its optimal matrix proportioning even for similar ones like vegetable fibres. Thus, in the asbestos cement industry the matrix is a cement paste without sand. Steel fibres have been used in concrete and mortar matrices. In using sisal fibres, for a sand-cement ratio by volume equal 1 there were an increase on flexural strength. Changing the ratio from 1 to 2 the sisal cement beams showed flexural strength with smaller values than plain mortar beams Guimarães (1987).

Coir fibres in matrices with sand-cement ratio equal 3 had showed an increase on flexural strength, a reduction on range of tests results and best fibre matrix bond.

2.4 Composite processing techniques

The processing technique used had a decisive influence on vegetable-cement composites properties due to the difficulty in mixing fibres and fresh matrix with a low water cement ratio, ocasionaly leading to fiber balling. It is necessary to assure a minimum matrix recovery of the fibres, good compactness and appropriate porosity.

For sisal cement composites three processing techniques were tested: a) compaction on two layers; b) casting pressure varying from 1.9 to 3.1 MPa; c) use of vibration table for 1, 2 and 3 minutes. Flexural test analysis showed an "optimum casting pressure" (2.2 MPa) in which a maximum flexural strength, a maximum compaction and a minimum absorption are reached simultaneously. The beams had cast by this process, at optimum pressure, holds higher flexural strength (6.2 MPa) than others cast with vibration (5.7 MPa)or by compaction (4.1 MPa). On the other hand, the variation of vibration elapsed time had no effect on flexural strength of the composites, Guimarães (1987).

2.5 Vegetable-cement composites dwelling components

In accordance THABA purposes dwelling components were cast by a very simple technique: fibre cement composites were cast into a wooden hollow frame as gauge over a plastic canvas (polyetilene). Even in fresh state the flat composite plate was raised by a plastic canvas and put over a mould in order to reach the final shape.

The flat plates were cast in a sandwich form (mortar-fibres-mortar) or in a randomly dispersed fibres into matrix mixing, Guimarães (1984),(1987).

The flat plates can be bonded together after curing with wire mesh and mortar on the edges making water tanks.

Figures 3, 4 and 5 show dwelling components casting using these techniques.

3 Durability

Lost of tensile strength and embrittlement in vegetable fibre cement composites are well known. In the tropical environment, deterioration is faster due to greater humidity and higher temperature. The reactions between fibre and alkaline pore water in concrete matrices is the reason for this behaviour, Gram (1983).

In order to improve the durability of these composites, four recognized approaches has been analysed: reducing the matrix alkalinity, sealing the matrix pore system, sealing surface of the components and protecting fibre by impregnating agents, Gram (1983) and Agopyan et al. (1989).

THABA's experiments were directed to evaluate the performance of composites using fibre impregnating agents. The results of these tests, performed until recently, are presented here although they are in progress and far to be completed.

The criteria to choose the impregnating agents are that they must be alkali-resisting,must have affinity to the fibre and the cement based matrices and have low cost and large availability.

Fig.3. Coir-cement 1.5 cm long roof tiles 3 years old, at CEPED, Brasil

Fig.4. Field tests on coir-cement flumes for drainage and irrigation purposes. In the first plane there is a water-tank made with sisal-cement in which their plates were bonded together after curing.

Two impregnating agents were tested to protect the vegetable fibres: the former, a liquid for demoulding adopted for fiber-glass applications, easily available in commerce as 7.5% weight polyvinylalcohol (PVA) aqueous solution; the last is polyacrylonitril diluted in dymethylformanide (DMF),as it were proposed by Tan (1988). The liquid for demoulding was diluted in water and the final impregnating agents concentrations are indicated on Table 2. The PVA impregnated fibres were heated up 200° C during 30, 60 and 90 minutes. There were, during the research, an expectancy that this treatment could cause cross linking of the polymer and turns it water-insoluble.

Fig.5. Coir-cement kitchen sink cast by simpliflied process.

Table 2. Impregnating Polymers

Polymer	Solvent	Concentrations (%)	Impregnating time at 25° C (min)
PVA	water	0.375; 0.750	1
		1.125; 1.500	1
PAN	DMF	2 ; 7.5	10 to 15

Sisal fibres were chosen to test the impregnating agent because their susceptibility to chemical alkali

Table.3. Sisal fiber tensile strength

Concentration Solution (%)	Impregnating Agent (name)	Heating Treatment (min)	Tensile strength (MPa) and Time of exposure of fibres under saturated lime (days)					
			0	28	56	84	112	140
Unipregnating (UF)		-	722	197	252	192	245	147
0.375	PVA	60	498	355	303	279	317	262
0.375	PVA	90	492	443	284	221	177	238
0.75	PVA	30	586	305	283	189	189	161
0.75	PVA	90	414	357	218	168	120	89.4
1.125	PVA	30	447	359	263	207	167	81.4
1.125	PVA	60	487	364	259	215	220	84.6
1.5	PVA	60	439	326	190	126	135	136
1.5	PVA	90	441	351	-	144	157	91.4
2.	PAN	-	491	374	228	233	244	229
7.5	PAN	-	477	234	280	146	129	213

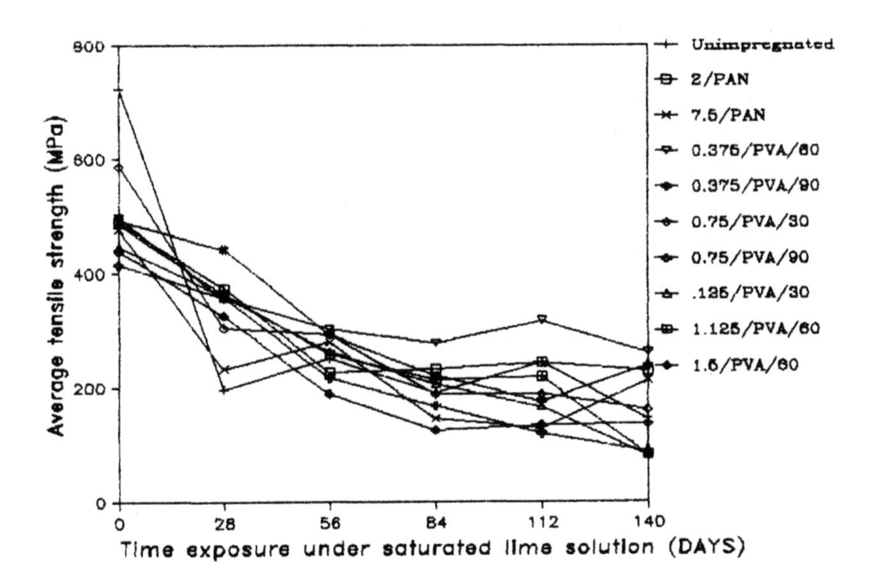

Fig.6. Sisal fibres tensile strength.

attack, Gram (1983). The alkali resistance test were carried out on impregnated sisal fibres after they had been stored under saturated lime solution for 28, 56, 84, 112 and 140 days under room temperature of +25° C. Ten tests from each impregnating type and age were carryed out but some of them were lost by fibre pull-out from glue in the mounting tab. For tests details see Guimarães (1984).

The results are presented in Table 3 and plotted in Figure 6. From the analysis from these results it could be observed that:

a) The impregnating agents as used in these tests didn't prevent the decrease in the tensile strength of the impregnating sisal fibres.
b) The tensile strength of the unimpreganated fibres had dropped drastically from 0 to 28 immersions days showing that the chemical attack occurs early in a few days or maybe in few hours.
c) After 28 days under exposure, the fibres had showed a stabilisation on tensile strength decrease, even for unimpregnated sisal fibres.
d) The impregnation 0.375% PVA aqueous solution heated during 60 minutes has presented the best results in protected sisal fibres, showing a tensile strength 78% greater than unimpregnated sisal fibres after 140 days under lime solution exposure.
e) The impregnation leads to a reduction on the tensile strength sisal fibres. This effect wasn't expected because PVA and PAN can be fiberized into resistant fibres up to 1150 MPa, see Studinka (1989) and up to 250 to 568 MPa, see Brandrup (1975), respectively. The expected result was that the fibres impregnation, the tensile strength should be increase but never a decrease as it happened.

Despite the obtained results up to this moment other impregnating agents will be tested varying the treatment conditions.

4 References

Agopyan, V. Cincotto, M.A. and Derolle, A.(1989) Durability of vegetable fibre reinforced materials, in **CIB - 89**, Paris.

Brandrup, J. and Immergut, E.H. (1975) **Polymer Handbook**, 2nd ed. John Willey and Sons, N.Y.

Gram, H.E. (1983) **Durability of natural fibres in concrete**. Svenska Forskningsinstitute, Stockholm, 230p.
Guimarães, S. da S. (1982) Utilização de fibras vegetais como reforço para argamassa de cimento, in **Seminário Latino-Americano sobre Construcion de Viviendas Econômicas**, Vol.2, Monterrey, pp. 181-206.
Guimarães S. da S. (1984) Experimental mixing and moulding with vegetable fibre reinforced cement composites, in **International Conference on Development of Low-Cost and Energy Saving Construction Materials**, eds. K. Ghavami and H.Y.Fang, Rio de Janeiro, pp.37-52.
Guimarães S. da S. (1987) Some experiments in vegetable fibre-cement composites, in **Building Materials for Low-income Housing In Asia and the Pacific**, organized by SCAP-RILEM-CIB, Bangkok, pp. 167-175.
Guimarães S. da S. (1987a) Fibra vegetal-cimento resultados de algumas experiências realizadas no THABA/CEPED, in **I Simpósio Internacional sobre Produção e Transferência de Tecnologia em Habitação: da Pesquisa à Prática**, Vol I, São Paulo, pp. 103-109.
Studinka, J.B. (1989) Asbestos substitution in the fibre cement industry. **The Int J. of Cement Composites and Lightweight Concrete**, Vol. 11 No. 2, pp.73-78.
Tan, J.K. (1988) Internal and personal communication from the polymer laboratory of CEPED.

12 LIMIT STATE OF CRACK WIDTHS IN CONCRETE STRUCTURAL ELEMENTS REINFORCED WITH VEGETABLE FIBRES

A. La TEGOLA and L. OMBRES
Department of Structures, University of Calabria, Cosenza, Italy

Abstract
The possibility of improving the mechanical characteristics of concrete through the use of reinforcement fibres has provoked in recent years a special interest for this new construction material, particularly in those areas where fibres, especially vegetable fibres, can be easily found and consequently have a low price.
The study of the behaviour of concrete reinforced with vegetable fibres is then necessary with a view to singling out and defining the most appropriate fields of application, both from the technical and the economic point of view.
The most remarkable consequences of the use of fibres in concrete are the increase of ductility and the presence of a residual resistence to traction even after cracking.
There is consequently a better behaviour as far as the deformation and the crack widths are concerned. In this work we analyse the positive contribution of vegetable fibres in ordinary concrete in the evaluation of the limit state of crack widths.
Keyword: Crack Width, Limit State, Reinforced Concrete, Vegetable Fibre

1 Introduction

In almost every area of civil engineering the use of fibre-reinforced concrete is becoming ever more widspread. Fibrous reinforcement leads to a noteable improvement in the characteristics of concrete, creating in effect a "new" material which shows excellent ductility, an improved tensile strength, improved resistence to shocks and to wear and tear and, above all, good results with regard to the limitation of cracking and deformation. Because of this, fibre-reinforced concrete can be used in many fields either as normal concrete, or as reinforced concrete in which traditional steel reinforcement is used side by side with fibre reinforcement. The latter type of reinforced concrete seems to be the most appropriate for structural purposes;

the fibres improve the material while the steel reinforcement guarantees resistence to tensile strain caused by external forces.

The type and nature of the fibres used varies; they can be metallic, synthetic or natural. However, it should be noted that first-quality concrete with superior mechanical characteristics(i.e. with metallic or synthetic fibres) is very expensive, whereas natural fibres with modest resistence are widely available from low-cost natural sources.

Furthermore, fibre-reinforcement of any type does not have an excessive influence on the resistence of concrete; there is, in fact,only a modest increase of resistence at collapse. This is due to the percentage of fibres in the conglomerate mix, which, for technical reasons, must be kept low.

Therefore it seems difficult to justify the use of high-cost fibres, especially when low-cost vegetable fibres are readily available. With these considerations in mind the use of natural fibres can be considered highly appropriate especially in the construction of certain types of structures such as reservoirs, pipes, floors and covers. It is necessary to safegard the behaviour of such structures when in use because the limit state when in use can, for pratical purposes, be considered equal to the final limit state.

Therefore structures strong enough to meet the above requirement guarantee an improvement in their own durability, an ever more important factor in structural planning. This paper analyses the advantages of vegetable fibres with regard to the service limit state; particular reference is made to the crack width limit state. Such an analysis can be useful in valuating deformations in flexural members since, up to almost the service limit state the sections can be considered totally resistant. So, as far as deformations are concerned, they behave like more rigid structures and are therefore less sensitive to imperfections. This has obvious implications with regard to problems of stability.

2 Mechanical characteristics of fibres and constitutive laws

Asbestos was the most widespread natural fibre; it has, however, been found to be poisonous and is virtually no longer used, having been substitued by vegetable fibres which in many cases possess mechanical characteristics very similar to asbestos, thereby offering an effective alternative. The most widely used vegetable fibres include sisal, coir, bamboo, banana, jute, pineapple, etc. The choice of which fibre to use depends largely on where the work is being carried out and the type of vegetation common in that area.

Vegetable fibres show a low value of the elasticity modulus with respect both to that of other types of fibre and to that of the concrete matrix. Moreover there is a tendency to absorb the water in the mix, and a certain susceptibility to action by corrosive agents. The effect of these factors has already been shown by other authors(Swift 1981a). Nonetheless, concrete reinforced by vegetable fibres offers all the advantages of concrete reinforced by other types of fibre; the constitutive law shows good ductility, good tensile strength and a residual tensile strength even after cracking.The resistence and deformation limits of the mixture depend on the type of fibre used, the percentage of fibre in terms of volume and the aspect ratio of the fibre.

The constitutive law of fibrous concrete is shown in the diagram below.

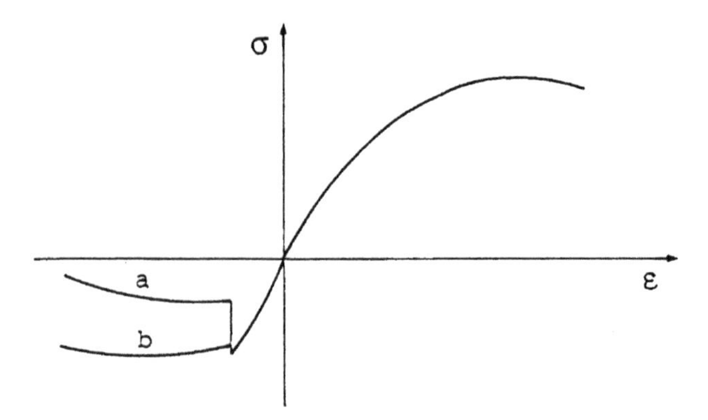

When considering compressions, the CEB parabola-rectangle diagram is used, while in calculating traction, two diagrams are brought into play to describe the post-cracking behaviour on the basis of the quantity of fibre V_f.

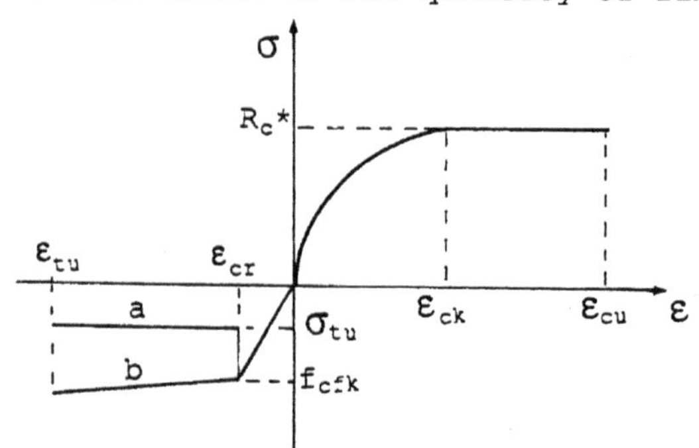

Bearing in mind the equations based on the theory of mixtures, the quantities which define diagram σ–ε are:

$\varepsilon_{cu}=3.5-4.0x10^{-3}$ last deformation in the compressed concrete;

$R_c{*}=R_{ck}/\gamma_c$ permissible stress of compressed concrete;

$\sigma_{cr}=\varepsilon_{cr}xE^{*}_{ct}$ tensile resistence of the concrete;

$\varepsilon_{cr}= f_{cfk}/E_{cm}$ deformation of the concrete in tension at the moment of cracking; f_{cfk} and E_{cm} represent the characteristic resistence and the usual elasticity modulus with the cement matrix in tension;

$E^{*}_{ct} =\eta_1\eta_o E_f V_f+E_{cm}(1-V_f)$ elasticity modulus of the fibre-reinforced concrete in tension; V_f is the percentage volume of fibre; E_f is Young's modulus of the fibre; η_o is the orientation factor; η_1 is the coefficient of the effective length;

$$\sigma_{tu}= 2\eta_1\eta_o V_f\ \tau_u\ l_f/d_f \qquad l_f<l_c$$

$$\sigma_{tu}=\eta_1\eta_o V_f[l_c/l_f(2\tau_u l_f/d_f)+(1- l_f/d_f)\sigma_{fu}] \qquad l_f >l_c$$

τ_u is the ultimate bond strength;
σ_{fu} ultimate strength of the fibre;
$l_c=0.5\ \sigma_{fu}d_f\ /\tau_u$ is the critical fibre length;
l_f, d_f length and diametre of the fibre.

3 Limit state of crack width in flexural members

With reference to beams made of concrete reinforced with vegetable fibres together with traditional steel reinforcement, the limit state of crack width caused by a constant bending moment is analysed. The analytic model used had already been developed for structures in ordinary reinforced concrete(La Tegola 1977, La Tegola et al.1984). In this analysis, using the constitutive law described above we obtain for a rectangular cross-section the following equations:

- **Moment of first crack m_{pf};**

$$m_{pf} = \frac{1}{3n_{ct}(1-s_{pf})}[s_{pf}^3+n_{ct}(1-s_{pf})^3+3n\mu(1-s_{pf}-\bar{\delta})^2+3un\mu(s_{pf}-\bar{\delta})^2] \quad (1)$$

$m_{pf}=M_{pf}/bH^2\sigma_{cr}$ being in adimensional form the moment of first crack;

b,H width and height of the section;

δ reinforcement covering;

$n=\frac{E_s}{E_c}$; $n_{ct}=\frac{E_{ct}}{E_c}$

$\bar{\delta}=\frac{\delta}{h}$; $\mu=\frac{A_s}{bH}$; $u=\frac{A_s'}{A_s}$

A_s' area of compression steel reinforcement;

A_s area of steel reinforcement in tensile zone;

$$s_{pf} = \frac{x_{pf}}{H} = \frac{n\mu(1+u)+n_{ct}}{(1-n_{ct})}\{-1+[1+\frac{(1-n_{ct})[n_{ct}+2n\mu(1-\bar{\delta}+u\bar{\delta})]}{[n_{ct}+n\mu(1+u)]^2}]^{\frac{1}{2}}\} \quad (2)$$

which is, in adimensional form, the value of the position of the neutral axis.

-Minimum distance between two cracks, l_{min}

The hypothesis of a uniform distribution of the bond stress between concrete and steel reinforcement gives:

$$l_{min}=(\lambda_{smax}-\lambda_{smin})\frac{A_s}{\tau_{ad}p} \quad (3)$$

where

$$\lambda_{smax}=\frac{\sigma_{smax}}{\sigma_{cr}} \ ; \ \lambda_{smin}=\frac{\sigma_{smin}}{\sigma_{cr}} \ ; \ p=\sum_{i=1}^{n}\pi\Phi_i$$

Φ_i is the diametre of steel reinforcement.

- **Maximum distance between two cracks, l_{max}**

$$l_{max}= 2\ l_{min} \quad (4)$$

- **Tension in the reinforcement within the cracks, σ_{smax}**

a- Totally resistant section

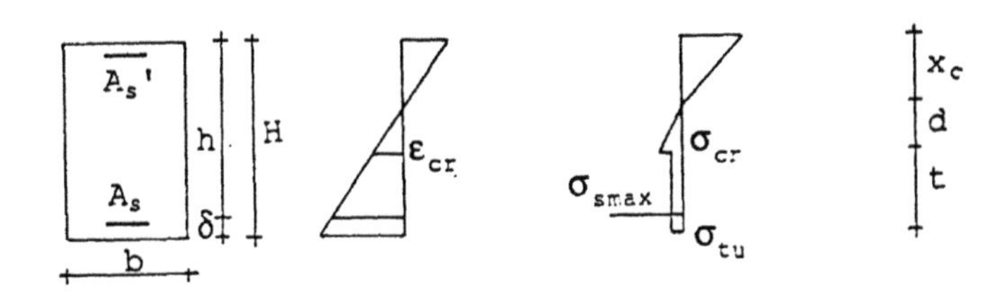

$$s_d = \frac{p_\sigma - n'_{ct}}{2p_\sigma - 1 - n'_{ct}} (1-s)\{1+$$

$$-\left[\frac{(2p_\sigma - 1 - n_{ct})\left[s^2 - n_{ct}n'_{ct}(1-s)^2 - 2\mu n\left((1-s-\bar{\delta}) - u(s-\bar{\delta})\right)\right]}{n_{ct}(p_\sigma - n'_{ct})^2(1-s)^2}\right]^{\frac{1}{2}}\} \tag{5}$$

$$m = \frac{1}{3n_{ct}s_d}[s^3 + (1 - \frac{3}{2}p_\sigma)n_{ct}s_d^3 + \frac{3}{2}n_{ct}p_\sigma(1-s)^2 s_d +$$

$$+ n_{ct}n'_{ct}(1-s-s_d)^2(1-s-\frac{s_d}{2}) + 3n\mu(1-s-\bar{\delta})^2 + u(s-\bar{\delta})^2] \tag{6}$$

$$\lambda_{max} = \frac{n}{n_{ct}} \frac{1-s-\bar{\delta}}{s_d} \tag{7}$$

In these equations the following values are adopted:

$$p_\sigma = \frac{\sigma_{tu}}{\sigma_{cr}} \; ; \; s = \frac{x_c}{H} \; ; \; s_d = \frac{d}{H} ; \; n'_{ct} = \frac{E'_{ct}}{E_{ct}} \; ; \; m = \frac{M}{bH^2\sigma_{cr}}$$

(in the case of the σ–ε law with traction as described in curve b we obtain $E'_{ct}=0$)

b- Partially resistant section

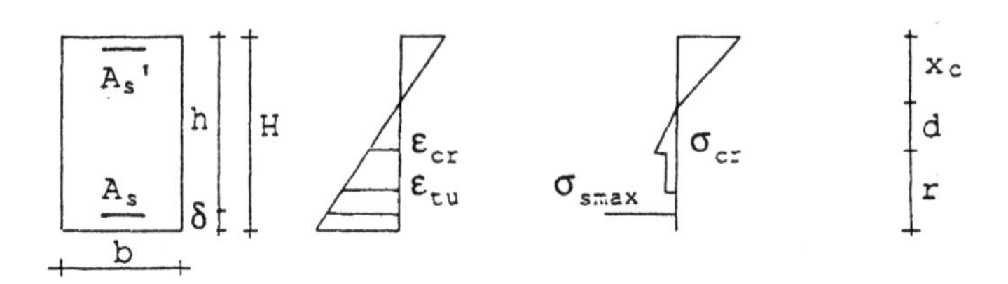

In this case we obtain

$$s_d=\left\{\frac{s^2-2\mu n[(1-s-\bar{\delta})-u(s-\bar{\delta})}{n_{ct}[1+2p_\sigma(\frac{\varepsilon_{tu}}{\varepsilon_{cr}}-1)+n'_{ct}(\frac{\varepsilon_{tu}}{\varepsilon_{cr}}-1)^2]}\right\} \tag{8}$$

$$m=\frac{1}{3n_{ct}s_d}\{s^3+n_{ct}s_d^3[1+\frac{3}{2}p_\sigma(\frac{\varepsilon_{tu}^2}{\varepsilon_{cr}^2}-1)+n'_{ct}(\frac{\varepsilon_{tu}^3}{\varepsilon_{cr}^3}-\frac{1}{2}\frac{\varepsilon_{tu}^2}{\varepsilon_{cr}^2}-2\frac{\varepsilon_{tu}}{\varepsilon_{cr}}+\frac{3}{2})]+$$

$$+3\mu n[(1-s-\bar{\delta})^2+(s-\bar{\delta})^2]\} \tag{9}$$

$$\lambda_{max}=\frac{n}{n_{ct}}\frac{1-s-\bar{\delta}}{s_d} \tag{10}$$

The solution of the system of equations relative to cases a) and b) can be obtained by successive iteractions; m is known, s is assigned a value, s_d is obtained and the correspondent $\bar{m}$ is calculated. The solution is reached when $m=\bar{m}$.

-Tension in steel reinforcement in the part where cracking is about to take place, σ_{smin}

In dimensional form the equations which resolve the problem are:

$$s^3-3(1-\bar{\delta})s^2-\frac{3n_{ct}}{1-n_{ct}}[2m+(6\mu\frac{n}{n_{ct}}u+1)(1-2\bar{\delta})]s+$$

$$+\{\frac{6n_{ct}}{1-n_{ct}}[m+\mu\frac{n}{n_{ct}}u\bar{\delta}(1-2\bar{\delta})]+\frac{n_{ct}}{1-n_{ct}}(1-3\bar{\delta})\}=0 \tag{11}$$

$$m=\frac{1}{6n_{ct}(1-s)}[6\mu nu(s-\bar{\delta})(1-2\bar{\delta})-s^3+3s^2(1-\bar{\delta})-n_{ct}(1-s)^2(1-s-3\bar{\delta})] \qquad (12)$$

$$\lambda_{min}=\frac{1}{n_{ct}(1-s)}\{\frac{1}{2\mu}[(1-n_{ct})s^2+2n_{ct}s-n_{ct}]+un(s-\bar{\delta})\} \qquad (13)$$

-Crack width, w_{max}

Using

$$\Delta_s=\int_0^{l_{min}}\varepsilon_s(z)\,dz \qquad (14)$$

to indicate the sudden lengthening of steel reinforcement in tension in the section of a beam between two cracks, and using

$$\Delta_{ca}=\int_0^{l_{min}}\varepsilon_{ca}(z)\,dz \qquad (15)$$

to indicate the lengthening of the concrete in tension at the level of the steel reinforcement, the maximum width of the crack is given by the relation

$$w_{max}=2(\Delta_s-\Delta_{ca}) \qquad (16)$$

a-Calculation of Δ_s

Using the above hypotheses we obtain

$$\Delta_s=\frac{\sigma_{cr}}{E_s}\,\frac{\lambda_{max}^2-\lambda_{min}^2}{2\tau_{ad}\,p} \qquad (17)$$

b-Calculation of Δ_{ca}

With reference to a generic cross-section part at l_{min} we obtain

$$\Delta_{ca}=\frac{\sigma_{cr}}{E_{ct}}\int_0^{l_{min}}\frac{\sigma_{ca}}{\sigma_{cr}}dz \qquad (18)$$

This quantity must be calculated by use of a numerical method, following the procedure show above in (La Tegola 1977, La Tegola et al.1984).

4 Moment-curvature relationship

The analysis of the behaviour of structures in fibre-strengthed reinforced concrete at the service limit state can be easily carried out using moment-curvature diagrams.These diagrams can be obtained analysing the tension and deformation state of the cross-sections when subjected to stress. They are influenced by the presence of the fibres.

In fig.1. the m-χ relationships are shown for the case of reinforced structures with rectangular cross-sections, with varying volume-percentages of fibres.

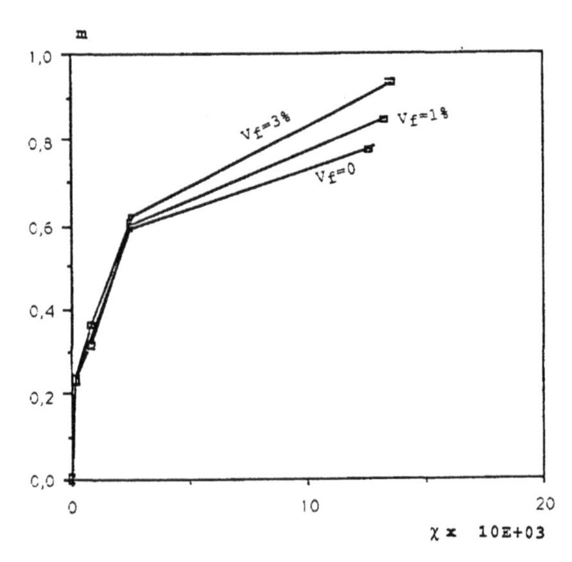

Fig. 1. **Moment-curvature curves** [Plain concrete(V_f=0); Sisal-concrete(V_f=1%, V_f=3%)]

5 Numerical analysis

In order to valuate the influence of vegetable fibres on the limit state of crack width, some numerical analyses were carried out relative to reinforced concrete beams with rectangular cross-section in flexure. The quality of the concrete and the steel, and the quantity of reinforcement were set as constant parametres; the type and the quantity of fibres,however, were variable.The results are shown in the following diagrams (Figs. 2., 3. ,4.):

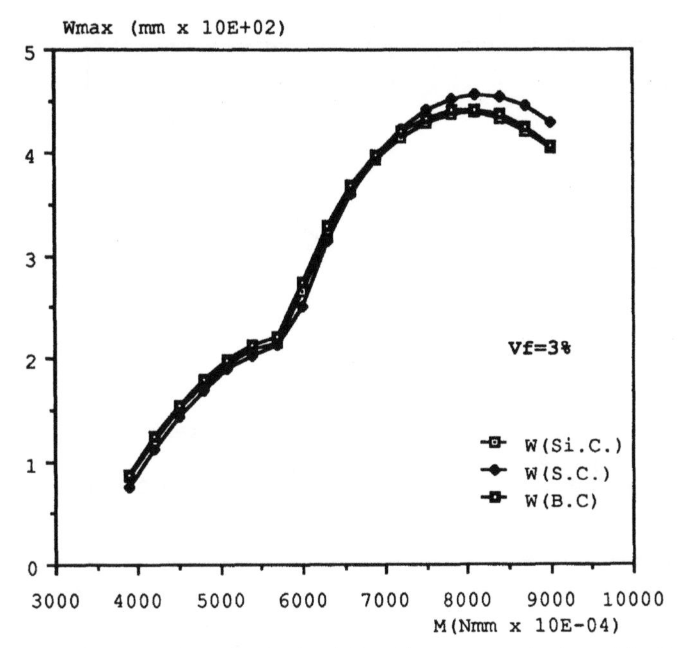

Fig. 2. M versus Wmax[S.C.(Steel-concrete); Si.C.(Sisal-concrete); B.C.(Bamboo-concrete)]

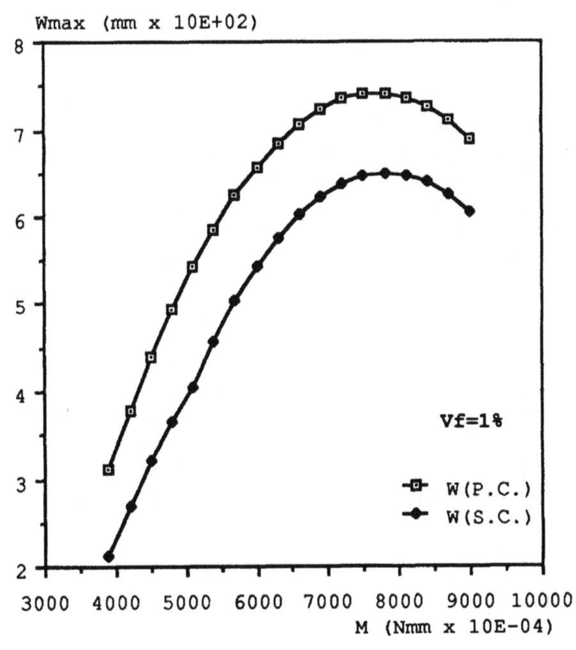

Fig. 3. M versus Wmax [P.C.(Plain concrete); S.C.(Sisal-concrete)]

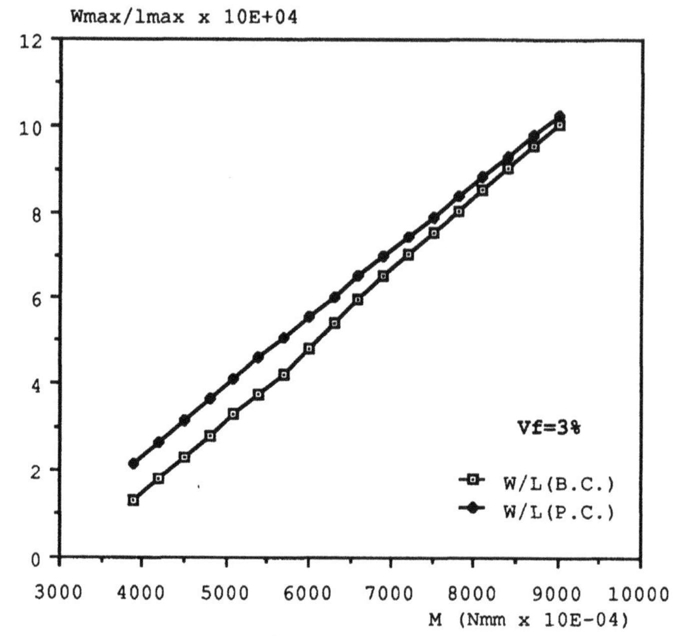

Fig. 4. M versus Wmax/lmax [P.C.(Plain concrete); B.C.(Bamboo-concrete)]

6 Conclusions

On the basis of the results of this numerical investigation the following conclusions can be drawn:

-reinforcement fibres produce more widspread cracking with a reduction in crack maximum width; the greater the volume of fibres, the more evident is this phenomenon;

-the effect of vegetable fibres is virtually the same as that of metallic fibres, especially for low quantities of fibre V_f;

-The moment-curvature graphs of fibre-reinforced concretes are in the initial phase, more markedly concave compared to concretes without fibres; this points to an improvement in the response of slender structural elements with regard to stability. The use of fibres, therefore, leads to an increase in the rigidity of structural elements, which are indeed less sensitive to imperfections. This guarantees greater margins of safety in the above-mentioned problems of stability.

7 References

La Tegola, A.(1977) Sulla fessurazione nelle travi inflesse in cemento armato. **Giornale del Genio Civile**,n°7-8-9,Roma.

La Tegola, A. Ombres, L. and Totaro, N. (1984) **Sulla fessurazione delle travi inflesse in cemento armato.Confronti teorici e sperimentali.** Giornale del Genio Civile, n°7-8-9, Roma.

Satyanarayana, K.G. Kulkanrni, A.G. Sukumaran, K. Pillai, S.G.K. Cherian, K.A. Rohatgi, P.K.(1981) On the possibility of using natural fibre composites, in **Composite Structures**(ed. I.H. Marshall), Applied Science Publishers, London, pp.618-632

Swift, D.G.(1981) The use of natural organics fibres in cement:some structural consideration, in **Composite Structures**(ed. I.H. Marshall), Applied Science Publishers, London, pp. 602-617.

Swift, D.G.(1985) Sisal-cement composites and their potential for rural Africa in **Composite Structures**(ed. I.H. Marshall), Applied Science Publishers, London, pp.774-787

13 POSSIBLE WAYS OF PREVENTING DETERIORATION OF VEGETABLE FIBRES IN CEMENT MORTARS

M.E. CANOVAS, G.M. KAWICHE and N.H. SELVA
Polytechnic University of Madrid, Spain

Abstract
Cement mortars reinforced with vegetables fibres have shown to be suitable for low-cost housing. However, the mineralization of the fibre by the cement alkaline with time makes the composite brittle. Experiences indicates that this embrittlement can be prevented using natural products only or combined with some other elements and still remain a low cost composite in all aspects.

This work contains an intensive revision of the structure of the fibre and the possible means of their mineralization in cement mortar including impregnants with possible means of blocking water penetration in the fibre and their reactions with alkalis.

The work end with some experimental results which prove this.

1 Introduction

Composites reinforced with fibres are as old as the creation of living multicells in the world and, based on this natural reinforcement the man conceived the idea of reinforcing brittle materials with organic matters long before Christ; that many years ago he reinforced clay with straws in order to increase the resistance obtaining adobes.

Technological development permited the creation of new materials which are partially reinforced with different industrial products with better results; for example concrete reinforced with steel. However, problems such as ambiental pollution originated in the elaborations, scarcity of prime materials, increase in the energy cost, etc., forced the research to develop new materials based principally on the use of local products of main availability and which needs appropiate technology.

Varios researches carried out so far in line with the above have shown that cement mortars and soil-cement reinforced with vegetable fibres can be used with success because they have initial mechanical properties almost similar to cement mortars reinforced with steel fibres, plastics, asbestos, etc. However, the major inconvenience which these composites present lie in the easy mineralization of the fibres by the alkalis present in the cement which make the product brittle with time and for that unsuitable for structural purposes.

At the present exist several researches directed to avoid or resist this effect, but such studies are limited on solutions based on the use of cheap products locally availables if possible and which can be applied with appropiate technology. In any case, the studies require deep knowledge of the nature and structure of the fibres, whose physical-mechanical properties are intensively affected because their nature is sensibly different from the matrix.

This paper is part of the investigation of the Doctor thesis carried out in the Department of Civil Engineering-Construction in the Polytechnic University of Madrid and whose objective is to contribute to the solution on the problem of vegetable fibres mineralization in the cement mortars. This work offers results of the various experiments carried out which show that the use of certain natural products contribute a great deal on solving this problem.

2 Fibre descriptions

Up to the moment, the vegetable fibres used in cement mortar reinforcement are the commercial types which are extracted by different method such as beating, bacteria desintegration in water, chemical disolution and mechanical scraping. Such methods produce generally some modifications in the fibre structure, disolves parcially the cementing polymers, relaxes the molecular bonds, etc., all of which increases the absorption and decreases their mechanical resistance. For this reason there are a great variations in the physical and mechanical properties of the fibres from the same plant.

Moreover, the fibres have variable characteristics which depend on the type, origen, age and species. Thus agaves are comprised of 20 genuses of approximately 700 species which differ much in their physical-mechanical and chemical properties, still only 3 of them have commercial value. Consistently, a knowledge of the fibre structure and composition is essential in the search for methods which can increase or improve their durability.

Vegetable fibres are complex materials formed by narrow and stretched cells which reinforce the plant stems and leaves; they are variable in size and form within the same fibre, run a long the length of the fibre in a form of well oriented fibrils made up cellulose and which act like a reinforcement of the cells.The fibrils are found oriented in complex layers in the cells and they have crystalline and amorphous regions. The content and volume of the cells, as well as the orientation angle of the fibrils and the densities of the crystalline and amorphous regions control the mechanical properties and the water absortion of the fibres. The cells and their fibrils are cemented by the following natural polymers: lignin, hemicellulose, pectin and extensin. These cementing materials are affected by miscellaneous factors like heat, alkaline and acid dissolutions, etc., which therefore reduce the durability of the fibres.

3 Influential actions on the fibre degradation

The principal factors which induce the fibre deteriorations are:

The mechanical actions of extraction which produce an increase

fragility and destruction of the internal and external structure of the fibre with a consequent loss of mechanical resistance. During their service in the matrix the fibres suffer similar damage due to dimensional changes of both the matrix and the fibres originated by environmental changes.
The thermic changes in the matrix induce tensional forces in the fibres which sometimes can exceed their limit of resistance.
Dry and wet cycles modifies the fibril orientation of the fibres which when combined with compressive stresses due to mortar shinkrage can cause a lot of damage to the fibres. Also the drying and welling cycles provoke continuos abrasion between the fibres and the matrix due to their differential coefficients of friction and expansion. This abrasion cause strain (fatigue), reduce the contact surface and produce cuts on the fibres.

All these actions which are present in cement mortars reinforced with vegetable fibres affects the fibre durability and continously reduces their resistance capacity, but also one can analyse the effects of the temperature and humidity on the fibre. These induce dimensional and volumetric changes in their structure beside chemical reactions.

The high temperatures produces structural changes because of the reorientation of the polymeric molecules to the new thermic conditions which result in an increase of unoccupied spaces, this way favoring more absorption of solutions of degradation to the fibres.

With respect of the humidity its principal effects are:

Increase the diameter of the fibre which provoke intermolecular and intramolecular disorder, moreover at the same time it increase their permeability, softening the fibres and reducing the temperature of softening.
The alkali solutions presents in the cement mortar poros and the various aggresive chemicals of the air cause serious damages to the fibres through their reactions with the cellulose and the cementing materials which forms the fibres.
Thus, the alkalinity and such solutions provoke solvatation of the hydroxil groups with cell swelling, polysaccharide disolution, alkaline hydrolysis of the glicosidic bonds, degradation of the dissolved hemicellulose, etc. All these reactions which are catalysed by temperatures weaken the fibres structures decreasing their mechanical properties remarkable.

With time the solar radiation produces progressive deterioration on the external layers of the fibres, as well as chemical changes refered to the reduction of the grade of polymerization of the lignin and hemicellulose.

Lastly, microorganisms (bacterias, fungi and mould) ferments and insects degrades also the fibres because they are carbohydrates.

4 Procedures to reduce the deterioration of the vegetable fibres

The factors analysed before including their actions on the vegetable fibres demostrate the complexity of the fibres degradation problem. Protection procedures which contribute to give some satisfactory solutions to this problem are:

Impregnating of the fibres by means of chemical compounds which impede the water and the matrix pore disolutions from penetrating into the fibre or which can react with the alkalis (particularly the calcium hydroxide) thus reducing partial or totally their concentrations.
Sealing of the matrix pore system using chemical compounds which can impide disolutions passing into the matrix.
Using quick hardening cements like aluminious cement which confines the fibres in a minimal diameter tunnel thus avoiding the water and disolution molecules from acting in the fibres with maximum intensity, that is to say, such cements decrease remarkable the fibre swelling when compared with portland cements.
Using neutral cements or cements with additives which reduce its alkalinity or partial substitution of cements with puzzolanic materials.
Combining all above mentioned procedures.

5 Vegetable fibres impregnation

In order to carry out an effective impregnation on the vegetable fibres which can prevent their mineralization is neccesary to take in consideration various factors in relation to the impregnanting agent. It is so because from the previous analysis fibres are heterogeneous materials of great complexity and as such their different components suffer different types of chemical reactions which affect them in various ways due to the actions of substances (ions and molecules) present in the mortars alkalis disolution.

Principal factors to consider in the sellection of the impregnants are related to their nature and structure, solubility and viscosity.

In this way and lacking enough research on the fibre studies carried out so far on the preservation of timber shows that the best results are obtaining by the use of organic compounds than with the inorganic ones because the first ones are insoluble in water, are good water repellent (damp proof), they have great attachment power to the cells of the fibres and they create bigger contact surface impregnant fibre. With an aim of obtaining a better penetration efficacy one ought to take care that its viscosity is not more than that of water which can be achieved by the use of appropiate dissolvents and relatively high temperatures, including adequate duration of treatment.

The impregnants used in our work and which are based on the above observations are: colophony (natural resin), tannin and clove oil; these compounds on top of being naturals have low cost and are economically availables.

The impregnation method used has been Hot and Cold Dipping, which consisted in dipping the fibres in a dip which contains the disolved impregnants in a predeterminates concentration and at a temperature between 80°C and 120°C., and for a certain period of time. Immediately the fibres are transfered to another dip which contain the same impregnants in the same concentration but at an environment temperature. This cycle is repeated several times so that through successives expansions and contractions any air ocluded in the fibres is eliminated thus improving the penetration of the impregnant.

6 Pore sealing and alkali reductions in the cement mortar

According to studies done by several researches the porosity reduction in the cement mortar requires the elimination or at least the limitation of the calcium hydroxide to the most possible because this compound is the one which mineralizes the vegetable fibres most. As such the sealing of the pores of the cement mortar reinforced with vegetable fibres is of great importance and the use of compounds which can react with the calcium hydroxide for its elimination is neccesary.

A bibliographical revision on the above requirement shows that the most adequate compounds for sealing the pores for the results obtained as well as for the economic availability are phosphoric acid, carbonated waters, fat acids, fats and oils, stearites, waxes, natural resins, etc., all have the tendence of reacting with $Ca(OH)_2$ to form insoluble compounds.

It is evident that a lot of more compounds have been tested on the same intention. In which ever case it is necesary to have in mind that in order to have efficient pore sealing the selection of adequate compounds only is not enough but furthermore it must include the physical-mechanical characteristcs of the mortar and the method adopted as such it is important to carry out preliminary studies in order to establish the cuantity of sealing material to employ.

Backed by the already existing informations on the effectiveness of organic additives in the construction materials and on the previous lines of improving the durability of cement mortars reinforced with vegetable fibres, studies were carried out to analyse the influence of the impregnantes used in the fibres in the portland cement mortars.

Materials used as impregnantes and then as additives in the mortar were colophony, tannin and wax. Results of these studies are already published by the authors. In such studies it was observed that the colophony has hydrofobic and hydrophilic characteristics which favorable improves the physical-mechanical and chemicals properties of the cement mortars. Moreover, all mensioned compounds single or combined reduce averagely the water/cement ratio by 20 per 100, absorcion by 60 per 100 and apparent porosity by 66 per 100, when compared with a control mortar of 1:2. In relation with the mechanical resistance they are acceptables especially the flexural resistance and when the samples were heat treated the mechanical resistance increased very much.

The last behaviour is advantageous when considering that the composite will be more employed in hot cimate areas where the high temperature will accelerate the reaction between the alkalis and the additives. Another observation made was that the addition of disolved colophony in turpentine (0.4 M concentration) in alkalis disolutions of calcium and sodium hydroxides with pH = 12 reduce the pH to 6.5 and 9 respectively.

7 Experimental procedures

The results presented in this paper are axial tensile resistance and water absorption of impregnated and unimpregnated sisal fibres, including also the flexural behaviour of portland cement mortar reinforced with impregnated and unimpregnated sisal fibres.

The sisal fibres used are commercial grade A from the Republic of Tanzania. The impregnants used were disolved in turpentine or alcohol The mixtures were agitated by magnet while being electrically heated up to its total disolution. In case of the colophony and tannin disolution it was necessary to filter them. The impregnation process was carried out as explained before submergening a bundle of fibres of 225 mm length into the disolutions.

One completed the impregnation and the drying, the axial tests and absorption were carried and the results are presented in table 1. The axial tensile tests are based on mean value of 10 fibres selected randomly while the water absorption is based on first weighing a small bundle of air dry fibres then submerging into water for 24 hours, thereafter surface drying and weighing them.

The results of the bending behaviour of mortar reinforced with sisal fibres are presented in figures 1, 2 and 3. Such results are based on speciments of 14x48x225 mm cut out from a sheet of 14x225x625 mm and made with portland cement type II-Z35 which complis with the Spanish Norm UNE 80 301, washed river sand of maximum size of 2 mm and water in a mixing ratio of 1:2:0.45 (cement, sand, water) in weight. The volume fraction of the fibres was 2 per 100. The fibres were laid in two layers uniformely distribuited on the 625 mm between thin layers of mortar well compacted by vibration such that each fibre is surrounded by mortar.
The speciments were cured in water for 28 days thereafter the rest were conditioned as indicated in figures 2 and 3.

The flexural test was with a central load increasing at the rate of 3 mm/min with a free span of 200 mm.

Table 1. Influence of impregnants used in the sisal fibres

Type of impregnants	Proportion in weight	Axial tensile resistance (N/mm^2)	Absorption of water (%)
Colophony + Turpentine	1:6	487.0	33
Clove oil + Xilene + Turpentine + Alcohol	13:3:30:1	530.0	14 + 20
Tannin + Alcohol + Xilene	1:30:1	407.5	39
Control	-	570.0	70

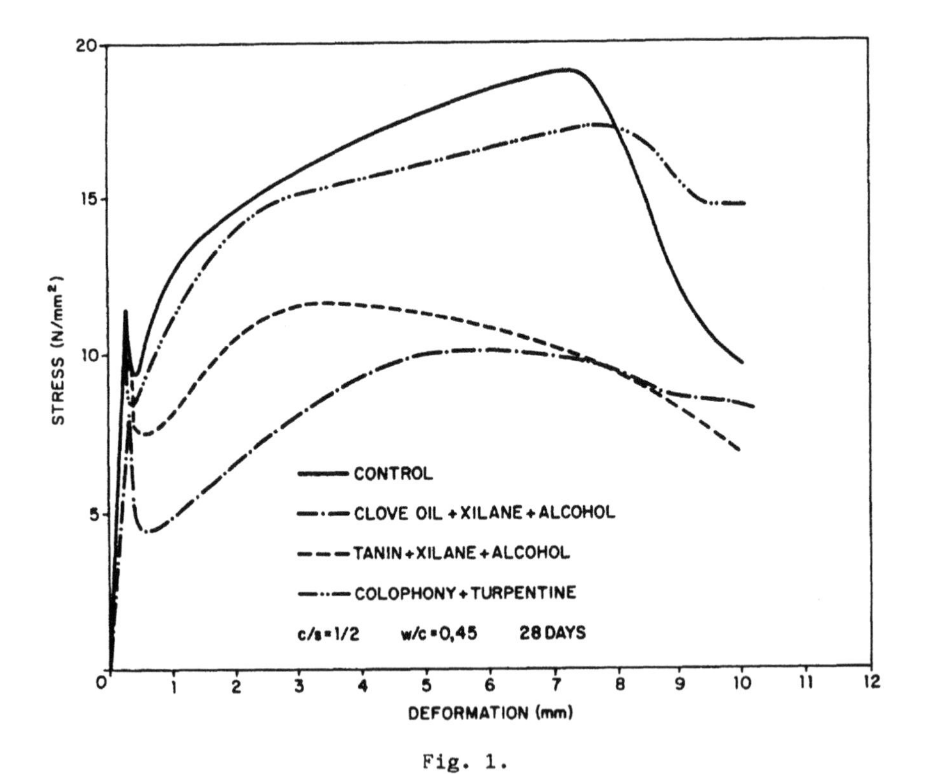

Fig. 1.

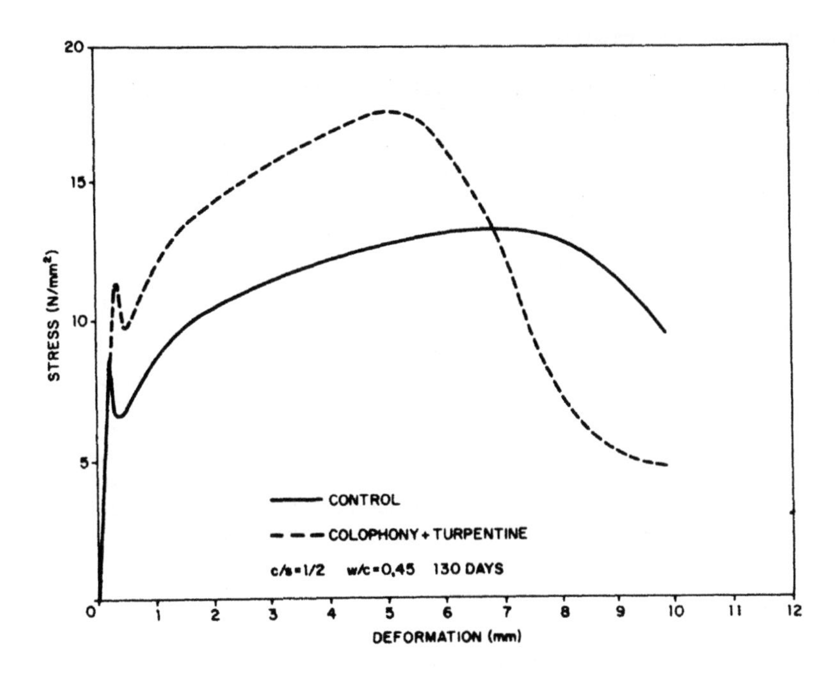
STRESS (N/mm²)
DEFORMATION (mm)
CONTROL
COLOPHONY + TURPENTINE
c/s=1/2 w/c=0,45 130 DAYS

Fig. 2.

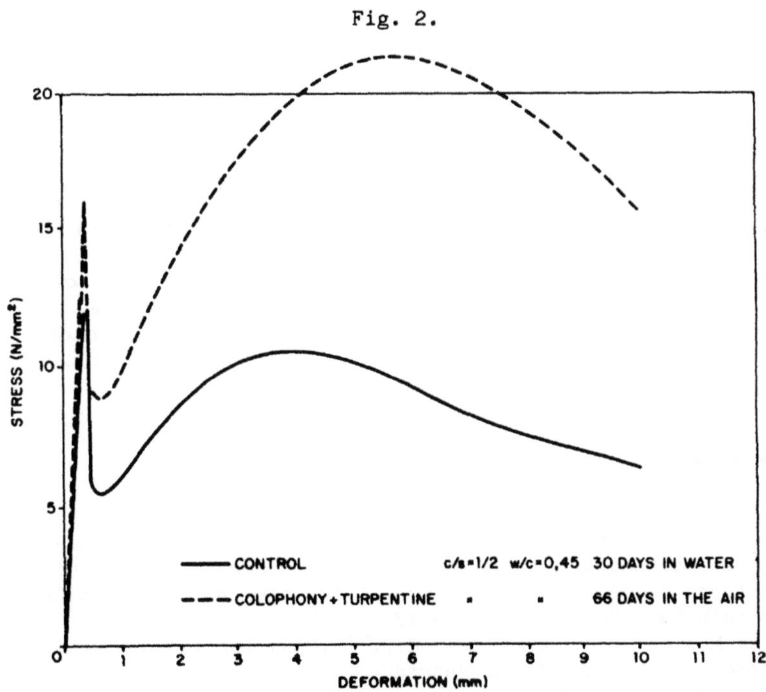
STRESS (N/mm²)
DEFORMATION (mm)
CONTROL
COLOPHONY + TURPENTINE
c/s=1/2 w/c=0,45 30 DAYS IN WATER
66 DAYS IN THE AIR

Fig. 3.

8 Discussion of the results

According with the impregnation results the water absorption of the fibres has been reduced bellow the 55 per 100 of the unimpregnated fibres. This is probably due to the reaction of certain metals present in the fibres with the resinic acids, fat acids, glycerin and others acidic residous of descomposition and present in the organic compounds used. Such reactions have a tendence of causing modifications which accelerate the formation of an elastic solid and impermeable thin film of the organic compounds. Also the metal soap formed is impermeable with soft and greasy nature and with fungicide characteristics.

Results of flexure resistance of mortar reinforced with colophony impregnation shows success on the effect of reducing mineralization of the fibres (Figures 1 and 2) when compared with mortar reinforced with unimpregnated fibres. The results are more surprising when the fibres are exposed to air during summer time where the resistance is even higher than the rest. This is due to the formation of alkali soaps which is accelerated by heat and because of the characteristics mentioned before the deterioration is greatly reduced.

9 Conclusions

Attemps to reduce mineralization of the vegetable fibres by the use of colophony as an impregnant has proved to be successful when compared with the mortar reinforced with unimpregnated fibres and the results obtained are as good as those obtained using others impregnants. This must be considered as the first attempt capable of being improved particularly controlling the content of all acids present in the organic compounds, etc.

The composition and the forms of reaction of certain metals with colophony to form metalic soaps is almost similar to that of tannin, fats, glycerin and other organic compounds with almost similar composition.

Studies carried out to find the influence of tannin and colophony to the reduction of the cement alkalis and the sealing showed great effect in both. The reduction of the permeability and the flexure resistance are increases with heating which indicates the reduction of the alkalis in the mortar.

10 References

Ambros, E.J.and Easty, J.N. (1.977) Biomolecular celular.

Cánovas, M.F. Kawiche, G.N. and Selva, N.H. (1.989) Influence on the physical-mechanical properties of portland cement mortars of admixtures of colphony and tannin. **Materiales de Construcción.** Instituto Eduardo Torroja. Madrid.

Cil, A.V. Mugica, M.G. and Ochoa, M.G. (1.969) Los taninos vegetales. **Ministerio de Agricultura.** Madrid

Gram, H.E. Durability of natural fibres in concrete (1.983) **CBI Research** fo 1-83.

I.I.C.E. Tratamiento y conservación de la madera. (1.976) **DL-M 24261**
Kirk, R.E.and Othmer, D.F. Enciclopedia de Tecnología Química (1.962)
Luis, E.W. and Edwin, C.J. (1.952) Wood Chemistry. **Reinhold Publ. Corp.** 2º Vol 1.
Moragues, A. Macias, A and Andrade C. Equilibria of the chemical composition of the concrete pore solution. Part. 1. Comparative study of synthetic extracted solution (1.987) **Cement and Concrete Research** Vol 17 Nº 2.
Mukherjee, P.S. and Satynarayana, K.G. (1.984) Structure and properties of some vegetable fibres. Part 1. Sisal fibres. **Journal of Mat. Science.**
Renata, M. and McGover, S.D. (1.970) Relation of crystaline dimensions and fibrillar orientation to fibre properties. **The physics and chemistry of wood pulp fibres. Special Tech. Assoc. Pub.** Nº 8.
Wegener, F.D. (1.984) Wood chemistry ultrastructure reactions. **Walter de Gruyter.** Berlín. New York.

14 MORTAR REINFORCED WITH SISAL – MECHANICAL BEHAVIOR IN FLEXURE

A.C. De C. FILHO
Federal Technical School of Pernambuco, Recife, Brazil

Abstract
Sisal fiber as reinforcement of cement mortar was researched. Three different fibers length ratios, 1.80 cm, 3.70 cm and 5.60 cm, were tested in different composites.
Samples with different mix proportion were compared to brittle matrix in which no sisal fibers were used. This comparison was done by the flexural strengths at the age of 7, 28 and 63 days. In addition the impact strengths was also determined. Composite were made with variable volume of fibers and constant length ratios; variable length ratios or humidity conditions of fibers (dry and saturated fibers) were used in composites with constant volume of fibers.
Flexural strength decreased in the composite compared to brittle matrix, but only in the absolute value. The composite showed an elasto-plastic behaviour after multiple cracking. A significant impact strength was noticed in all cases.
Keywords: Sisal fiber, Cement mortar, Flexural strength, Impact strength, Brittle matrix, Elasto-plastic behavior, Cracking.

1 Introdution

The incontrollable raise of prices of basic material necessary for developing housing projects makes impracticable policies of poor country governments. As the high deficit of housing usually reaches the low income populations, the solution comes forth with the adaption of alternative technologies which, if not innovating, they bring back old and forgotten ways of building.

Thus, the divulgation of new materials identified with traditions and usages, as well as adapted to climate and socio-economical conditions of the region would help the low income families to build their own houses with costs compatible with their monthly incomes and without exposing the minimal patterns of housing and perfomance.

If there is decision and investment for research, the solution will be based in the next points:

a) Usage of raw material from natural and renewable sources.
b) Implementation of the usage of industrial wastes which reach low marketing prices.

It's important to underline that these solutions will only be successful if they come from technologies easily taught to non-specialized workmanship and comunities.

2 Reinforced materials with vegetable fibers. Sisal fibers

The natural fibers, mainly the vegetable ones, are produced in many countries of the third world. Most of them require a low degree of industrialization in their processing and in their improvement. The amount of energy required for their cost are very low, if compared with the strength improvement which is gained.

The use of fibers in cement composites aim to modify certain properties (Agopyan,1983), such as:

- .Increasing the flexural strength.
- .Increasing the impact strength.
- .Controlling the rupture cracking and modifications giving improvement in the ultimate tensile strength.
- .Modifying the material rheological caracteristics.

Many factors act on the properties of composites reinforced with fibers. The diversity of factors, allied to their constituents, has identical responsibilities on the properties influenced by them, not only for the fresh state, but also in the hardened state.

AZIZ(1981) lists, according to Table 1 some of these factors and the development of technologies suitable for identifying researches will contribute for them to be solved.

Table 1. Factors affecting the natural fiber reinforced materials.(AZIZ,1981).

Factors	Constituents
1.Type of fiber	Coconut, sisal, sugar-cane, bamboo, jute, elephant-grass;
2.Fiber geometry	Length, diameter, cross-section, conformation;
3.Fiber shape	Monofilament, multifilament;
4.Fiber surface	Rough, glossy, overlapped;
5.Matrix property	Type of cement, type and granulometry of aggregate, type of additive;
6.Mix proportion	Water/cement ratio, consistence, fiber content;
7.Mixing method	Type of concrete mixer, fiber additions sequence, mixing time;
8.Molding method	Vibration conventional method, sprinkling method, extruding method, projecting method, pressing method;
9.Curing method	Conventional, special.

The sisal fiber reinforced materials show feasibility for usage in some kinds of parts and components for civil engineering with their capacity improved by fibers, the composites can resist to flexural effort. Associated to the way the cracking are controlled, their use is possible in several fields of binding construction, permitting different kinds of components to be made, such as: panels, smooth and corrugated tiles, waters tanks, wash basins. (BAHIA,1982;BAHIA,1984; GUIMARÃES,1987;ANDRADE et alii,1987;HINGIRA,1987).

3 Sisal fiber disposibility

Sisal is a tropical plant of the "AMARYLLIDACEAE" family. The "SISALANA PERRINE" species is the only in the "AGAVE" genus to be successfully cultivated in Brazil.

Originated from the tropic, this caudex plant has rigid, smooth, bright-green leaves, about 10 cm wide and 150 cm long. It doesn't usually frutify. It flourishes from 5 to 8 years of age and then it dies.

Its culture is successful in places where the temperature is high most of the year, in permeable silicious-argillaceous soil, relatively profund, about pH 5,5 (IPT,1983).

Excepting Bahia, Paraíba leads all other Brasilian States, producing over 30% of the total national production. The other states contribute with less than 10% of the total. Ceará has the smallest percentage of the total national production - about 0,1%.(IBGE,1988; CARVALHO FILHO,1989).

In Paraiba, the cropping area according to IBGE'S information(1988) is approximately 95,000 Ha, comprising 71 cities in Sertão, Agreste, Brejo, Litoral of the state.

The Curimatau region, in Paraíba's sertão, has been responsible for a production of 31,2% with a fiber average efficiency of 851 Kg/Ha. This value can be considered normal for the national productivity patterns but it lies below the levels of African producing countries where the productivity reaches 2.000 kg/Ha.(BAHIA,1984).

4 Program of study

In the program of study elaborated for this research, three main points were defined as give emphasis to verifying the flexural mechanical behavior of the sisal fiber reinforced composites. They are:

- .Comparing with a matrix mortar, composites of different volumes of addicted fibers, the average length of the fiber being kept.
- .Comparing with a matrix mortar, composites of a constant fiber volume and varying the length of the addicted fibers.
- .Comparing with a matrix mortar, composites where the fiber volume, as well as the average fiber length is kept constant, but in their initial condition, dry or saturated fibers would be used. For that, they should be kept in water for 24 hours before being used.

Thus, the test specimens were elaborated with different mix proportion as to satisfy the studying points identified above, according to Table 2 that follows.

Table 2. Samples with different mix proportion used in the research.

Samples	Composition cement (kg)	sand (kg)	sisal fibers (kg)	W/C ratio (l/kg)	Fiber content (volume %)	Density (g/cm^3)
Mix 1	12.0	24	-	0.46	0	2.16
Mix 2	13.0	26	0.297	0.50	1.12	2.01
Mix 3	12.5	25	0.438	0.50	1.70	2.11
Mix 4	12.0	24	0.574	0.58	2.20	1.99
Mix 5	12.0	24	0.734	0.64	2.70	1.94
Mix 6*	12.0	24	0.421	0.54	1.66	2.10
Mix 7*	12.0	24	0.421	0.58	1.62	1.99
Mix 8+	12.0	24	0.421	0.58	1.62	2.01

Obs:
(*)Mix 6 and mix 7 were obtained the length of their fibers. Fibers 1.80 cm and 5.60 cm long respectively were used.
(+)Mix 8, the fibers were previously saturated, lent in the original ratio w/c, the water, carted by the fibers into the composite was considered.

For making a comparison among the different situations in which the prepared samples were shown, two evaluating points were stablished. First, flexural test were to be made as to obtain the flexural strength for the different mixtures at varied ages. Second, to evaluate the behavior of the same material, in the same samples, when under impact charge effects.

With the results of the flexural and impact test, to analyse the improvement factors which the fibers bring to the composites and to the methodological influences of the mix proportion, mixture condesation and cure used in the process, evaluating the adapted steps so that the results could be used as recomendation to be followed in the future when components for civil building is to be prepared.

5 Flexural test

The flexural test applied to the test specimens aims to stablish the flexural strength of the samples when under a continuous gradative changing. Thus, the influencing factors such as: fiber critical volumes, fiber critical length, water/cement ratio, fiber/matrix adherence and other factors inherent to the composite will probably be detected.

The method that was adapted for the test was based in what was stablished by ASTM C-683-82 (ASTM,1982), the charges would be applied to the sample though 4 cutlasses. As the test specimens used in the work are planteshaped and not prismatical-rafter shaped as indicated by the method, added to the behavior next to the ductile state that the material would have and, for that, subject to great deformites before the rupture, there was necessity to choose a test condition in which the supporting elements wouldn't take the assembly to hiperstaticity, once another variable will appear in the experimented test specimen: its stiffness.

The solution was to adpt a flexural test device, according to the above mentioned method, changing one of the curlasses. For that, a kind of support was project as to permit two degrees of freedom to the acting efforts, i. e., it would allow only vertical reaction and no opposition to the horizontal and tension efforts.

The projected supporting device is made up of three cylindrical rollers, bounnd thoung axes to two lateral plates following a triangular distribution so that two of them would be place in the lower part working as a basis and the other one, on the upper part where it will be supported by the upper face of the test specimen.

This way, the friction effects would be minimized and the equipment efficiency would be secured.

5.1 Results

Table 3 summarizes the results of the flexural test for the different mix proportion suggested in the program of study.

Table 3. Summary of the results in test of flexural strength.

Samples	Flexural strength (MPa) Age of the test specimens (in days)		
	07	28	63
Mix 1	4.2	4.8	6.0
Mix 2	3.8	5.0	5.6
Mix 3	4.2	4.7	5.4
Mix 4	3.3	4.0	4.6
Mix 5	2.9	3.5	3.5
Mix 6*	3.6	4.6	5.2
Mix 7*	3.5	3.7	4.2
Mix 8+	3.7	4.2	5.0

From the moment we can keep fixed the volumes of the fibers and vary their length for the various composites (mix 3, 6 and 7) the flexural mechanical strength losses, in relation to the matrix (mix 1) manily concernig the lengths 1.8 and 5.6 cm (mix 6 and 7). For the length 3.7 cm (mix 3), the strength has values which are very close to the average values of the matrix composite strength, being the same on the seventh days and having a loss not greater than 2% in the 28th days. For the age of 63 days, its loss is about 10% in relation to the matrix, but the results obtained the dispersion as to the average values was lower than the fiberless composite.

These results make us believe that we are very near the fiber critical length in the suggested composite and it shows what was researched in bibliography about the compatibility of the fiber average length related to certain volume of it without the risk of balling.

Considering that the matrix had been elaborated with a water/cement ratio 0.46 and that the composites, according to their workability, had had a rather greater water/cement ratio, the final result shows a real additions in the strength of composite with fibers, the length of which is not above 4.0 cm. It also shows that during the mixing, the fibers had had a random distribution. Possibly, greater lengths would make the fiber distribution inside the mortar more difficult and thus, the dispersion would be much greater than the one which was found.

For the ages of 28 to 63 days an acceptable performance occurred also for the fibers of average length 1.8 cm (mix 6). Since there was a smaller dispersion in its average length 3.7 cm(mix 7) and also that the water/cement ratio used was 0.54, what causes loss of strength, the lengths 1.8 cm and 3.7 cm show acceptable perfomance to the composite studied.

When we report the second condition of the research, one can observe a better strength in the samples whose volumes of incorporated fibers are between 1.12 and 1.7% (mix 2 and 3). Thus, for the average length of fibers fixed in 3.7 cm, the composite shows better strength at the age of 7 days for a volume of 1.7% of fibers.

However, for the age of 28 do 63 days, the better perfomance refers to the volume 1.12%, although a greater dispersion is noticed. Probably the fiber critical volume acceptable by the composite is located in that range.

It must be emphasized that, when we mention a better strength, we don't refer only to the absolute value obtained in the test machine. There's another non-registered value which concerns the material behavior after the matrix cracking where the piece breakdown doesn't occur, once the fibers make their true work contibuting for an additional strength to the composite, concerning the capacity of absorbing deformations. Even after the piece cracking, there's no separation of the broken pieces.

Concerning the variation of the initial condition of the fiber humidity, either dry or saturated, the results show that when the fibers are previosly saturated, losses of strentgh occur. It's not possible to say if the initial swelling, that must occur when the fibers acquire a saturation condition, influence the fiber/matrix linking, but it seems that there are probably losses as for the physical linking due to the retraction that the drying up will bring to them, as well as the superficial saturation condition must contibute strongly for a humidity concentration in the interface, and consequently, a differentiated water/cement ratio, which would also damage the linking strength.

The results obtained show that the influence caused by the initial state of the fiber humidity doesn't reflect a very sensible variation as to characterize an obstacle to their use. It was evident, though, during the preparation period, that there was a greater difficulty to make the composite homogeneous.

6 Impact test

For obtaining additional information about changes occured in the matrix mortar when the sisal fibers were added, impact tests were made in the same test specimens already tested as to flexion, since the increase of the impact strength is the most observed property was: as soon as the flexural test was finished the test specimens already broken were marked in the remaining part so as the result would be a plate measuring 150 mm x 150 mm x 31.7 mm.

For the impact test we chose to check the material performance when subject to the action of impact with a body of a known mass and the energy was supplied by the body to provoke the appearing of the first visible cracking as well as the necessary to open the same cracking to 1.5 mm.

We can't forget, however, that the real energy that provokes the visible damage to the test specimens suffers consecutive impacts and, then, the cracking, as well as the opening of these cracking are due to energies intermittently received. But as it's a comparative test, we believe that the results have acceptable values and don't invalidate the methodology used.

The equipment used for the test was developed by another researcher of EPUSP, Engineer Holmer Savastano Jr. who has used it for his studyies about the use of vegetable fibers - coconut fibers - as modifier of the fragile matrixes behavior. (Savastano Jr.,1988).

6.1 Results

The results obtained in the impact test carried on in the test specimens at the age of 7, 28 and 63 days, have showed strength improvements due to the matrix reinforced with the sisal fiber, independently of the length or volume, raised the energy to values near 150% of the one found for the matrix, when done for the appearing of the first visible cracking. and above 195%, when the opening of the cracking came to 1.5 mm.

7 Comment on the results

The flexural strength shown by fiberless matrix, always superior to that of the fibered composites, is mainly due to the low water/cement ratio used in the mixture so that it becomes necessary to improve the studies about the total amount of water involved in the making of the material, since a greater volume of water necessary to assure the workability. That's an intrinsic exigence of vegetable-fibered materials. Thus, it will need a greater attetion in searching a method that evaluates the material rheological behavior in better levels and that makes it possible to reach the improvement of the amount of water necessary and so, to minimize its effect on the loss of strength of composites.

Another observation made in the results, and that deserves considerations in that, although there was decay of the composite strength values as compared to the fiberless matrix for different conditions tested, those variations are in an interval considered normal since the results obtained reach variations not above 20%.

In general there were raises of strength as the time passed what made it possible for us to reduce the supposed influencing conditions of the chemical attack on fibers. Also, we must consider the importance of fibers on material modification before the flexion solicitations, since they did their job adequately, allowing the after multiple cracking without collapsing the composite - ductile behavior - which was guarante in all situations tested.

In verifying the impact strength, the values found were very significant. They prove that the material has a satisfactory performance when under dynamic charge effort and, consequently, with a good behavior that permits its use in civil engineering. From what was observed we can think it is necessary to keep the behavior in all cases tested, but with a decay in the resisting energy value. This may be due to a loss of adherence in the fiber x matrix interface for the passing of the time.

8 Final considerations

The material studied has been very promising to be used in civil engineering; however we know that, in other studies made in Brazil and abroad, a durability of no more than 2 years was found. This can and must be solved with the development of researches aiming to find more economical forms than the ones already known and that could be used to minimize the alkalinity attack of the matrix on the lignine that makes up the fiber.

However, even though the fiber durability is one of the compromising factors in selecting the material to be used in civil engineering, we believe that, as for the disposibility of fibers in the producing areas, as for the possibility of improvement in the mechanic characteristics, and as for mixture easiness and piece-shaping, and, since it is easy to substitute the damaged components in a constructive system, adapted to the material. Its use will be quite attractive for building houses for the low income populations.

9 Acknowledgments

To CAPES, for awarding a scholarship thoungh the PICD plane.

To Professor Dr. Vahan Agopyan for orientations during the research and the M.Sc. at "Departamento de Engenharia de Construção Civil" of "Escola Politécnica da Universidade de São Paulo".

10 References

Agopyan, V. (1983) Materiais com fibras para a construção civil. Revista Politécnica., 182, 66-70.

Andrade, L.E.Z.M, Neves, C.M.M. (1987) Alternativas de componentes habitacionais: pias e telhas de argamassa reforçada. in Habitec/87 pp.31-38.

ASTM., American Society for testing and materials., Flexural strength of concrete (Using simple beam with third-point loading) C78. in anual book of ASTM standards. Philadelphia.1982. pp.40-42.
Aziz, M.A. (1981). Prosppects for natural fibre reinforced concretes in construction. The international journal of cemente composites and lightweigth concrete. 3(2). pp. 123-152
Bahia, CEPED.(1984) Utilização de fibras vegetais na construção civil. Relatorio 94/84. 61p.
Bahia, CEPED.(1982) Utilização de fibras vegetais no fibro-cimento e no concreto fibras. BNH/DEPEA. 72 p.
Carvalho Filho, A.C. (1989) Argamassa reforçada com fibras de sisal - comportamento mecânico à flexão. Dissertação de Mestrado. Departamento de Engenharia de Construção Civil da EPUSP. 161p.
Guimarães, S.S.,(1987) Fibra vegetal-cimento- resultados de algumas experiências realizadas no THABA/CEPED. in Habitec/87. pp 103-109.
IBGE,(1988) Produção agrícola municipal - 1986. Culturas temporárias e permanentes - região norte-nordeste. vol 13, tomo 1.
IPT, Instituto de pesquisas tecnologicas de São Paulo,(1983) Fichas de caracterização de materiais têxteis - tópicos texteis, 5(1).pp. 7-11.
Hingira, J.,(1987) Sisal-cement sheets & burnt clay brinks-cheap and durable building materials. in Habitec/87. pp 133-142.
Savastano, J.,(1987) Fibras de coco em argamassas de cimento portland para a produção de componentes de construção civil. Dissertação de mestrado. Departamento de Engenharia de construção civil da EPUSP.

15 APPLICATION OF SISAL AND COCONUT FIBRES IN ADOBE BLOCKS

R.D.T. FILHO
Department of Agricultural Engineering, Federal University of Paraiba, Campina Grande, Brazil

N.P. BARBOSA
DTCC, Federal University of Paraiba, João Pessoa, Brazil

K. GHAVAMI
DEC, Pontifical University of Rio de Janeiro, Brazil

Abstract
Adobe houses built of earth in North-east of Brazil, called "taipa" usually have a short lifetime. The average life of the walls of the taipa houses is estimated to be several years and mostly present large cracks after the construction. To overcome these problems, physical and mechanical properties of the abundantly available sisal, coconut fibres and bamboos have been studied and are incorporated into the adobe blocks. The primary results of the investigation for the production of the fibre-adobe blocks carried out at the Federal University of Paraiba are presented in this paper. This is part of a continuing research program which started at the Pontifícia Universidade Católica do Rio de Janeiro, since 1979, and implemented at the Federal University of Paraiba since 1988, to find low-cost and low-energy materials for the solution of housing problems both in the slums of the big cities and the rural housing.
Keywords: Fibre, coconut, sisal, adobe blocks.

1 INTRODUCTION

In recent years, many researchs around the world have began to explore the use of low-cost and low-energy substitute construction materials. Among the many possibilities for such substitutions, bamboo, which is one of the fastest growing plant, coconut fibres and sisal have got a great economic potential. These materials have been used in constructions of bridges and houses for thousands of years in Asia.

With the production of new materials, such as steel, concrete and, in general, factory manufactured materials, at the begining this century, these low-cost and low-energy construction materials were used less and less. Very few scientific works were carried out in universities and research institutions to study the behavior of locally available materials, such as bamboo, sisal and coconut fibres.

However, the drastic cost increase experienced during the

last two decadas using so called conventional building materials, has encoraged the researchers to explore the use of natural fibre in civil construction. The application of bamboo, sisal, coconut fibres for the construction of low-cost housing become more attractive when we look at the population increase and ecological problems caused by the use of timber for the construction of houses.

Since 1979 several research programs into the use of bamboo and coconut fibres have been carried out at Pontifícia Universidade Católica do Rio de Janeiro, PUC-RJ, and have shown very promissing results [1-4]. Two research programs into the use of bamboo, sisal and coconut fibres are in progress at the Federal University of Paraiba.

This paper presents the results of the investigation carried out to find the physical and mechanical properties of sisal and coconut fibres in Paraiba. Two types of water repellent treatments were used to improve the water absorption of the fibres studied [5]. The properties of the local soils used in the construction of taipa houses also were studied. The stress-strain behaviour of the adobe blocks produced manually without and with sisal and coconut fibres, considering different water/earth percentage, also have been investigated. The results of these findigs are also discussed.

2 Physical and mechanical properties of sisal and coconut fibres

Sisal cultivation originated in Mexico and has spread to tropical regions throught the world. In general Brazil, Indonesia and East African countries are the main producers of the sisal fibres in the world. In Table 1 the production of sisal in Brazil and in some Northeast states of Brazil is given [6]. It can be noted that the main production factories of sisal in Brazil are situated in Northeast states of Brazil. Most of the producers in this part are looking for new ways to use their product. Moreover the increase of the sisal production has got an ecological advantages as it has proved to be a mean of arresting desertification of the lands.

Table 1. Sisal production in Brazil and some of northeastern states, in 1000 tonnes.

Year	Brazil	BA	%	PB	%	RN	%	PE	%
1980	235,0	133	56,6	81,0	34,5	13,7	5,8	6,9	2,9
1981	239,2	158	57,7	80,2	33,5	14,4	6,0	6,2	2,6
1982	251,3	150	59,6	81,1	32,3	14,2	5,7	5,8	2,3
1983	180,8	75	41,4	88,5	49,0	12,4	6,7	4,6	2,6
1984	224,7	119	52,9	83,3	37,1	16,1	7,2	5,8	2,6
1985	290,9	190	65,3	78,3	26,9	17,8	6,1	4,6	1,6

The coconut fibres in Brazil as a whole are considered to be an agricultural residue. Except a small amount of the coconut fibres which are used for the production of carpets, brushes and some objects of handicraft, the rest are burned or dumped as rubbish. No official data have been found to give the global and local production of coconut fibres in Brazil or in the Northeast of Brazil.

2.1 Physical properties

Physical properties of sisal and coconut fibres of the state of Paraiba have been studied. The average length and diameter of the sisal fibres, which were obtained from the town of Monteiro, are 870mm and 0,15mm and that of cleaned coconut fibres were 140mm and 0,35mm, respectively. For each type of fibres, 10 specimens were chosen randomly from the sample bulk. The cross section of the fibres were measured at several length along the fibres using paquimeter of the 0.01mm precision. Their mean values then were considered as the diameter of each fibre.

To find the natural humidity, the fibres were air dried for 5 days and then these fibres were dried in oven with 105°C temperature during 24 hours. Their weights were measured using electronic balance of 0,01 gram precision. The normal humidity, "H" was calculated using equation 1.

$$H = \frac{P_d - P_o}{P_o} \cdot 100 \quad (\%) \qquad (1)$$

Where **Pd** is air dried weight of the fibre and **Po** is the oven dried weight of the fibre.

The specific weight of the air dried fibres also were established using equation 2.

$$\gamma = \frac{Pd}{V} \qquad (2)$$

The volume of the fibres "V" corresponds to the volume of displaced water after the immersion of a bulk of fibres into the drinking water. The calculated results are given in column 2 and 3 of table 2.

Table 2. Natural humidity, specific weight and water absorption of sisal and coconut fibre.

Type of fibres	Natural humidity (%)	Specific weight (kN/m³)	Water absorption (%) in hours			
			4			
1	2	3	24	48	72	96
Sisal	14,6	7,6	193	199	220	215
Coconut	14,3	10,6	109	110	118	128

2.2 Water absorption of the fibres

One of the main problem in using the sisal and coconut fibres as a matrix with earth is their water absorption during first 96 hours when the soil is drying. Therefore the water absorption capacity of the naturally dried fibres under study was measured. For this purpose, the air dried fibres were immersed in drinking water. Then their weights were measured with the electronic balance after each 24 hours, during four days. The average percentage of the water absorption is given in table 2 and was calculated using equation 3.

$$W = \frac{Ph}{P_h - P_d} \quad 100 \quad (\%) \qquad (3)$$

Where **Ph** is the weight of soaked fibres. The results are, also, plotted in Fig.1. It can be seen that water absorption of sisal after 72 and 96 hours is 87 and 68 percent more than that of coconut fibres. Therefore to produce a good bonding between the earth and these fibres, some type of impermeable material should be used.

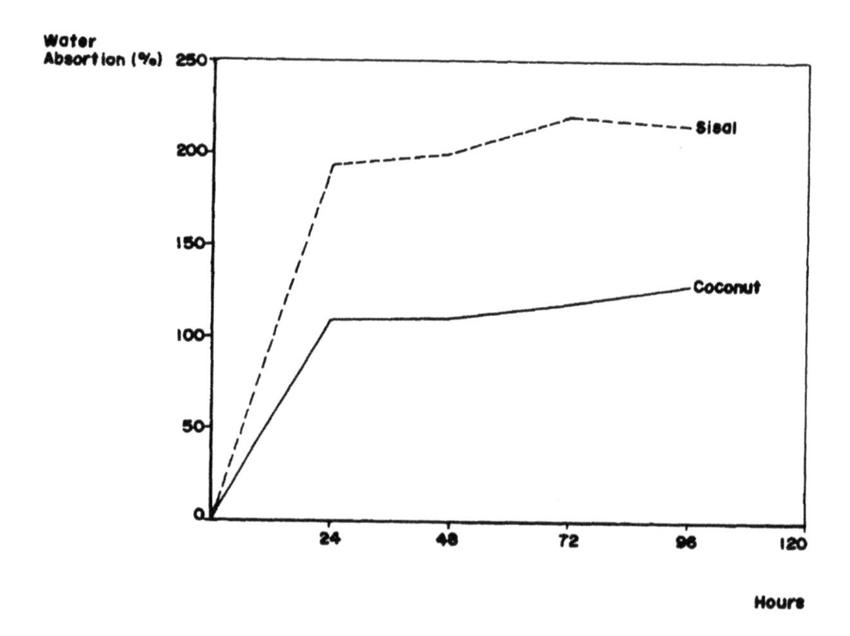

Fig. 1. Water absorption of sisal and coconut fibres.

The transversal and longitudinal changes of the fibres during these predetermined periods also were measured. The dimensional changes are given in Table 3.

Table 3. Dimensional changes of sisal and coconut fibres due to water absorption.

Type of fibres	Direction	Dimensional Changes (%)				
		24h	48h	72h	96h	168h
Sisal	Longitudinal	1,2	1,2	1,2	1,2	1,6
	Transversal	29,7	35,1	35,1	35,1	33,1
Coconut	Longitudinal	1,2	1,2	1,2	1,2	1,2
	Transversal	11,7	11,7	11,7	11,7	13,9

It can be noted from the obtained results that the longitudinal changes of fibres due to water absorption is not significant. The main changes indeed occur transversely and this change for sisal fibres is much higher when compared with that of coconut fibres.

2.3 Mechanical properties of the fibres

In order to establish the mechanical properties of the sisal and coconut fibres, various factors such age, method of separation, and storage condition should be considered. Due to lack of data on the effect of each of these variables, several batch of freshly separated sisal and coconut fibres from the local producer were obtained.

To find the stress-strain "σ-ε" relationship, ultimate tensile strength, 55 samples for each type of fibre have been selected. The tensile tests were carried out in a tensile test machine with the maximum capacity of 200N, fabricated by Branco-Marcallo (Varese). The load was applied at the speed of about 1mm/sec. Through this research work, the cross-section of the fibres were considered as circular. As the length of the tensile tests on these types of fibres as yet is not standardized, the influence of fibre length on the ultimate tensile strength was studied.

For coconut fibres, two different length of 65 and 120mm have been chosen. As the length of sisal is higher than coconut fibres, an additional sample of 500mm also was chosen. The average values of the obtained results are given in table 4.

Table 4. Tensile strength and strain of sisal and coconut fibres.

Type of fibres	Specimen length(cm)	Diameter (mm)	Tensile strength(MPa)	Strain at failure(%)
	6,5	0,13	535	7,7
Sisal	12,0	0,13	539	3,6
	50,0	0,12	470	1,6
	6,5	0,25	140	19,9
Coconut	12,0	0,31	112	15,7

As can be noted in Table 4, the length of the test specimen has a significant effect on the tensile stress and strain behaviour. The longer the specimens length are the lower the stress and strain. The obtained results, compared with those available in the literature proves this trend. The average tensile strength and strain of 460MPa and 4% were abtained on specimen of 450mm of sisal and 180MPa stress and 28% strain were measured on 50mm length of coconut fibre by CEPED [7]. Mukherjee and Satyanarayama [8] have carried out tensile test on 50mm length of sisal fibres presented a mean tensile strength and strain of 560MPa and 4,2% respectively. The tensile tests carried out by Ghavami and Veloso [4] on 100mm length coconut fibres were in the range of 35 to 121 MPa and maximum deformation of 32%. The average tensile strength and strain of 366MPa and 5,2% and 107MPa and 35% for sisal and coconut fibres respectively were obtained by Agopyan [9]. The length of the tensile specimen is not given in his paper.

Comparison of the results for these two different fibres shows that the tensile strength of sisal fibres is much higher and its strain is much lower when compared with those of coconut fibres. However, the surprising results is the water absorption of the sisal which is much higher than that of coconut fibres. This is probably related to the impermeability properties of the coconut fibres.

3 Water-repellent treatment

One of the major weaknesses of the sisal and coconut fibres as matrix with earth is their water absorption. As the fibres absorb water they expand and when the earth is drying they lose water and their dimensions change to original size. Therefore they lose the bonding with the filler material. In order to overcome this handicap two different water-repellent agents which are cheap and available locally were used. These water repellents are Betume based material and are used in civil construction as impermeable material. They are: a) Piche, which is a liquid betume and fabricated by Betumat, Bahia; b) Cipla which is also liquid and produced by Cia Industrial de Plásticos, Joinvile, SC.

The percentage of water absorption of the treated fibres with piche and cipla, measured after each 24 hours during four days is given in Table 5. These are mean of five measurements for each type of water repellent. All the specimens were approximately of 100mm length. Before application of the water repellent, the samples were dried for two days. Then the fibres were kept inside the liquid water repellent for some seconds. The weight of the treated specimens was measured by an electronic scale of 0,01 gram precision before immersing them completely into drinking water at room temperature. The same precedure was repeated at the predetermined intervals while taking out the specimens from water and measuring its weight for the determination of the moisture absorption.

Table 5. Water absorption of treated fibres with water repellent.

Type of fibres	Water repellent	Water absorption (%)			
		24h	48h	72h	96h
Sisal	Piche	34,5	67,9	66,7	65,3
	Cipla	61,8	78,6	81,7	99,3
Coconut	Piche	60,4	76,8	79,8	87,2
	Cipla	61,6	84,3	76,7	97,1

As it can be seen from Table 5 and Fig. 2 for both type of the fibres, the piche produced better results. It is interesting to note that although the natural water absorption of the sisal is much higher than the coconut fibres the treated fibres showed the reverse results. The water absorption of sisal treated with piche after 96 hours is 25 per cent less than that of coconut fibres. This could be attributed to the better penetration of the water repellent into the sisal fibres.

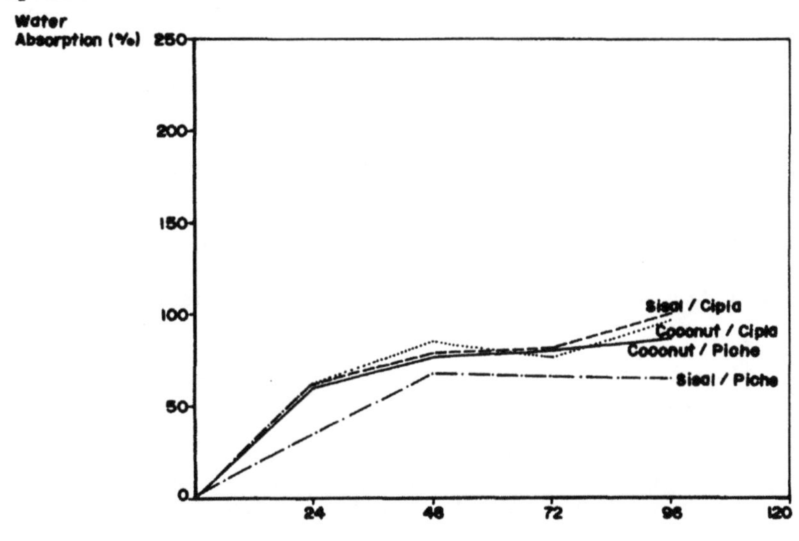

Fig. 2. Water absorption of treated fibres.

4 Properties of the soil

To produce the adobe blocks, the soil which is commonly used for the construction of taipa houses from different regions of the state of Paraiba was selected. The soil was taken three regions called Taperoá, Areia and Campina Grande. The particle size distribution of the material are shown in fig. 3.

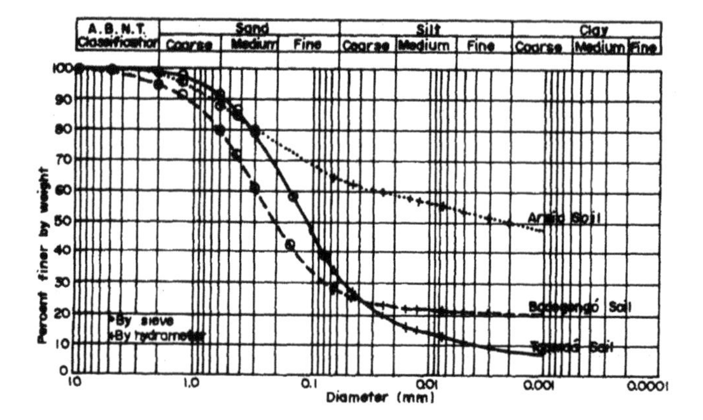

Fig. 3. Particle size distribution curves of the soils.

The next step was to find the Atterberg limits and consistency indices of the soils. These were: Liquid Limit "LL"; Plastic Limit "LP"; Plasticity Index "IP"; Shrinkage Limit "LS" and Activity of the chosen soils. The results of the tests are given in table 6. It was found that Taperoá soil presents characteristics of a practically stabilized soil while Areia and Campina Grande soil presented higher plasticity limit. Therefore these soils should be used in conjuction with some type of soil stabilizer. The dry specific density and ideal humidity, Wót, of the soils also were measured and are given in Table 6. It was found that Taperoá soil has got the highest dry specific density.

Table 6. Atterberg limits, consistency indices, humidity and dry specific density of the soils.

Soil type	LL (%)	LP (%)	IP (%)	LS (%)	Activity of soil	Wót (%)	Dry specific density (kN/m³)
Taperoá	23,5	17,1	6,4	22,4	0,48	14,5	18,8
Areia	54,0	30,4	23,6	31,7	0,48	27,0	14,6
C. Grande	45,6	21,4	24,2	20,8	1,07	15,9	17,4

4.1 Stress-strain behaviour of the soil

To investigate the stress-strain behaviour of Taperoá soil with and without fibres and to establish the optimum soil/water percentage, the compression cylindrical tests on 9,38 x 12,34cm specimens were carried out. The tests were conducted in compression testing machine, Pavitest, fabricated by Contenco. The load was applied through electric motor driving the mechanical ram. The applied pressure was measured using a load dial gauge transducer so that the vertical stress could be measured directly to 0,1MPa. Mechanical deflection dial gauges were used to monitor the relative vertical deformations of the samples. The dial gauge could be read directly to 0,01mm. The apparatus which was used to find the stress-strain behaviour of the soil is shown in Fig. 4.

Fig. 4. Test machine used in compression tests of the soils.

The stress-strain behaviour for different cases is presented in Fig. 5. The curves are generally similar in shape, being composed of initial portion where the stress-strain is linear up to the first peak load. The soil specimens showed a very abrupt and fragile behaviour when the maximum load was reached. Four percent fibre-soil specimens after achieving the first maximum load followed a plateau where the stress increased very smoothly with large displacement andattained a peak stress at 20% strain.

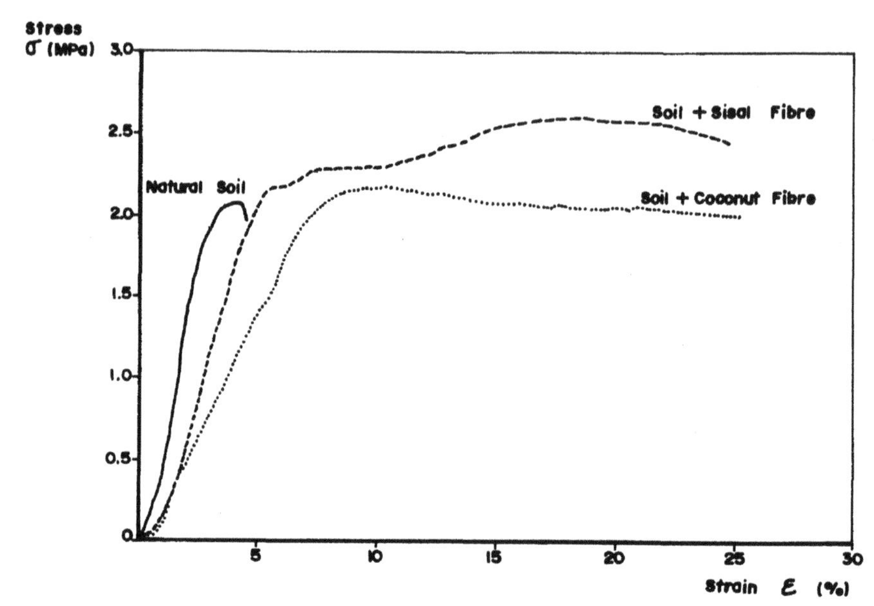

Fig. 5. Stress-strain curves of soil and 4% fibre + soil.

5 Concluding remarks

The authors realize that the low-cost housing construction both in rural and urban areas is a multidisciplinary problem, the solution of which demands a dynamic approach. However one of the most important aspect for the solution of this important problem in our time is the use of all locally available material, especially in developing countries such as Brazil.

The results of the investigation reported here have shown the great potentials that sisal, and coconut fibres have for the improvement of the adobe houses. The application of 4% of sisal and coconut fibres have improved the brittle behaviour of adobe blocks drastically.

Different percentages of sisal and coconut fibres, and various length of the fibres needs to be studied to establish their optimum values in the adobe bricks. There are several other variables such as stabilizers, and other local waterproofing agents which need to be considered for further research.

The development of simple process for the production of the adobe blocks is needed so that the rural people can produce and use these blocks with facility.

The soil stabilizer, like fly ash if locally available and or lime should be investigated in conjunction with the use of the soil.

6 Acknowledgment

The authours would like to acknowledge the contribution and cooperation of Mrs. Soênia Marques de Souza and Mr. Francisco Batista dos Santos in execution of the experimental works.

Further they would like to thanks CNPq for the financial support given for this project.

7 References

1 - Culzoni, R.A.M. - Caracteristicas dos bambús e sua utilização como material alternativo no concreto . Tese de Mestrado DEC-PUC-RJ, Rio de Janeiro, 1986.

2 - Ghavami, K; Van Hombeeck, R. - Application of bamboo as a construction material. Simpósio Latino Americano de Racionalização da Construção. IPT, São Paulo, out/81.

3 - Ghavami, K; Van Hombeeck, R. - Application of coconut husk as a low cost construction material. Conference on development of low-cost and energy saving construction materials and applications. Rio de Janeiro, jul/84.

4 - Ghavami, K; Veloso, R.F.R. - Análise Microscópica do bambú e comportamento mecânico de fibras de coco. Relatório DEC-PUC-RJ nº 031/85. Rio de Janeiro, 1985.

5 - Toledo Filho, R.D. - Utilização de materiais não convencionais nas construções rurais. Relatório Técnico, DEAg, UFPb, Campina Grande, Jan/90.

6 - Souza, M.L. - A produção de sisal na Paraiba. O município de Cuité: Um estudo de caso. Tese de Mestrado, UFPb, Campina Grande, 1987.

7 - CEPED-Ba - Utilização de fibras vegetais no fibro-cimento e no concreto-fibra. BNH/DEPEA, Rio de Janeiro, 1982.

8 - P.S. Mukherjee, K.G; Satyanaryama - Structure and properties of some vegetable fibres. Journal of Material Science nº 19, 1984.

9 - Agopyan, V.; Cincetto, M.A.; Derolle, A. - Durability of vegetable fibre reinforced materials. IPT, São Paulo, 1986.

16 THE USE OF COIR FIBRES AS REINFORCEMENT TO PORTLAND CEMENT MORTARS

H. SAVASTANO Jr
Academia da Força Aérea, Pirassununga, Brazil

Abstract
The production of coir fibre reinforced cement mortar is presented. The coir fibres are by-products of carpet industry without any commercial value. A semi-industrial production methodology is tried. Mortars with three different water-cement ratios are tested reinforced with dry or satured fibres. The consistence of the mixes is determined. At the age of 28 and 90 days, the compressive strength, the tensile strength in bending tests and impact strength are presented. The use of coir fibres as a reinforcement of cement mortars increases the impact strength of the matrix up to 220% and the tensile strength up to 175%. The compressive strength is reduced.
Keywords: Fibre Reinforced Mortar, Coir Fibre, Composite Strength.

1 Introduction

The interest of this present study consists in drawing attention to the production of fibre cement composites in a semi-industrial level with the utilization of local raw-materials and of the vegetable fibres available in the market and answering to a fixed standard of quality. The use of vegetable fibres in portland cement mortars should be subject to a technological control not only of the composites produced but also of the fibre, so as the physical and mechanical characteristics must be, as much as possible, constant for the resulting composite.

The vegetable fibre reinforced cement mortars have their application turned towards the production of panels for vertical load-bearing walls, roofings, floors and other components where the ductility is an important characteristic. In the case of coir fibre, that is a vegetable fibre used in the present research, it is known that its use in internal panels, that are not subject directly to climatic conditions, makes the time life of these panels very much longer. However, it is necessary to state that, among the vegetable fibres, the coir fibre is one of the most resistent to the alkalinity and the variations of moisture in the matrices [Gram(1983), Ramaswamy et al.(1983)].

The basic principle of reinforcement of a brittle matrix with fibres is to get fibres that will sustain and distribute the tensile stresses in the composite by way of their anchorage in the matrix. It is, then, an idea similar to the functioning of the reinforced concrete, although it is not possible to establish a theoretic anology between the two cases as the fibres are thin, of short length and distributed randomly and homogeneously in the matrix.

Analyzing the deformation of a fibrous composite required for tensile stress, there is a pre-cracking stage where the elastic shear bond is intact followed by a post-cracking stage of the matrix giving origin to an elastic and frictional bond combined. In the third and last stage, the cracked matrix is useful only for the fibres anchorage, that are now the only part responsable for composite strength [Aveston et al.(1971)].

In the case of fibres of low elasticity moduli, it can be noted that the growth of stresses in the post-cracking stage tends to be small. Even so, these fibres, because of their high capacity of deformation and by the effect of vibration of these (relative slip along the fibre matrix interface), give to composite an elevated ductility, being recommended their use to resist accidental and dynamic or impact loadings.

Under a dynamic loading, the composite is required in three direction stresses and also the fact of having or not microcracking alters very much its strength. Because of this, it is very difficult to get the theoretical composite impact strength, seeing that the principles and suppositions proposed for static stresses, with axial tensile, for example, are not valid for the dynamic case.

During the composite loading (tensile load), the energy dissipation occurs specially due to the following reasons: elastic energy absorption by the fibres and matrix in the uncracked matrix; and, in the multiple fracture stage, energy dissipation due to the fracture surface work in forming a crack in the matrix and also due to the debonding and slipping at the fibre matrix interface [Agopyan(1982)].

2 Coir fibres - application in cement composites

Among the coir fibres suitable to the production of composites with brittle matrices of mortars, the main idea should be the utilization of by-products, once that the longer fibres already have their market garanted for the production of ropes. Being cheaper, these shorter fibres require, nevertheless, studies to make possible their use in production of fibrous composites.

The vegetable fibres have their physical and mechanical

properties very affected by the variabilities of the climate, quality of the soil where they are produced and the time of year that they are reaped at. Even the process of obtaining the fibre, if more sophisticated or rudimentary, using chemical processings or not, influence the final quality of the fibre. Thus, comparing characterizations about the Brazilian coir fibres and fibres from other countries made by Guimarães(1984), Cook (1980) and Das Gupta et al.(1978),there happens to be differences that can be superior to 50%. Also coir fibres from different Brazilian regions certainly have variations in their physical and mechanical characteristics.

The vegetable fibres have a high tensile strength and low elasticity moduli when compared with common cement mortar. In the case of coir fibre, its tensile strength is approximately 100 times larger than the cement matrix strength. In the Table 1 there are some physical and mechanical properties of some vegetable fibres compared to the steel and glass fibres.

Table 1. Physical and mechanical characteristics of some Brazilian fibres [Guimarães(1984)]

Fibre	Elasticity modulus (10^3MPa)	Tensile strength (MPa)	Elongation at failure (%)	Aspect ratio (l/d)
Coir	2.8	180	29.2	35.0
Sisal	15.2	458	4.3	152.6
Bamboo	28.8	575	3.2	170.5
Piassava	5.6	143	5.99	-
Sugar-cane fibre	5.0	181	5.11	66.1
Glass	55.6	1328	3.77	-
Steel	200	100-300	3-4	-

The water absorption of vegetable fibres is very high, specially for the coir, sisal and jute fibres reaching more than 100% in only one hour of immersion in water [Tezuka et al.(1984)]. Analyzing the use of these fibres in the cement matrix, if the fibre is dry, it tends to draw towards itself the hydration cement water, hindering the fibre matrix bonding, seeing that at the interface the cement has its hydration deficient. Now the use of wet fibre for the production of composite needs a careful study of the influence that this causes on the water cement ratio of matrix, affecting the strength of the last.

The coir fibre is obtained from the fruit of coconut palm, that is a palm tree of smooth cilindrical surface, erect or curved, with up to 30 m in height and 0.6-0.7 m in diameter. Its leaves are petiolate, with up to 6 m in length, and inflorescence in ramous spadix, with white flowers. The fruit is egg-shaped , with 0.3 m in length and 0.25 m in diameter approximately, mesocarp is fibrous and rough - region from where the fibres are extracted - and hard endocarp with a large central cavity which contains a sweet liquid (coconut water) [IPT(1983)].

The coir fibres have an increase in their brittleness when they are subject to alternated situations of wetting and drying. This weakness, however, is very inferior to the one that occurs with the sisal fibres in cement mortar composites. The chemical composition and morphological properties of coir fibres certainly give more protection against deterioration. Gram(1983) states that high temperatures (around 50°C) accelerate the weakness of the fibres because the chemical process catalysation.

However, it can be said, in light of the results shown above, that, for an application of coir fibres in composites with portland cement mortars in no structural uses, there is no need of any treatment of the fibre or diminution of the alkalinity of the matrix, seein that the weakness of the fibre is small in time.

3 Testing methodology

The materials used in the elaboration of composites were ordinary portland cement CP32, fine silica sand (fineness modulus of 2.20) dried in a hot stuffy and coir fibres without chemical treatments. These fibres are by-products of a carpet industry, found in two average lengths: 38.2 mm and 9.7 mm (type A and B fibres, respectively).

The composite mix was made in an inclined axis mixer following a sequence of introduction of materials so as to avoid composite intermesh. The compaction was handmade with a socket and also using a vibrating table.

In the fresh composite, it was found the consistency index (NBR 7215) and also specific gravity with the intention of checking the interference of the fibres in these two properties of the matrix.

There were produced 20 series made up of 12 specimens each one, varying from one series to other the type of fibre used (type A or B), the conditions of humidity in which the fibres were used (dry or satured fibres), the fibre fractions introduced to the cement mortar matrix and also the water cement ratio used for the matrix.

Among the produced specimens, 6 are of dimensions (100x100x20) mm^3 for the impact test, 3 are of dimensions (300x150x50) mm^3 for the bending test and 3 specimens of cilindrical shape of diameter 50 mm and 100 mm height for the compression test.

The specimens are kept in the curing room until the date of the test which occurs at the 28th or 90th day of age.

The impact test apparatus measures the specimen strength

by the fall of a steel ball (mass 639.98×10^{-3} kg) from different heights.

The present test consists in submiting each specimen to a series of impacts, being the height of the fall in the first impact fixed at 400 mm and the subsequent growing at 50 mm each time. At each impact it is checked in the inferior face of the specimen the formation of crackings. The height of the fall of the steel ball that causes the first visible crack (fvc) is taken note of, and from there on, with the help of measurer of crack openings (cracking measurement ruler), it is taken note of the maximum opening that is obtained with each impact until they reach the width of 1.5 mm, situation in which the test is finished. If there happens to be the occurence of cracks in the superior face of the specimen or the complete rupture of it, these facts are taken note of. It must be made clear that the crack read throughout the procedure explained above is residual, being its opening inferior to the one that occurs on the moment of the impact of the metallic ball on the specimen.

The results obtained from each series are compiled for the interpretation of the behavior of the composites of cement mortar with coir fibre when submitted to impact test. This interpretation is made determining the medium number of impacts needed for the apparition of the first visible fissure or maximum opening of a crack of 1.5 mm. For such it is found the average number of impacts for the 6 specimens of each series.

It is also determined for each series the average absorbed energy applied both in the situation of apparition of the first visible crack as in the major opening of crack correspondent to the number of impacts in each case. It is assumed that the absorbed energy in the impact is the same as the potential energy of the metal ball before it is allowed to a free fall.

The bending test equipment has 4 points, being 3 of these fixed articulations and 1 of these a movable articulation. The objective is to avoid the testing machine to be an indetermined structure, which would include an extra variable in the specimen: its stiffness. Then, it is obtained an average tensile strength on bending for each series made up to 3 specimens.

Finally, the axial compression test is done in the same procedure as the NBR 7215.

4 Results and discussion

Once the measures of specific gravity of the composites are made on fresh state for each fibre fraction in the series that were produced, it is noted that the specific gravity tends to diminish in the same proportion as the fibre fraction grows, due to the larger air contents incorporated by the presence of the fibres.

Also the consistency of the fresh composite grows in the same proportion that grows the fibre fraction, according to the flow table test.

The growth in the fibre fraction in the composite leads to an increase in the impact strength and a decrease in the compression strength. The tensile strength on bending test arrives at a maximum value for an optimum fibre fraction of 3.2% in volume (increase of tensile strength of up to 165% in relation to the brittle matrix).

Due to the presence of fibres, there are increases of more than 4 times the impact strength of the composite when compared to the brittle matrix (fibre fraction of 4.4% in volume). It can be checked this through the figure 1 where are correlated applied energy to 1.5 mm opening cracks versus the volume fraction of short fibres (type B).

The fact of a fibre being long or short, type A or B

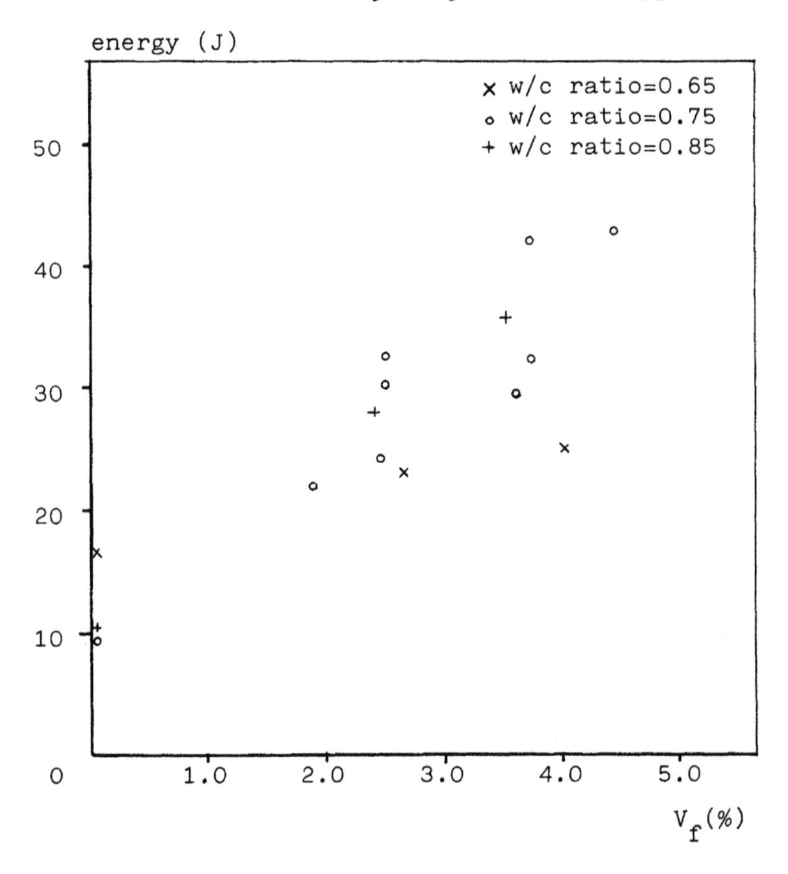

Fig.1. Applied energy to 1.5 mm cracking opening x fibre volume fraction.

respectively, does not leave a very clear influence on the impact strength but, on the bending and compression tests, the best results where obtained in composites with type A fibres (longer fibres). The fact that the fibre used is dry or satured does not leave a clear effect on the results. The water cement ratio considered ideal for the materials used is 0.75 because it causes a more homogeneous composite and at the same time not very affected by the air content left by the water in excess that evaporates.

It is advisable that future studies should use a sand not as fine as the one used in this research, with fineness modulus of around 2.70 for example, so as to obtain a good homogenization of the composite keeping the high fibre fractions here used and diminishing the water cement ratio of 0.75 to about 0.60. Following these advices, it will be possible to obtain mechanical strengths still superior to the produced composites.

The specimens tested with 90 days of age and with high fibre fractions (up to 4.4% in volume) show a tendency of decrease of strength, probably because the effect of degradation of the fibres is stronger due to the higher quantities used.

5 Other considerations

The type B fibres show a very satisfactory performance in the matrix, comparable to the type A fibres. The impact strength and tensile on bending obtained by the composites using type B fibres are very superior to the brittle matrix and, when inferior to the results obtained with type A fibres, the differences are always smaller than 4%.

The short length of the type B fibres (l=9.7 mm), so much inferior than the critical length (shorter anchorage length) of the coir fibre in the cement mortar matrix, diminishes the anchorage of these in the matrix, also diminishing the composite strength. However, the fact of being short allows the volume that is going to be added of type B fibres to be up to 4 times larger than the maximum volume of type A fibres possible to be used without causing composite intermesh. The possibility of using higher fibre fractions in its turn, causes an increase of the mechanical properties of the composite obtained.

Another factor that is favorable to the use of type B fibres is the fact of being the more abundant residue of the carpet industry, originating from the process of leveling of the carpet hairs. The only necessary act before using them is to seive the existing powder of fibres.

By the results obtained, it is seen that the application of coir fibres reinforced composites for the production of panels subject mainly to impact and tensile stresses is perfectly possible. It can be recommended studies relative to the panels production not only about non-load panels but also load-bearing ones. It is of great importance also to define the mix, moulding and cure procedures in the case of production of bigger bits for vertical walls, roofs and floors, for example.

6 Acknowledgements

This is a part of the research done in Escola Politécnica, University of São Paulo, to get the Master of Engineering degree. The author wishes to acknowledge Dr. Vahan Agopyan by the Orientation of this research and the financial support given by the Fundação de Amparo à Pesquisa do Estado de São Paulo.

The author also wishes to acknowledge the support given by the Academia da Força Aérea - Ministério da Aeronáutica do Brasil - and the Coordenadoria para o Aperfeiçoamento do Pessoal de Ensino Superior.

7 References

Agopyan, V. (1982) The Preparation of Glass Reinforced Gypsum by Premixing and its Properties under Humid Conditions. University of London, PhD Thesis, London.

Aveston, J. Cooper, G.A. and Kelly, A. (1971) Single and Multiple Fracture, in The Properties of Fibre Composites (ed National Physical Laboratory) pp. 15-26.

Cook, D.J. (1980) Concrete and Cement Composites Reinforced with Natural Fibres, in Concrete International: CI 80-Fibrous Concrete, Lancaster, pp. 99-114.

Das Gupta, N.C. Paramasivam, P. and Lee, S.L. (1978) Mechanical properties of coir reinforced cement paste composites. Housing Science, 2, 391-406.

Gram, H.E. (1983) Durability of Natural Fibres in Concrete. CBI Forskning Research, Stockholm.

Guimarães, S.S. (1984) Experimental mixing and moulding with vegetable fibre reinforced cement composites, in International Conference on Development of Low-Cost and Energy Saving Construction Materials, Envo Publishing, Rio de Janeiro, v.1, pp. 37-51.

Instituto de Pesquisas Tecnológicas do Estado de São Paulo (1983) Fichas de caracterização de materiais têxteis. Tópicos Têxteis, 5, 7-12.

Ramaswamy, H.S. Ahuja, B.M. and Krishnamoorthy, B. (1983) Behavior of concrete reinforced with jute, coir and bamboo fibres. The International Journal of Cement Composites and Lightweight Concrete, 5, 3-13.

Tezuka, Y. et alii (1984) Behavior of natural fibers reinforced concrete, in International Conference on Development of Low-Cost and Energy Saving Construction Materials, Envo Publishing, Rio de Janeiro, v.1, pp. 61-77.

PART FOUR

BUILDING COMPONENTS WITH VEGETABLE FIBRE COMPOSITE MATERIALS

17 COMPARISON BETWEEN GYPSUM PANELS REINFORCED WITH VEGETABLE FIBRES: THEIR BEHAVIOUR IN BENDING AND UNDER IMPACT

R. MATTONE
Turin Polytechnic, Italy

Abstract
Whit the aim of exploring possible applications of vegetable fibre reinforced gypsum, the behaviour of thin panels of gypsum reinforced with either sisal or coconut fibres was investigated. The test pieces were produced through a vacuum process to reduce the water/gypsum ratio, increase matrix compaction and improve the bond between the fibres and the matrix so as to obtain high performance composites. Bending tests were performed on test pieces measuring 30x40 cm and bending and impact tests on panels sized 80x80 cm. The behaviour of these composites was compared to that of traditional building materials used for similar purposes.
Keywords: Composite materials, Vegetable fibres, Sisal, Coconut fibre, Gypsum, Panels, Flexural strength, Impact strength.

1 Introduction

In recent years, possible applications of vegetable fibres in the production of building materials and components for low cost housing have been the subject-matter of numerous studies and investigations. The mechanical characteristics and physical properties of these materials have been examined together with their ductility and the durability of the matrices, mostly consisting of cement reinforced with different kinds of natural fibres, such as sisal, coconut, jute, flax, bamboo, wood, etc.

This report deals with research work done in this field: based on feasibility studies carried out earlier, a testing programme was performed in order to evaluate the possible applications of gypsum as a matrix of composites reinforced with sisal or coconut fibres.

The choice of such a matrix was prompted by various considerations: gypsum is readily available, suitable for matrix production in small runs, even through handcraft processes and using a reduced amount of energy.

In order to arrive at a more accurate definition of the performance characteristics of these composites, it was deemed necessary, at this stage:

- to integrate the tests carried out during the previous testing programmes with additional experiments, designed to obtain a more thorough understanding of such characteristics;
- to compare the performance of panels made of gypsum reinforced with vegetable fibres with the behaviour of test pieces prepared with gypsum reinforced with glass fibres (which possess very good mechanical properties and are also widely used in other fields) and that of gypsum-board (an industrially produced, commonly used material).

This made it possible to assess the utilization potential of such panels for the production of interior partitions, false ceilings and, provided they are suitably protected against moisture beforehand, external curtain walls.

2 Experimental work

2.1 Materials

This section describes the materials used for the tests.

2.1.1 The matrix

The tests were carried out on gypsum kindly made available by a local gypsum manufacturing company. This material, produced by means of a rotary kiln, had a rather coarse grain size, selected so as to simulate as closely as possible the characteristics of gypsum as can be obtained from an unsophisticated manufacturing process.

Gypsum slurry was produced with a 1:1 W/G ratio: this rather high ratio was deemed necessary in order to facilitate the mixing in of the fibres.

The mechanical characteristics of the gypsum used in the testing programme, as assessed from RILEM tests conducted on 4x4x16 cm prisms, are listed in Table 1 below.

Table 1. Mechanical characteristics of the matrix

W/G	Flexural strength (MPa)	Compressive strength (MPa)
0.8	2.32	6.10

2.1.2 The fibres

Sisal, coconut and glass fibres were employed. The lengths adopted for each individual type of fibre are listed below:

- Sisal fibres: these were used in both chopped (l = 5 cm) and natural form (l = ~90 cm);

- Coconut fibres: these were used in their natural form (length between 10 and 30 cm) as well as curled (a treatment these fibres are subjected to for industrial purposes);
- Glass fibres: type E fibres were tested in chopped form (l = 2.5 cm).

The different kinds of fibres were added to the matrix in the amounts that previous experiments had shown to be ideal in terms of workability and strength of the composite.

In particular, vegetable fibres were added to the gypsum slurry with a volume fraction V_f = 5,6%, and glass fibres with V_f = 4%.

2.2 Test specimens

All the different kinds of panels employed were produced by means of a steel formwork (measuring 90x90 cm) equipped with a vacuum mat connected to a vacuum pump. Before the casting, the mat was faced with a filtering layer so as to enable excess water to be drained out during the vacuum treatment (0.8÷0.9 bar).

Owing to the variations in fibre length, the panels had to be manufactured according to two different procedures. The panels made with long sisal fibres and coconut fibres were obtained by placing a layer of fibres on the vacuum mat and pouring onto it the gypsum slurry in very fluid form (W/G = 1); then the surface was smoothed with a calibrated wooden template so as to ensure even thickness and the vacuum treatment was applied. This arrangement enabled the matrix to penetrate in between the fibres giving rise to uniform impregnation. Some difficulties were encountered only when dealing with long sisal fibres arranged in two layers at right angles to one another: the gypsum slurry was not always able to cover perfectly the surface of the fibres in contact with the filter. This drawback, however, had no significant repercussions on panel performance.

Short sisal fibres (l = 5 cm) and glass fibres, instead, were added directly to the water-gypsum matrix and mixed in with great care; after that, the composite was poured into the formwork, smoothed and vacuum treated following the same steps as for the other panels.

The vacuum treatment was seen to be very effective: it made it possible to reduce the amount of water present in the mix - which initially, as mentioned above, was very high, so as to facilitate the mixing of the composite - to W/G = 0.4, resulting in a considerable improvement in the mechanical characteristics of the matrix.

Table 2 lists the various types of panels produced and the characteristics of the reinforcing fibres of each.

The specimens were cured in the laboratory, at an average temperature of 20°C and 55 ± 5% R.H.

Table 2. Characteristics of the panels

Panels	Nominal thickness (mm)	Type of fibre	Length of fibres (cm)	Arrangement of fibres	V_f (%)	Bulk density (kg/m^3)
A	15	long sisal	90	2 cross layers	5.6	1120(*)
B	15	short sisal	5	random	5.6	1250
C	15	straight coconut	10-30	random	5.6	1100
D	15	curled coconut	10-30	random	5.6	1100
E	15	E glass	2.5	random	4	1310
F	15	gypsum-board	==	==	==	925

(*) N.B.: the casting process left a few areas not fully impregnated.

3 Evaluation of test results

Each type of panel was subjected to bending tests (on 30x40 cm specimens and 80x80 cm panels) and to impact tests (on 80x80 cm panels).

3.1 Bending tests on 30x40 cm specimens

For this type of test, 4 test pieces were obtained from each of the individual panels manufactured with the different materials to be tested, according to the scheme shown in fig. 1.

Four test pieces of the same size were also obtained from a panel made of gypsum-board, of the type available on the

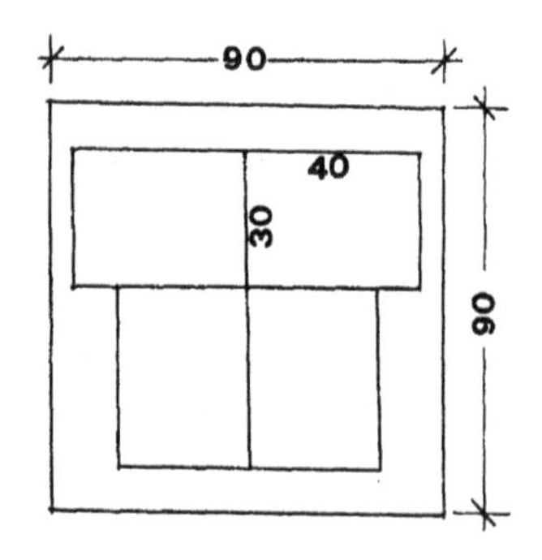

Fig.1. Test piece production scheme.

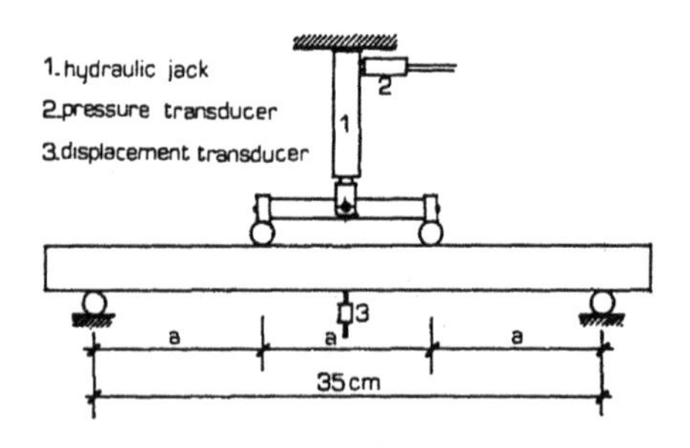

Fig.2. Testing device.

market, of a thickness comparable to that of the panels specially produced for the testing programme.

All test pieces were subjected to bending tests under monotonic loading cycles. Fig. 2 shows the testing apparatus, figs. 3, 4, 5, 6, 7 and 8 give the most significant diagram for each type of panel.

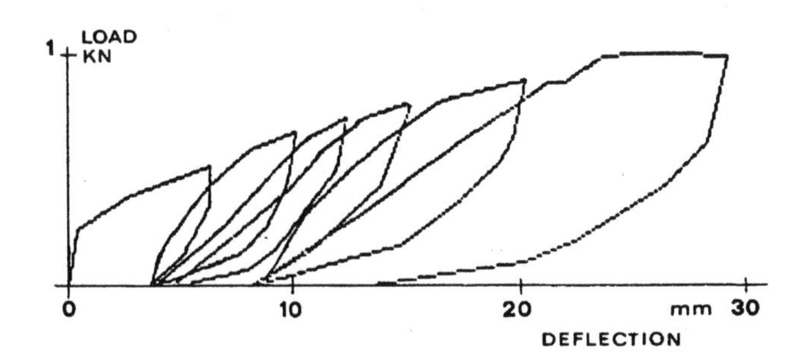

Fig.3. Load-deflection diagram: long sisal fibres (Panel A).

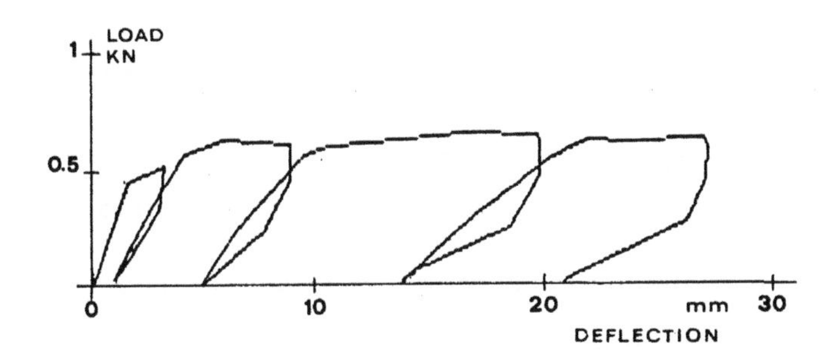

Fig.4. Load-deflection diagram: short sisal fibres (Panel B).

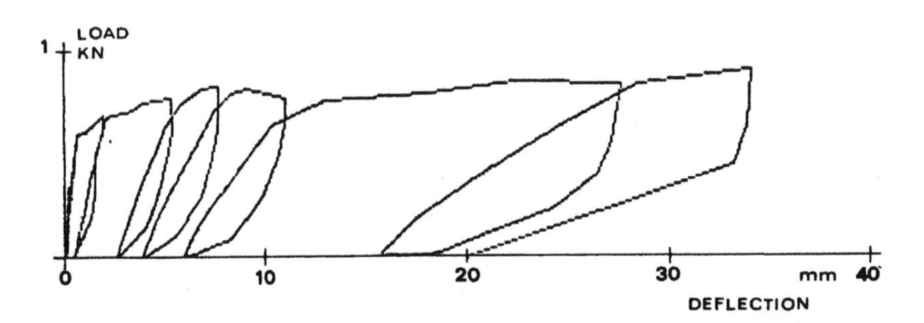

Fig.5. Load-deflection diagram: straight coconut fibres (Panel C).

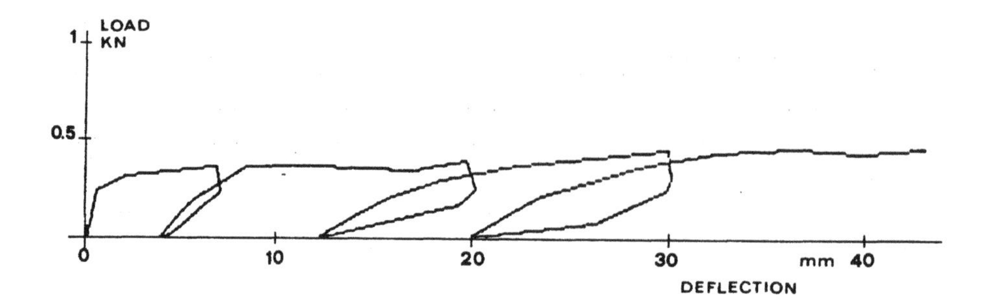

Fig.6. Load-deflection diagram: curled coconut fibres (Panel D).

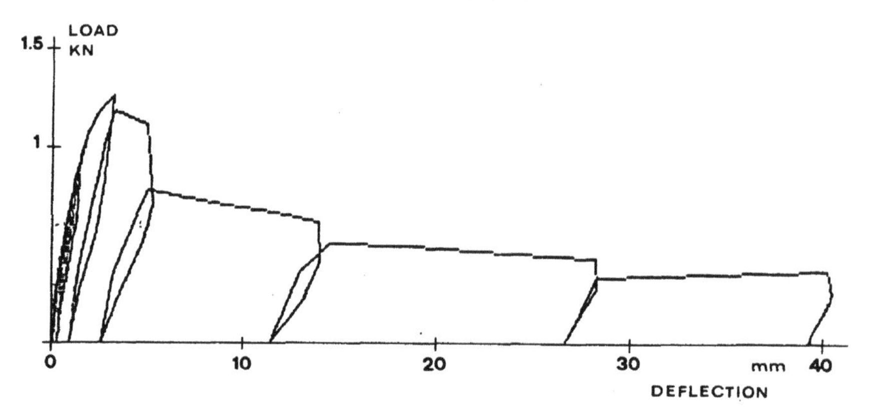

Fig.7. Load-deflection diagram: glass fibres (Panel E).

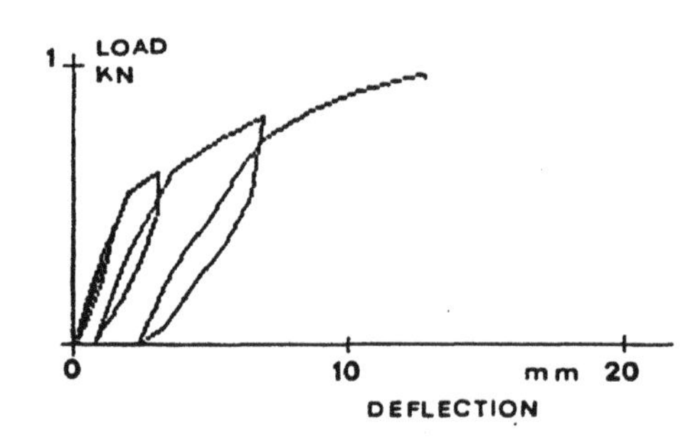

Fig.8. Load-deflection diagram: gypsum-board (Panel F).

Table 3 gives the mean values of failure stresses for the different kinds of panels tested and their E modulus.
From an examination of these values we can make the following statements:
- gypsum test pieces reinforced with long sisal fibres were seen to possess the most satisfactory strength

characteristics of all;

- gypsum tests pieces reinforced with coconut fibres (either straight or curled) and short sisal fibres have poorer strength characteristics compared to gypsum-board specimens and glass fibre reinforced test pieces; however, their strength values are still compatible with the intended applications.

Table 3. Flexural strength results obtained on 30x40 cm specimens

Panels	Effective thickness (mm)	Failure σ (MPa)	Young Modulus (MPa)
A	17,7	6,99(*)	3300
B	15,5	2,89	2700
C	17,8	3,07	3500
D	16,3	3,06	3300
E	16,7	4,98	5300
F	15,0	4,96	3500

(*) Mean values of flexural strength as determined for the different arrangements of the fibres in respect of the edge in tension.

The results obtained in earlier investigations on specimens made with the same type of long sisal fibres had been even more satisfactory. This was due to two factors: the unidirectional arrangement of the fibres and a fully manual impregnation method (the fibres were dipped into the gypsum slurry before putting them into the formwork). This procedure had been adopted, with excellent result, for the realization of gypsum-sisal structural elements (conoid elements, beams, slab members), but it cannot be employed for the production of flat panels, not even in a modest industrialization process.

3.2 Bending tests on 80x80 cm panels

From the original panels, sized 90x90 cm, test pieces measuring 80x80 cm were obtained.

These panels were rested, all along the edges, onto a sturdy steel frame and - after interposing a Ø 7 cm steel disk - loaded centrally until failure. In this case too, the panels were subjected to monotonic loading cycles; the values of the loads and the corresponding strains were recorded and plotted. Figs. 9 to 14 show the load-deflection diagrams obtained and indicate the maximum load applied for each type of slab.

As can be seen from these diagrams, the tests performed on 80x80 cm panels confirmed the good performance of long sisal fibre reinforcement and revealed the rather uniform behaviour of all the other types of vegetable fibres tested. Compared to the results obtained on smaller-sized test pieces, the behaviour of the composite reinforced with glass fibres is seen to remain essentially unaltered, whilst the carrying capacity of gypsum-board is seen to decrease.

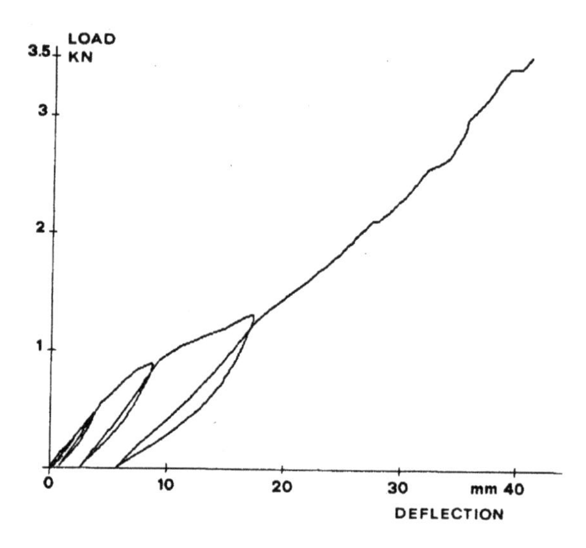

Fig.9. Load-deflection diagram: Panel A.

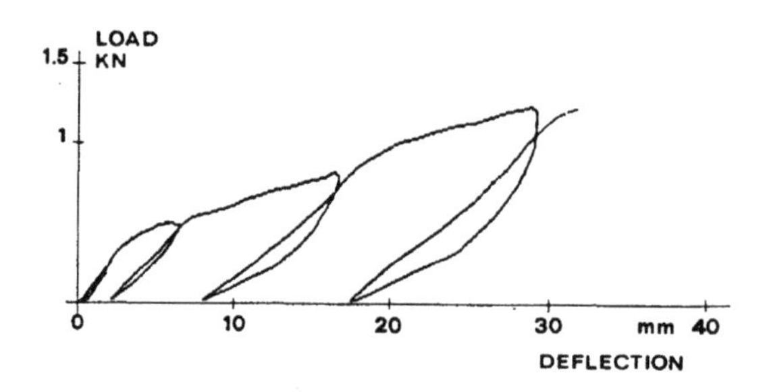

Fig.10. Load-deflection diagram: Panel B.

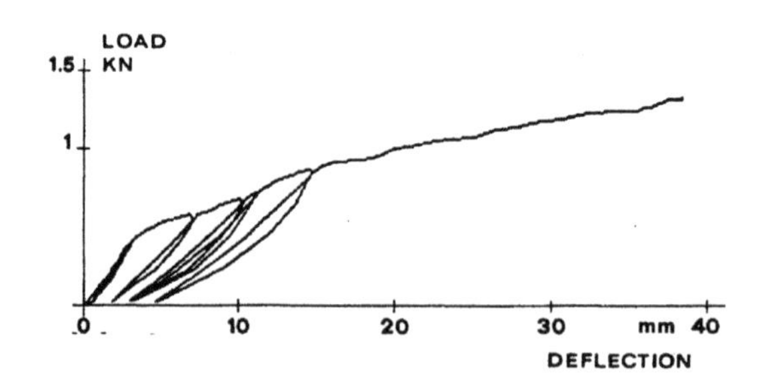

Fig.11. Load-deflection diagram: Panel C.

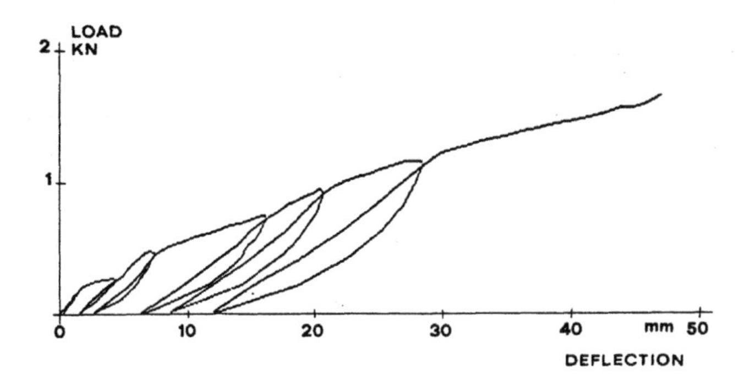

Fig.12. Load-deflection diagram: Panel D.

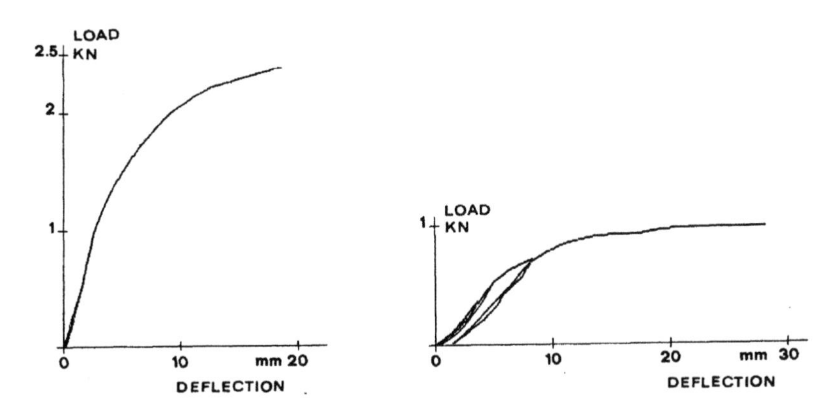

Figs.13-14. Load deflection diagrams: Panels E-F.

3.3 Impact tests

The panels employed for these tests, also measuring 80x80 cm, were rested at the edges on a sturdy steel frame fastened to a supporting masonry structure. They were impact tested by means of a 1 kg metal ball dropped onto the center of the panels from a progressively increasing distance. A Ø 7 cm steel disk was placed at the point of contact between the metal ball and the panel and, underneath it, on the opposite side of the panel, an inductive displacement transducer, connected to a digital oscilloscope and a P.C., was applied. The height from which the ball was dropped was increased in 50 cm steps up to a maximum of 3.5 m. Maximum and residual strains were recorded for each different height.

The diagrams shown in figs. 15-16 summarize the test results; furthermore, they indicate, for each type of panel, the time-deflection diagram relating to the impact corresponding to the maximum fall distance.

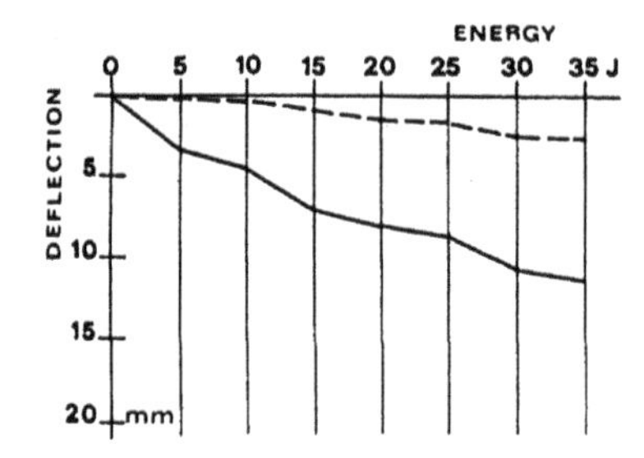

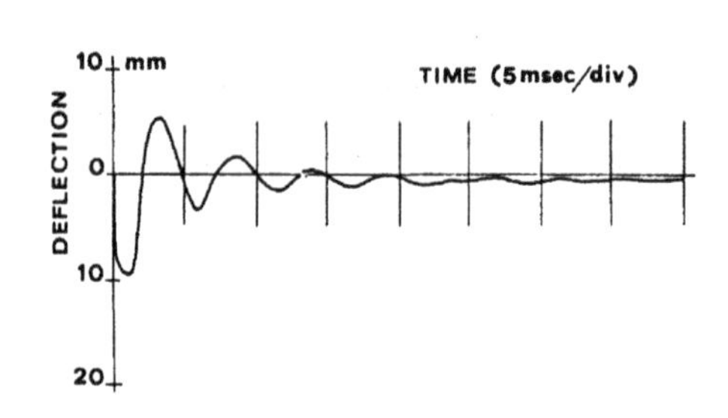

Panel A

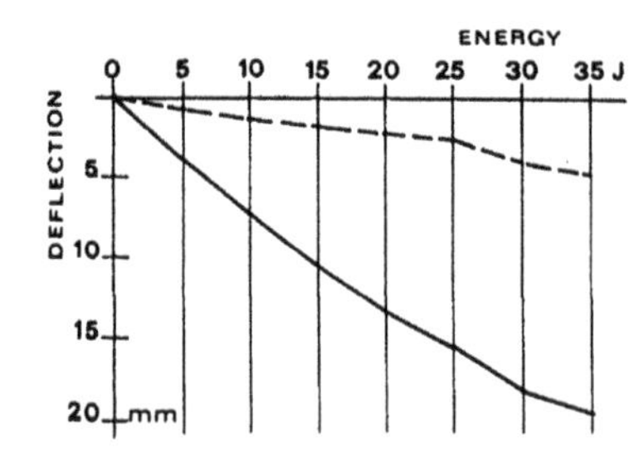

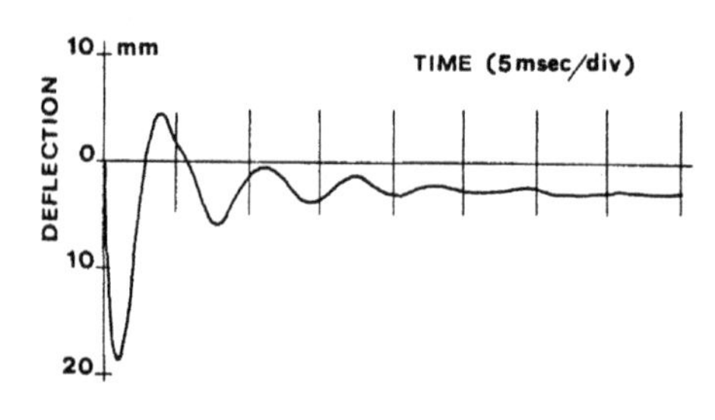

Panel B

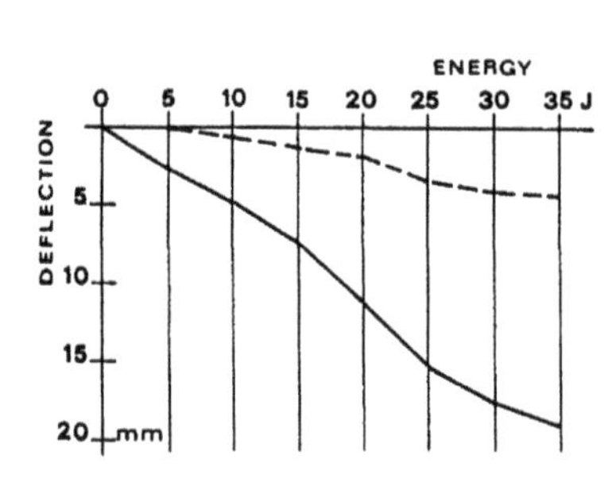

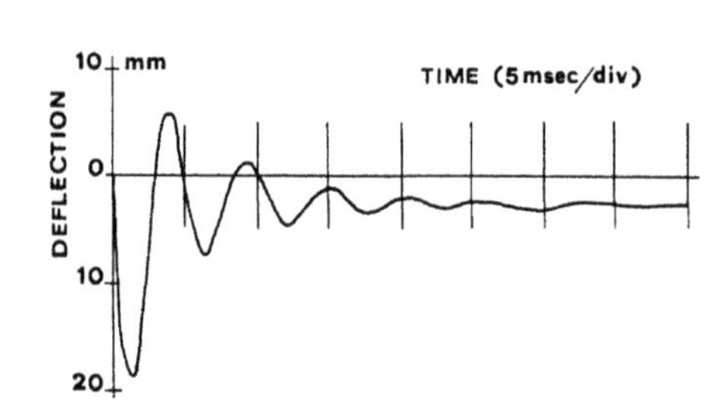

Panel C

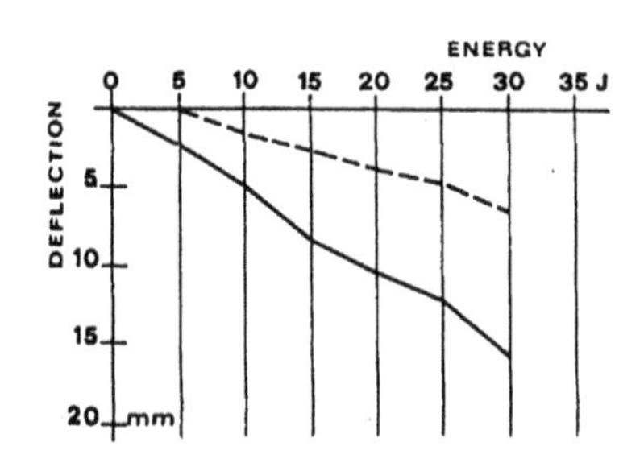

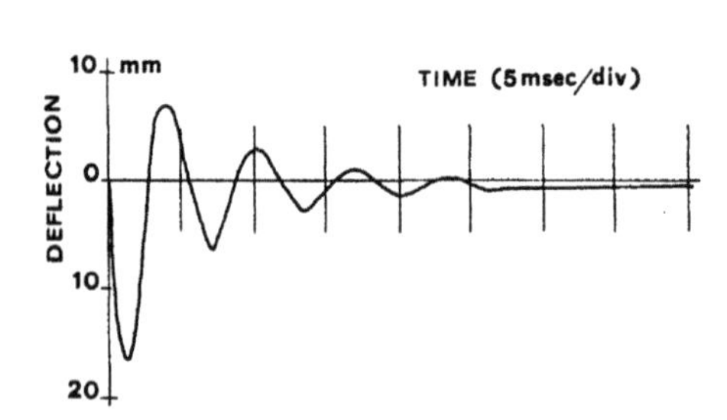

Panel D

Fig.15. Impact tests: Panels A, B, C, D.
(—— Maximum central strain; — — Residual central strain)

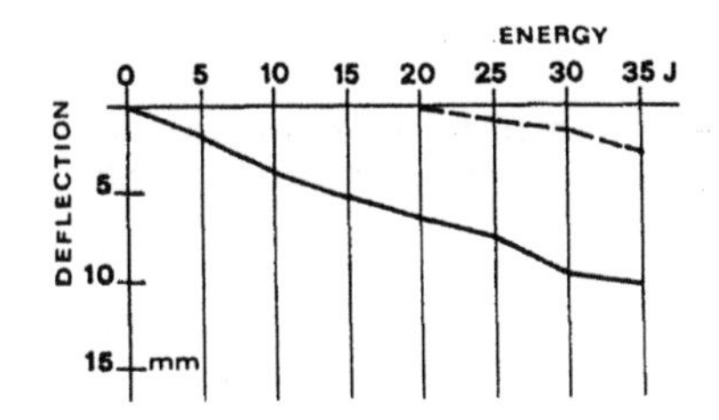

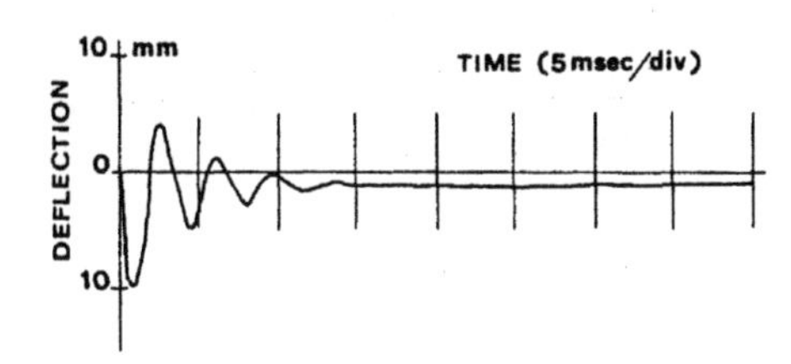

Panel E

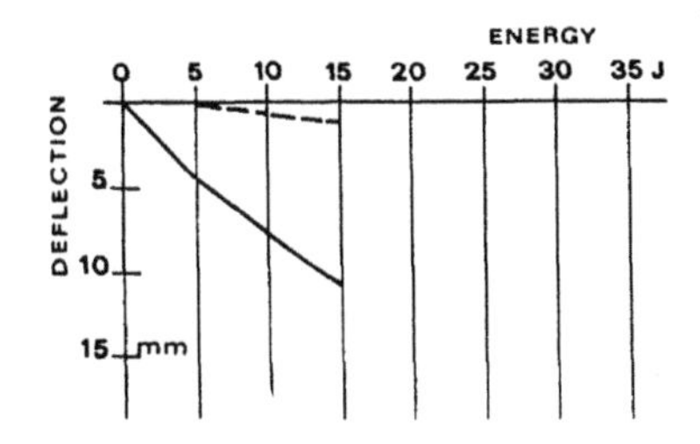

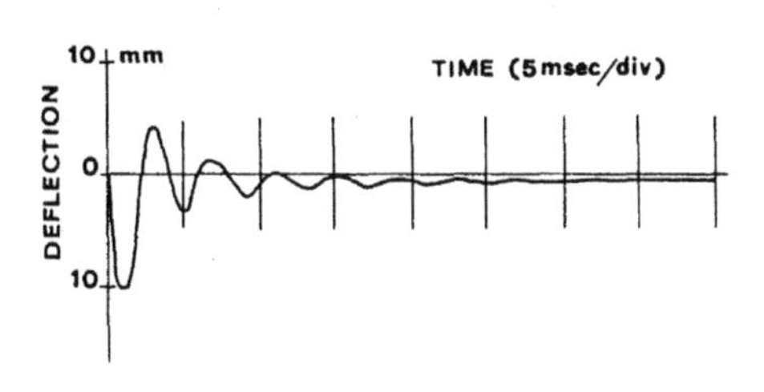

Panel F

Fig.16. Impact tests: Panels E, F.
(——— Maximum central strain; – – – Residual central strain)

As can be inferred from an examination of these diagrams, the behaviour of the composites reinforced with vegetable fibres falls somewhere in between the behaviour of gypsum-board and that of the gypsum-glass fibre panels. In particular, the evolution of strains recorded in gypsum panels using short sisal fibres was found to be very close to that observed in gypsum-board test pieces; however, while the latter were totally damaged by an impact corresponding to a fall distance of 2 m (20 J), the panels reinforced with short sisal fibres were able to withstand, with considerable strains but limited damages, the impact produced by the ball falling from a height of 3.5 m (35 J).

The behaviour of the panels reinforced with long sisal fibres was very similar to that of glass fibre reinforced gypsum panels.

The results obtained on composites employing straight and curled coconut fibres were better than those obtained on gypsmn-board, but less satisfactory than those relating to long sisal fibres.

On all the panels tested, save for gypsum-board specimens, visible cracks began to appear on the lower face for falling distances between 1,5 and 2 m (E = 15-20 J). Such lesions, however, covered a very limited area in the immediate proximity of the impact point.

4 Conclusions

The testing programme revealed the good performance characteristics of gypsum composites reinforced with vegetable fibres.

Gypsum panels produced under vacuum with the addition of sisal and coconut fibres can be used as a viable alternative in lieu of gypsum-board or panels reinforced with fibres possessing very good mechanical characteristics, such as glass fibres.

Furthermore, the use of coarse grain size gypsum confirmed that these panels could be produced without difficulties even in environments, such as those of the developing countries, where it might be hard to obtain high quality matrices through handcraft manufacturing processes.

Acknowledgements

The author expresses his gratitude to the Persano Company of Turin that kindly supplied the gypsum used for the tests.

This investigation was subsidized by the Ministry of Education.

5 References

Aziz, M.A., Paramasivam, P. and Lee, S.L. (1981) Prospects for natural fibre reinforced concretes in construction. **The International Journal of Cement Composite and Lightweight Concrete**, 3, 123-132.

Coutts, R.S.P. (1983) Flax fibres as a reinforcement in cement mortars. **The International Journal of Cement Composites and Lightweight Concrete**, 5, 257-262.

Mattone, R. (1987) Operational possibilities of sisal fibre reinforced gypsum in the production of low-cost housing building, in **Symposium on Building Materials for Low-income Housing**, Bangkok, pp.47-56.

Mattone, R. (1981) Il gesso rinforzato con fibra di vetro: possibilità applicative messe in luce da recenti esperienze. **L'Industria Italiana del Cemento**, 9, 601-606.

Nilsson, L. (1975) Reinforcement of concrete with sisal and other vegetable fibres. **Swedish Council for Building Research**, document D14, Svensk Byggtjanst, Stockholm, 1-68.

Ramaswamy, H.S., Ahuja, B.M., Krishnamoorthy S. (1983) Behaviour of concrete reinforced with jute, coir and bamboo fibres. **The International Journal of Cement Composites and Lightweight Concrete**, 5, 3-13.

18 SISAL-FIBRE REINFORCED LOST FORMWORK FOR FLOOR SLABS

H.G. SCHAFER
University of Dortmund, Federal Republic of Germany

G.W. BRUNSSEN
Failure Analysis Associates, Dusseldorf, Federal Republic of Germany

Abstract
Low cost housing projects in developing countries often suffer from both lack of structural timber as well as skilled workers to manufacture the shuttering for R.C. floor slabs. In this contribution a formwork system and its load-carrying behaviour is described by which structural timber can be substituted through cementitious "lost" formwork elements reinforced with sisal fibres. Among various forms arch-shaped units (sectroids), placed in between prefabricated beams have turned out to be most favourable.

The production technique of the sisal fibre reinforced units is simple so that unskilled workers can be employed. The units are strong enough to carry the weight of the fresh concrete and all loads which may occur during the casting of the slab itself. An appropriate bond quality between formwork units and in-situ concrete can easily be obtained with the help of a simple tool. Due to the high cement content the surface of the units is nearly gas-tight, so that it guarantees a high degree of corrosion protection. The lost formwork system as described fulfills the main requirements of an "appropriate technology" technique: simplicity and utilization of natural national resources.
Keywords: Sisal fibres, reinforced concrete, floor slab formwork, appropriate technology.

1 Introduction

The author named first had the chance of working for some years in Tanzania where he was confronted with the problems of building in countries of the 'Third World'. Building is especially difficult where without external help small scale development projects are to be realized in rural regions .- 'up-country'. There is an enormous shortage of skilled workers and building materials. Even where project managers succeed in securing cement and reinforcing steel the difficulty of manufacturing the shuttering still remains. Often neither suitable structural timber nor 'fundis' (Kiswahili: skilled workers) are available to produce the formwork.

Here a form of 'appropriate technology' is required, which makes use of native resources, simple tools, and which does not need more than basically trained staff. In nearly all developing countries mowadays there are cement factories and in all tropical countries high-strength organic fibres are available in abundance (sisal, coir, hemp etc.), which, in industrialized countries, have been substituted by synthetic fibres in the meantime.

For many years efforts have been made to use sisal as a substitute for reinforcing steel. These efforts, however, were due to fail because organic fibres as sisal are not resistant against alcalinity, Nilsson (1975). In developing countries there is nearly no chance of impregnating the fibres efficiently and durably, Gram (1983) and Ziraba et al. (1985).

For a short period (up to about 1 year, depending on the climate), however, organic fibres are quite useful, as they provide a concrete member a sufficient load-carrying capacity and ductility (Fig. 1). It seems, therefore, to be promising and reasonable to apply organic fibres for transient purposes only, so, for example, as reinforcing material in cement based 'lost formwork' units.

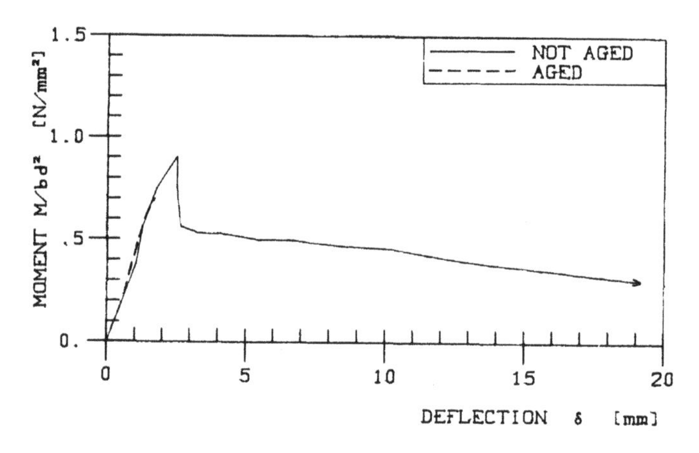

Fig. 1. Load-deflection-behaviour of sisal-cement specimens in a bending test

2 Investigations

2.1 Tests on materials

As mentioned by other authors long fibres (l_f = 900 mm) are much more efficient than short fibres (l_f = 30 mm) with regard to the tensile strength of the sisal-cement.

Due to the low Young's modulus of the fibres the stiffness of the fibre reinforced matrix is reduced considerably when cracking occurs. Therefore, in contrast to the behaviour of high-modulus-fibre reinforced matrices, large deformations are required in order to reach the tensile strength of the fibres. To achieve a good serviceability behaviour of the sisal-cement elements they should remain uncracked under service loads.

Fig. 2 informs about suitable percentages of reinforcement. Tests on specimens exposed to accelerated ageing reveal that the degradation of the fibres does not influence the cracking strength of the matrix.

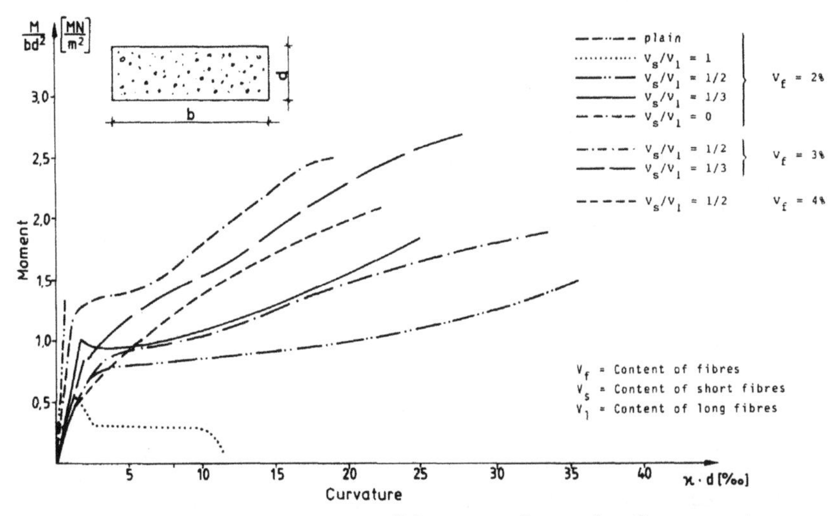

Fig. 2. Moment-curvature-diagram for sisal-cement specimens with different fibre content

Bending and compression tests show that both the Young's modulus of the sisal-fibre reinforced matrix as well as its strength are decreasing with the fibre content (Table 1).

Table 1. Strength values of sisal-cement

Fibre content [% w/w]	Young's modulus [MPa]	Compression strength [MPa]	Bending strength [MPa]	Remarks
0.0	29.3	41.2	8.1	plain
1.0	23.6	25.3	5.9	l_f = 30 mm
2.0	14.9	11.8	4.8	l_f = 30 mm

These reductions are mainly caused by the reduced workability and compactability of the fibre reinforced matrix. By applying superplasticizer it should be possible to reach a better performance.

2.2 Production technique for formwork elements

The formwork units which have a thickness of 10 to 20 mm are preferrably produced with a technique described by Parry (1985). The cement mortar is applied with a trowel on a plain wooden sheet covered by a plastic foil. If needed, long fibres are placed in between thin mortar layers. After compacting on a vibration table the plain sisal-cement unit can be placed on a form either being arch-shaped or trapezoidal.

This technique requires a developable surface; small deviations are allowed. In arch-shaped elements the overlapping depicted in Fig. 3 may be dropped. The trapezoidal elements, however, need overlappings at both endings; from this point of view arch-shaped elements are more suitable.

This manufacturing procedure is very labour-intensive and, of course, inappropriate for high-wage countries. This property, however, makes is particularly attractive for developing countries, where labour is practically the only "commodity" which is available in abundance. As the mortar is cast on a plastic foil the soffit of the units looks enamel-like and is nearly gas-tight.

2.3 Tests on formwork elements

Arch elements and trapezoidal elements can be utilized as formwork units (Fig. 3). Arch elements placed in between simple self-produced truss girders or light-weight precast beams proved particularly suitable (Fig. 4). The load carrying capacity of arch elements loaded on the vertex is in the order of 7 kN.

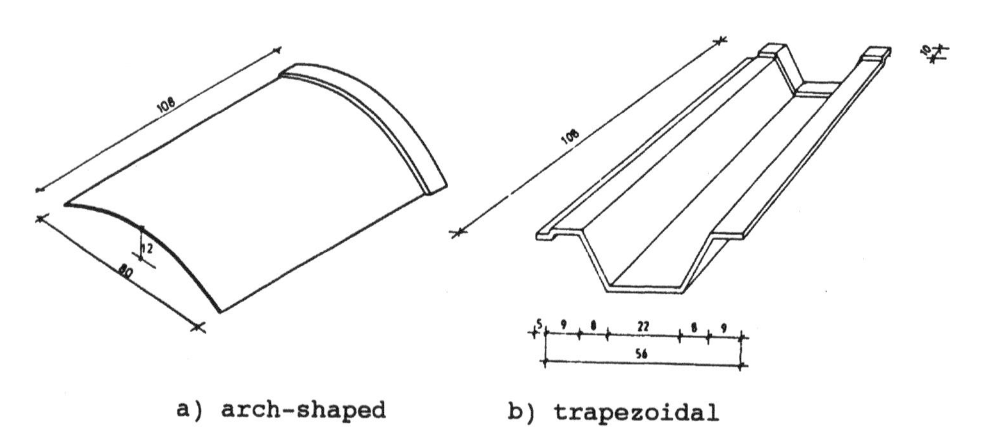

Fig. 3. Formwork elements

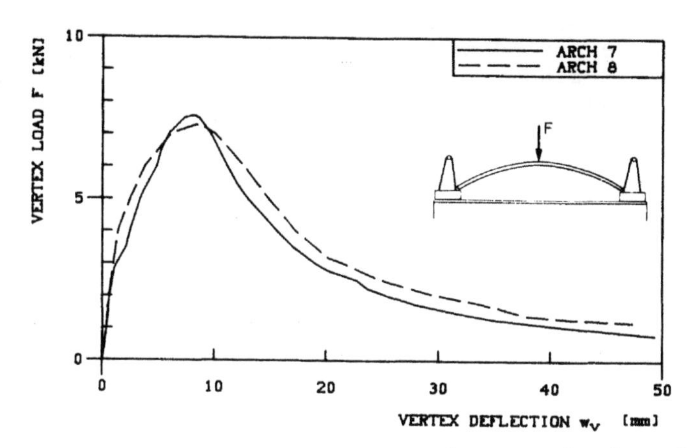

Fig. 4. Load-deflection behaviour of an arch-shaped element

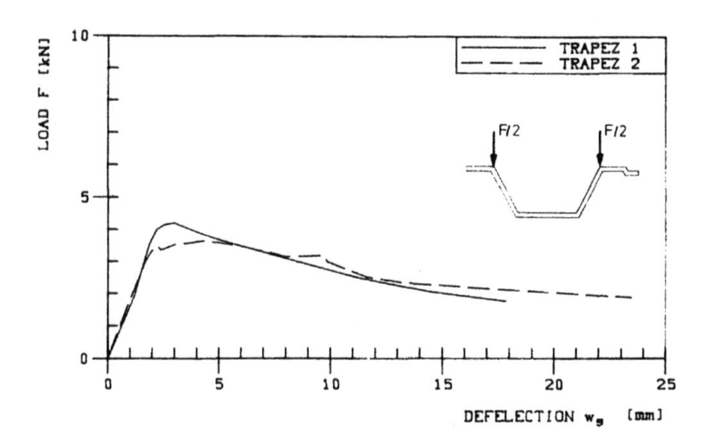

Fig. 5. Load-deflection-diagram for a trapezoidal element

Trapezoidal elements form the ribs of a ribbed slab, as they are oriented in the direction of the span. Precast girders, therefore, are not required. For the erection of the shuttering a transient support, of course, is needed in the form of timber beams and props.

2.4 Bond between sisal-cement unit and in-situ concrete

To obtain the full tensile and shear strength in the interface it is necessary to remove the cement slump on the units. Furthermore the surface has to be treated in such a way that it becomes rough. A suitable roughness can be obtained by treating the surface with a simple nail-roller about 4 hours after casting of the matrix, or by strewing gravel on the freshly cast surface. The gravel can be removed after a period of about 4 hours, leaving behind a castellated interface.

Table 2 shows results of tensile tests on specimens with different methods of treating the surface.

Table 2. Tensile tests on drilled cores

Specimen	Tensile strength [MPa]	Treatment
1	0,63	no treatment
2	0,84	roughened after 0 hours
3	1,59	roughened after 4 hours
4	1,42	gravel, grain size 2/4
5	1,80	gravel, grain size 4/8

Shear tests on beams without shear-reinforcement manufactured with sisal-cement elements confirmed the quality of the methods of treatment mentioned above.

2.5 Protection against corrosion

As mentioned before the lower face of the sisal-cement elements is enamel-like and nearly gas-tight. Hence the thickness of the lost formwork units can be taken into account to the concrete cover of the reinforcing steel bars. The carbonation depth is extremly thin due to the high cement content needed to manufacture the formwork elements.

Measurements on half year old specimens showed no measurable depth of carbonation at the lower surface of the elements. Furthermore tests on a **nine year old building** on the premises of Parry Associates in Cradley Heath, UK, which was built with sisal fibre reinforced elements produced comparable results. While other concrete surfaces of the building showed carbonation depths in the order of one centimeter the carbonation depth on the sisal-cement surface was found to be not greater than one millimeter only.

2.6 Tests on "two-aisle-slabs"

The final R.C. slab is a one-way spanning "sectroid slab" (Fig. 6); a two-aisle test slab under a concentrated load on the middel web revealed a good load distributing behaviour although no transverse reinforcement was applied.

Fig. 7 shows the portion of the load carried by the middle web of the slab versus the applied total load. For a comparison the load portions of an isotropic slab as calculated and a sectroid slab are plotted. It can be seen

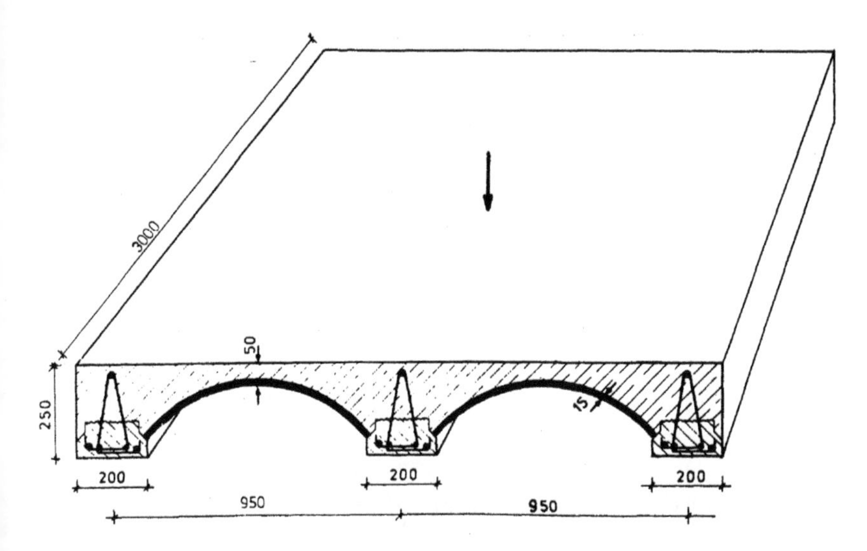

Fig. 6. Two-aisle test slab under middel-web-load

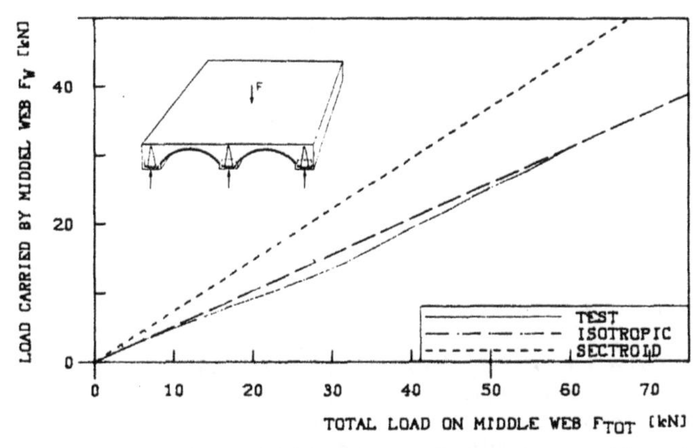

Fig. 7. Load distribution behaviour of a two-aisle sectroid slab under one concentrated load at the mid-span of the middle web

that the quality of the load distributing behaviour of the test slab corresponds perfectly to that of an isotropic slab.

It can be seen from Fig.7 that the load distribution ratio remains nearly constant over the whole loading range. Up to the ultimate load the slab does not show any cracks parallel to the vertex.

The most unfavourable way of loading a two-aisle sectroid slab is to apply a concentrated load at midspan of each of the vertices. In this case the arches are weakened so that the load distribution in transverse direction is

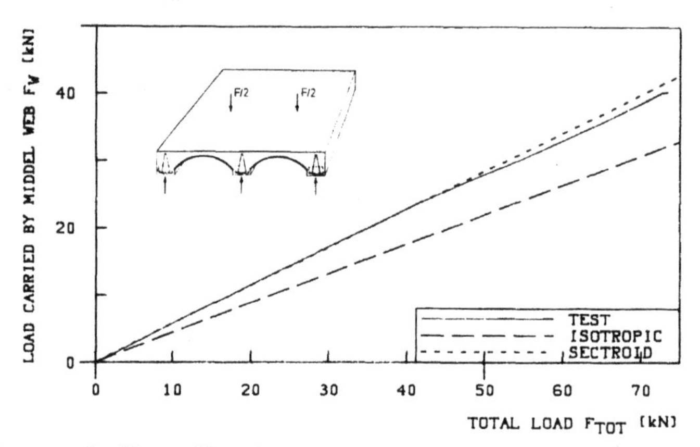

Fig. 8. Load distribution behaviour of a two-aisle sectroid slab under two concentrated loads at midspan of the vertices

reduced. As a consequence the middle web has to carry a higher load portion. A similar ratio, however, is obtained when the load distribution for a sectroid slab is calculated (Fig.8).

Up to the service load of the slab all crack widths in both tests remained smaller than 0.2 mm.

3 Conclusion

The investigations prove that sisal-fibre reinforced elements are suitable to be applied as components of a formwork system.

Due to the favourable cracking behaviour and the small carbonation depths it can be allowed to take account of the sisal cement for the concrete cover of steel rebars.

Due to the relatively low modulus of elasticity and low compression strength sisal cement in the compression zone of a slab should not be considered as load carrying.

If the surface of the sisal-cement elements is treated as described a sufficient bond between element and in-situ concrete is given.

According to the test results the arch-shaped elements are to be preferred. Due to the high load carrying capacity and the simplicity of the production technique these elements appear most suitable for a formwork system.

They are more economical although they require prefabricated beams. This is no disadvantage because they contain already the final reinforcement; hence additonal rebar laying on the construction site can be avoided. They should be light-weight so that they can be placed and moved by hand.

As a rule, on construction sites of developing countries cranes are not available.

The production technique of arch-shaped elements is simple as they can renounce of long fibres. The load carrying capacity of an arch-shaped element under a concentrated load is about 7 kN provided the support is horizontally fixed. This can easily be done by placing the elements into a thin mortar bed on the prefabricated beams. A failure load of 7 kN guarantees a high safety factor for all loading cases which may arise during the casting and construction process.

4 Acknowledgement

The authors wish to thank the 'German Research Association' (DFG) for supporting this idea by a subsidy. We wish to express our gratitude to Mr. J.P.M. Parry for allowing us to carry out some investigations on his premises in Cradley Heath.

5 References

Gram, H.E. (1983) **Durability of natural fibres in concrete.** CBI-Research No. 1-83, Swedish Cement and Concrete Research Institute, Stockholm

Mwamila, B.L.M. (1984) **Low modulus reinforcement of concrete - special reference to sisal twines.** Swedish Council for Building Research, Document No. D-10:1984, Stockholm

Nilsson, L. (1975) **Reinforcement of concrete with sisal and other vegetable fibres.** Swedish Council for Building Research, Document No. D-14:1975, Stockholm

Parry, J.P.M (1985) **Fibre concrete roofing.** IT Workshops, Cradley Heath

Persson, H., Skarendahl, A. (1978) **Sisal fibre concrete for roofing and other purposes.** Swedish International Development Authority, Report No. 1978-08-15, Stockholm

Schäfer, H.G., Brunssen, G.W. (1989) **Sisalfaser-bewehrte verlorene Schalelemente.** Interim report to the German Research Association (DFG), unpublished (in German)

Swamy, R.N. Ed. (1988) **Natural fibre reinforced cement and concrete.** Concrete technology and design Vol. 5, Blacky, Glasgow, London

Swift, D.G., Smith, R.B. (1979) The flexural strength of cement-based composites using low-modulus (sisal) fibres. **Composites** Juli 1979 pp. 145-149.

Ziraba, Y.N. et al. (1985) Use of plasticized sulphur in sisal-fibre concrete. **Durability of Building Materials** 3/85, pp. 65-76

19 STABILISATION D'UN TORCHIS PAR LIANT HYDRAULIQUE

(Stabilization of Adobe with Hydraulic Binder)

F. BUYLE-BODIN, R. CARBILLAC, R. DUVAL
and W. LUHOWIAK
Paris University X, Laboratoire LEEE, IUT, Cergy, France

Résumé
La terre est un matériau de construction traditionnel, en particulier en Europe , où il est réétudié scientifiquement depuis les années 80. Pour en faire un matériau économique et utilisable sans compétences particulières, il paraît intéressant d'en limiter l'usage au remplissage des constructions à ossature bois et de le renforcer avec des pailles. Un matériau traditionnel de ce type est largement répandu en France, le torchis. Partant d' études des techniques traditionnelles d'élaboration et de la faisabilité d'un produit industriel, les auteurs ont étudié l'opportunité de stabiliser le torchis. Cette technique conduit à une forte utilisation de végétal dans la construction, par l'ossature et par les fibres de renforcement de type paille.
Mots clés:Torchis, Terre, Paille, Stabilisation, Ossature Bois

1 Introduction

Dans l'éventail des techniques de construction en terre, le torchis, mélange de terre et de fibres végétales, occupe une place à part. Son rôle de matériau de remplissage des constructions à ossature bois impose des propriétés mécaniques très particulières et assez différentes de celles des autres matériaux de construction à base de terre. Sa stabilisation et sa protection sont également conditionnées par son mode de mise en oeuvre, ce qui explique l'intérêt d'utiliser des fibres végétales pour cela.

Dans une première phase de notre recherche sur une meilleure connaissance de ce matériau économique, nous avons essayé de stabiliser le torchis par adjonction de liant afin de pouvoir se dispenser de l'enduit de protection. Nous livrons ici des résultats partiels et provisoires sur ce matériau plusieurs fois millénaire et dont les nombreux avantages devraient amener un renouveau, auquel nous espérons contribuer par notre travail de recherche.

2 Les matériaux de construction en terre

2.1 La terre en France

Le renouveau de la terre comme matériau de construction date en France de la fin des années 1970, initié entre autres par l'équipe Craterre (1979). L'intérêt que présentait ce matériau pour le tiers monde a conduit à un regroupement des actions de recherche au niveau national sous l'égide du Plan Construction en 1981 (le Plan Construction est une émanation du Ministère de l'Equipement et du Logement français). Ont suivi un ensemble de colloques et séminaires en 1982, 1983, 1984 et 1987 où d'un point de vue technique la connaissance des méthodes traditionnelles et l'amélioration par stabilisation du matériau ont fait d'importants progrès.

Si la construction en terre est présente sur 20 % du territoire français, le torchis ne se trouve de façon significative que dans le Nord-Est et le Nord-Ouest de la France. Le végétal le plus utilisé semble être la paille d'orge pour des raisons de disponibilité et de résistance mécanique. Le torchis remonterait au néolithique et des vestiges du II et IIIe siècle ont été découverts en Picardie (Michel et al. 1987). Ce matériau a été pourtant jusqu'à maintenant relativement peu étudié en France.

Nous exposons dans un premier temps Les travaux les plus récents sur une meilleure connaissance du torchis en France, et sur la mise au point de procédés industriels d'élaboration suivant les recettes traditionnelles.

2.2 Le torchis traditionnel en France

Le torchis est un matériau de remplissage d'une construction à pans de bois, c'est à dire dont la structure primaire est une charpente. Celle-ci peut varier notablement suivant la région, en taille des éléments comme dans leur disposition ou dans les essences utilisées. Une structure secondaire d'accrochage, également en bois, va permettre de supporter et de rigidifier le remplissage (figure 1).

Celui sera préparé et mis en oeuvre suivant des techniques extrêmement variées suivant les lieux et les époques (figure 2).

La teneur en eau du mélange est élevée (20%, 30%, dans la technique "mille feuille", jusqu'à 40% dans la technique "à la torche"). Le pourcentage en volume de la paille est classiquement 50%-50% avec la terre mouillée, ce qui amène à 3% de fibres en poids. Les fibres pourront être longues (longueur naturelle d'environ 50 cm) ou courtes (coupées à la serpe à la longueur d'une main).

Le mur est ensuite recouvert d'un enduit à base de chaux grasse, de terre tamisée additionnée de sable, et de fibres animales ou végétales fines et courtes.

Le produit résultant présente en moyenne, d'aprés Calame (1985):

- par rapport à un béton dosé à 300 kg de ciment, 1/20 de résistance à la rupture en compression et 1/5 de résistance à la rupture en traction
- un module d'Young entre 200 et 300 MPa qui permet au torchis de suivre les déformations naturelles de l'ossature bois.
- une conductivité thermique voisine de 0,7 W/mK (mesurée à la sonde à chocs thermiques -Laurent et al.1984) .

La pathologie du torchis traditionnel présente plusieurs aspects

- fissuration et retrait excessifs, souvent dus à une trop forte présence d'argile (activité trop élevée)
- décollement du torchis , mal accroché à un support trop lisse ou trop serré.
- décollement de l'enduit par défaut d'accrochage ou manque d'élasticité par rapport à la souplesse du torchis (enduits à base de ciment)
- pourrissement des fibres végétales par manque d'échanges hygriques (air et humidité), à cause d'un enduit trop imperméable ou de l'usage d'un plastifiant.

Néanmoins des torchis datant du XVe siècle parfaitement conservés ont été retrouvés en Picardie, ce qui prouve la durabilité du matériau, à condition qu'il soit à l'abri de l'eau.

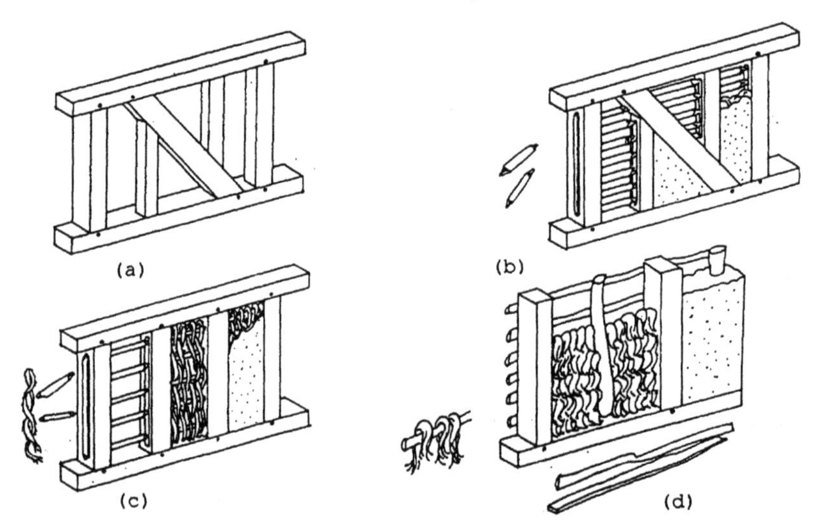

Figure 1 (par F. CALAME)
a: Module de base d'un pan de bois - système Picard
b: torchis à bois apparent - système champenois et picard
c: torchis à bois apparent - système Landais
d: torchis à bois apparent à l'extérieur posé en torche - Nord

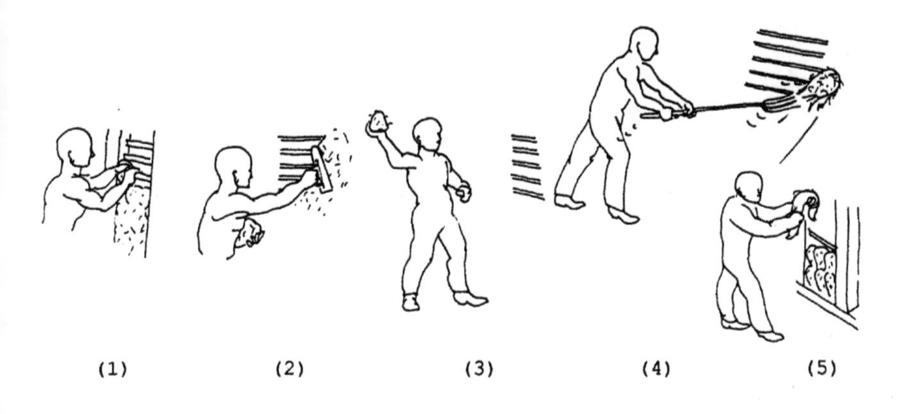

Figure 2 (par F. CALAME)
- pose du torchis à la main (1), au platoir (2), à la volée (3), à la fourche (4), à la torche (5)

2.3 Présentation d'un torchis industriel dérivé des techniques traditionnelles

La nécessaire réparation du bâti existant, combinée à un renouveau de la demande de matériaux traditionnels et liée à un goût prononcé de la population pour des constructions respectueuses de leur environnement, y compris culturel, ont conduit à diverses expériences sur la mise au point de torchis industriel (voir Calame 1985).

Il faut comprendre par industriel un produit mélangé en usine à sec à partir de terre de carrière et de pailles sélectionnées et livré ensuite sur le chantier en vrac ou en sac. La granulométrie peut ainsi être corrigée, en particulier la teneur en argile doit être abaissée en dessous de 18 % (valeur qui semble être une limite maximale) par adjonction de sable. Cette caractérisation de l'activité de la terre peut être menée par l'essai au bleu de méthylène (norme NFP 18592 juillet 1980) et donne, sur le passant à 400 microns, des valeurs de bleu qui doivent être inférieures à 3g/100g pour éviter de grands retraits, mais ne doivent pas descendre trop bas car le manque d'argile rend alors le matériau difficile à poser.

Le végétal peut être du foin, à tige fine et résistante mais relativement onéreux. Parmi les pailles, celle d'escourgeon (variété d'orge) est préférée à celle de blé dont la rigidité, due à un diamètre plus important, rend le torchis plus délicat à mettre en oeuvre.

Le matériau est préparé en usine (briquetterie) dans un malaxeur planétaire par voie humide à une teneur en eau de 30 % à 40 %. Il doit ensuite reposer un à deux jours pour perdre l'excès d'humidité. L'enduit est également produit industriellement. C'est un mélange : chaux aérienne éteinte (1/5), divers sables pour coloration (2/5), argile (1/5), fibre végétale (balle de lin) (1/5).

3 Vers un torchis moderne

3.1 Paramètres scientifiques, techniques et économiques à préciser pour développer un torchis moderne

L'étude de la pathologie des torchis traditionnels et la recherche d'un produit industriel équivalent ont permis de dresser la liste des paramètres à préciser pour développer un torchis moderne.

Pour le torchis, il faut :

- caractériser par des essais simples la teneur en argile
- préciser l'amendement nécessaire par adjonction de sable
- étudier l'influence du diamètre, de la longueur et de la nature des fibres sur le produit résultant
- rechercher une éventuelle stabilisation par adjonction de produits adaptés.

Pour le support, il importe de recherche l'influence de l'espacement, de la largeur et de l'état de surface des lattes,ou de l'opportunité d'un éventuel autre support (treillis, grillage)

Pour l'enduit, nous pouvons dresser la liste des caractéristiques requises :

- protection aux intempéries
- élasticité (souplesse)
- perméabilité à l'eau
- résistance aux ruissellement
- résistance aux chocs thermiques
- adhérence au support
- esthétique.

Enfin plus globalement, le procédé doit être économique, facile à mettre en oeuvre, durable, et assurer le respect de contraintes diverses, culturelles, esthétiques ..(Lahure 1987)

3.2 Recherche d'une stabilisation

Le problème de la protection du matériau terre n'est pas simple (Dayre 1983). L'enduit est le point faible du procédé, et il lui est souvent préféré une imprégnation de surface ou une stabilisation en masse qui permet au matériau de résister directement aux intempéries.

La stabilisation d'un matériau terre consiste en l'amélioration de ses caractéristiques mécaniques tout en réduisant sa sensibilité à l'action de l'eau. Pour cela il existe des méthodes mécaniques, physiques ou chimiques .

Pour des raisons liées à sa mise en oeuvre, le torchis doit être posé à l'état plastique, et toute stabilisation mécanique classique par compactage ou dessication est donc exclue. De plus les améliorations ainsi obtenues ne sont pas irréversibles pour les éléments fins, qui sont présents en plus grande proportion dans le torchis que dans les autres matériaux en terre.

La stabilisation physique est à notre sens assurée par l'adjonction de fibres. Le torchis est donc par essence un produit stabilisé par la présence des pailles, qui réduisent considérablement la fissuration de retrait (lors du séchage du produit), et assurent par la suite une bonne résistance à la rupture et une bonne tenue aux grandes déformations du matériau. La sensibilité à l'action de l'eau existe néanmoins toujours.

Seule la stabilisation chimique permet de résoudre le problème, par adjonction de produits variés. Le plus étudié est actuellement le plus répandu dans nos pays, le ciment.

3.3 La terre armée de fibres végétales

Diverses études ont été menées sur la stabilisation des terres par adjonction de fibres végétales. Plusieurs terres à proportion d'argile variée, renforcées de diverses fibres végétales (à l'exception des pailles) de différentes longueurs et mise en oeuvre par voie sèche, semi-sèche ou humide ont été étudiées (Martin 1984). Les résistances au cisaillement et à la compression, la résilience, la sensibilité à l'eau et la durabilité ont été mesurés sur des briques compressées statiquement à la presse, ce qui rend la transposition difficile au torchis.

Le mélange terre-paille a été particulièrement étudié et mis en oeuvre sur des chantiers expérimentaux en France (Simon 1984). Le produit diffère ici encore notablement du torchis par le mode de mise en oeuvre et les proportions d'eau et de végétal.

Le rôle du renforcement par fibres végétales est en conclusion reconnu comme très positif car il confère au matériau des propriétés de souplesse et de résistance appréciables à moindre coût. Mais l'insensibilité du matériau à l'action de l'eau n'est toujours pas assurée. Nous devons essayer d'y remédier sans atténuer les qualités fondamentales du torchis.

4 RECHERCHES ACTUELLES SUR UN TORCHIS MODERNE

4.1 Stabilisation par liant hydraulique (ciment)

Nous avons cherché dans un premier temps à stabiliser durablement le torchis par adjonction de ciment. La présence d'une forte proportion d'argile dans le torchis pouvait faire craindre des conflits entre cette variété de minéral et le ciment. Des recherches multiples sur cet aspect particulier de la stabilisation des terres sont menées depuis dix ans. Citons par exemple M. Laquerbe (1981) M. Albenque et al. (1982) et F. Buyle-Bodin et al. (1989).

Nous partons donc d'une terre corrigée respectant le fuseau optimal de composition d'un torchis moderne (figure 3) à teneur en eau initiale proche de 30 % . La terre humide est malaxée seule 6 mn, puis le ciment et la paille sont rajoutés et le malaxage poursuivi 3mn (malaxage par voie humide).

Des teneurs en ciment CPA 55 de 5 %, 10%, 15%, et 20% ont été essayées. La valeur de 5% est trop faible pour que le mélange soit homogène et que la stabilisation soit effective.

ANALYSE GRANULOMÉTRIQUE

TAMISAGE PAR VOIE HUMIDE/SECHE
SEDIMENTOMETRIE
ECHANTILLON
FUSEAU TORCHIS

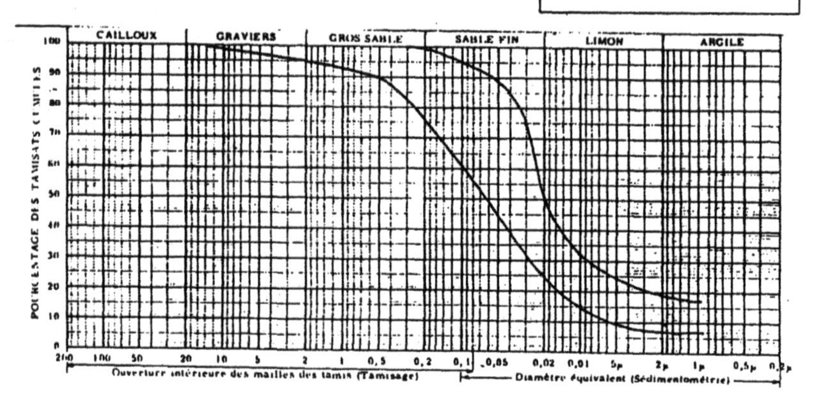

Figure 3 (par ENTPE)
fuseau optimal de composition d'un torchis moderne

Pour les trois autres dosages, nous donnons les résultats:

-de l'essai de prise à l'aiguille de Vicat (a)
-de la mesure du retrait et de la dilatation à l'eau (b)
-de la résistance à la compression (c)

à plusieurs âges en conservation à 20°, 50% d'humidité (figure 4).
Les essais de traction-flexion demandent quelques compléments.
Les résultats sont comparables à ceux donnés par d'autres expériences (Moussa et al. 1984). L'insensibilité à l'eau parait donc acquise, avec un renforcement notable des caractéristiques de résistance. Mais il semble que cela se fasse au détriment de la souplesse. Le début de prise au bout de 6 heures devrait également poser des problèmes de mise en oeuvre du torchis stabilisé.

4.2 Etude de l'influence des fibres.

L'ajout de paille semble avoir quelque influence sur la résistance mécanique à la compression, comme l'indique le tableau suivant :

tableau 1. Résistance à la compression à 28 jours en fonction de la composition (teneur en ciment C et en pailles P)

composition	10%C	15%C	20%C	10%C + 2%P	15%C + 2%P	20%C + 2%P
Rc en MPa	1	1,4	1,6	1,6(+60%)	2,1(+50%)	2,2(+37%)

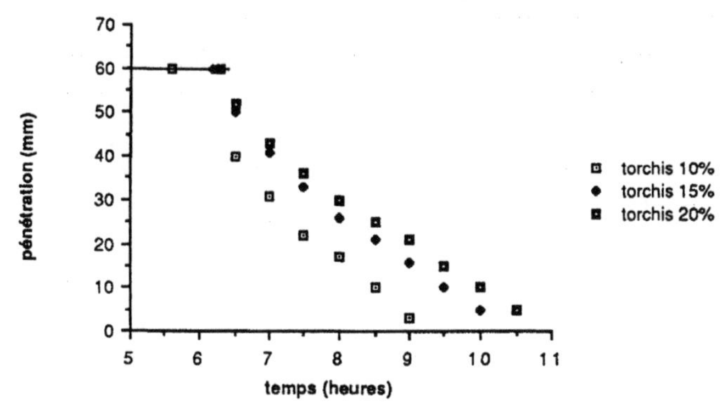

(a) essai de prise à l'aiguille de Vicat

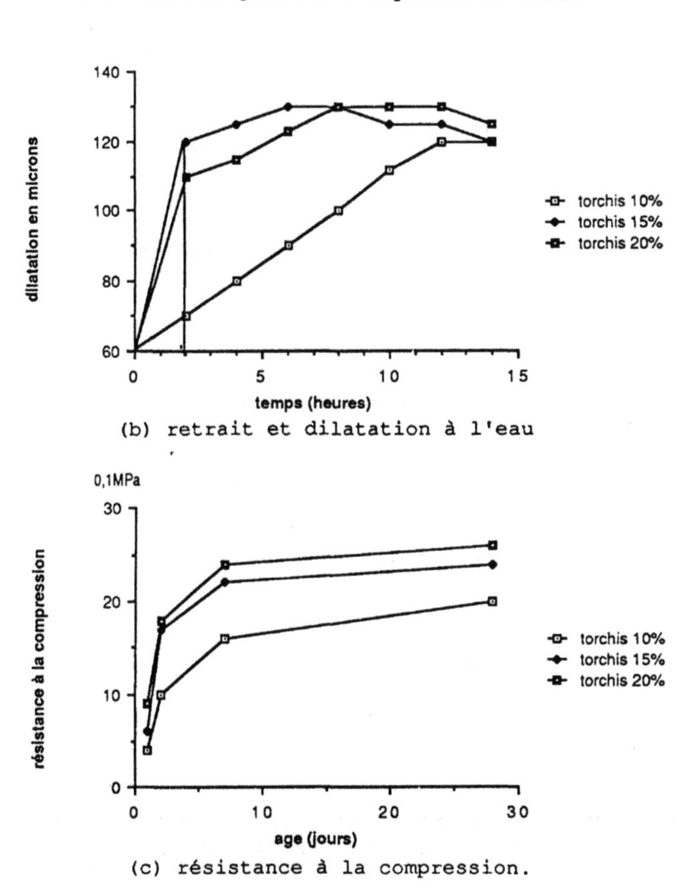

(b) retrait et dilatation à l'eau

(c) résistance à la compression.

Figure 4: résultats d'essais sur un torchis moderne

Cette influence est moins bénéfique quand la proportion de ciment augmente. Il est à noter que sans paille la rupture est de type fragile, avec paille nettement plus ductile.

Pour des raisons de malaxage, la paille ajoutée est coupée à une longueur d'environ 5 cm. Des essais sur matériau préparé à la main avec pailles entières n'ont pas semblé donner de résultats différents, mais la collection de mesures est trop hétérogène pour être fiable. Nous envisageons une nouvelle série d'essais en jouant sur les paramètres suivants :

- pourcentage de pailles : recherche de l'existence d'un optimum.
- longueur L et diamètre D des pailles : recherche de corrélation entre L/D et les caractéristiques mécaniques du produit résultant

L'ajout des fibres végétales donne un produit nettement plus performant et permet donc d'envisager une économie de liant. Mais le produit ainsi obtenu parait s'éloigner du torchis désiré au départ.

4.3 Etude des paramètres perméabilité à l'air et souplesse

Il est bon de rappeler ici que le torchis est au départ un matériau de remplissage. Donc sa résistance à la compression est un paramètre secondaire en regard de la souplesse et de la perméabilité à l'air.

La souplesse est caractérisée par le module d'Young, qui est de 200 à 300 MPa pour le torchis non stabilisé, contre classiquement 500 Mpa pour une terre compactée. La fibre végétale joue donc un rôle essentiel dans la souplesse du torchis simple.

Notre produit terre + ciment + paille présente un module compris entre 1 000 MPa et 2 000 MPa, comparable à celui d'une terre stabilisée au ciment. On peut constater que la paille ne joue plus aucun rôle dans ce cas et que la déformation admissible sans fissures est réduite de quelques pour cent à 0,5 pour mille.

Pour la perméabilité à l'air, des essais sont programmés. On peut néanmoins remarquer par un essai de reprise d'eau que le torchis simple absorbe 20 cm3 /10mn et que le torchis stabilisé absorbe si peu que nous tombons sous le seuil de sensibilité de la mesure. La porosité ouverte du matériau doit donc être quasiment nulle et laisser craindre une perméabilité à l'air très faible.

5 Conclusion

Nous pouvons conclure en proposant deux directions bien distinctes d'application du produit terre armée de fibres végétales du type paille.

La première est celle du torchis matériau de remplissage des ossatures bois. Des impératifs de souplesse et de perméabilité excluent l'usage des liants hydrauliques comme stabilisateurs. La durabilité du produit repose donc intégralement sur l'enduit superficiel et sur une bonne protection architecturale contre l'action de l'eau.

La composition de la terre et les principales caractéristiques du torchis sont maintenant bien connues. Des recherches complémentaires sont envisageables pour mieux fixer le rôle de la taille des fibres, mieux contrôler la perméabilité et la souplesse du torchis, améliorer l'adhérence du matériau sur son support et celle de l'enduit , perfectionner les moyens industriels de production et de mise en oeuvre d'un produit moderne fiable et durable.

L'autre direction de recherche concerne l'adjonction de pailles à la terre stabilisée au ciment, pour économiser ce dernier et obtenir ainsi un produit compétitif qui serait du genre bloc manufacturé. Nous développons actuellement un nouveau type de presse à blocs par compactage dynamique et recherchons systématiquement à cette occasion le gain en liant apporté par l'adjonction de fibres végétales.

6 Bibliographie

Albenque M. Collon L. Lorec S. Bombled J.P. Mortureux B. Regourd M. et Volant J. (1982) Briques de terre cuite et briques en argile stabilisée. **Ciments,bétons,plâtres,chaux,** 737, 217-226 et 738, 291-296.

Buyle-Bodin F. Cabrillac R. Duval R. et Luhowiak W. (1989) The influence of clay minerals over the elaboration of products made of soil stabilized by hydraulic binders. **Materiales y tecnologias para la construccion de viviendas de bajo costo - Tercer simposio CIB-RILEM**.ed. Infonavit Mexico pp 207-216.

Calame F. (1985) **Eléments d'un cahier des charges pour la réalisation d'un torchis moderne** ed.Plan Construction, Paris France.

Craterre (1979) **Construire en terre** ed. Alternatives Paris France

Dayre M. (1983) La protection du matériau terre. **L'habitat économique dans les pays en développement**.ed. Presses de l'ENPC, Paris, pp 37-42.

Lahure F. (1987) La réhabilitation du patrimoine terre en haute Normandie. **Le patrimoine européen construit en terre et sa réhabilitation**.ed. ENTPE, Lyon, pp 61-80.

Laquerbe M. (1981) La cristallisation des argiles à froid. **Recherche française et habitat du tiers monde** ed. Plan Construction, Paris, pp 73-79.

Laurent J. Quenard D. et Sallée H. (1984) Caractérisation thermique du matériau terre. **Modernité de la construction en terre**.ed. Plan Construction, Paris, pp 67-88.

Martin M. (1984) Etude du renforcement de la terre à l'aide de fibres végétales **Modernité de la construction terre**.ed. Plan Construction, Paris,pp 49-66.
Michel P. Poudru F. (1987) Le patrimoine construit en terre en France. **Le patrimoine européen construit en terre et sa réhabilitation**.ed. ENTPE, Lyon, pp 529-551
Moussa K. Prin D. Didier G. (1984) Valorisation des sols tropicaux **Modernité de la construction terre**.ed. Plan Construction, Paris,pp 93-110.
Simon D. (1984) Un mélange original terre paille:synthèse des connaissances. **Modernité de la construction terre**.ed. Plan Construction, Paris,pp 125-134.

20 PRELIMINARY WORK TO PRODUCE PAPYRUS-CEMENT COMPOSITE BOARD

K.S.J. Al-MAKSSOSI and W.A. KASIR
College of Agriculture and Forestry, Hammam Al-Alil,
Mosul, Iraq

Abstract
This research was designed to evaluate a preliminary work of papyrus-cement composite board. The research examined the influence of papyrus particle size and the board density on the mechanical and physical properties of papyrus-cement composite board. The boards were classified based on the distribution of two levels of two variables. The variables were: two sizes of papyrus and two densities.

To determine the mechanical and physical properties of the papyrus-cement composite board, four types of test were done. These tests include bending, tensile strength perpendicular to the surface (internal bond), compression strength parallel to the surface and water absorption and thickness swelling. Testing was essentially according to D-1037-78 of the American Society for Testing and Materials.

The observed trends were for higher test values for MOR, MOE, Spl in bending test gave better results with high density, never the less, internal bond and thickness swelling, regardless of density, highest values were exhibited with small particles.

The combination of small particles and high density gave the best value for maximum crushing strength.
Keywords: Cement, Papyrus, Calcium Chloride,Static Bending, I.B., Compression Strength, Water Absorption, Thickness Swelling.

1 Introduction

The concept of combining cement with some cellulosic materials to produce composite board dates back more than five decades. Dinwoodie and Paxton (5) mentioned that a light - weight wood wool board under the early 1930's. Also, there was a report in 1947 stating that a cellulosic materials such as wood, straw or baggesse were mixed with protland cement to produce a light - weight board for use during world war II (6).

In the United states, Elmendorf. Inc. developed a high -

density wood - cement board for which several patents was developed further in Switzerland by producing a new building board under the name of ((Duripanal)). The development was due to further improvement work done by Dorisol Villmegen AG. leading to the erection of a plant in 1968-69 at Dietikon, near Zurich, Germany (3,5,11).

As a matter of fact, papyrus has never used as a useful material to produce such a product in Iraq before. So, an attempt has been made to manufacture and examine a multi purpose construction board for the first time in Iraq. The effort was made for several reasons: first to utilize a natural resource self renewable, papyrus, which grown over a wide range of southern and mid parts of Iraq. Second, both materials, papyrus and cement, have been used in this product were very cheap. Finally, the board can be used as a construction material where the risk of fire, humidity and insects and fungi to the particleboard to be avoided. In regarding to preceding, development of useful products from papyrus that is often wasted, should certainly be encouraged.

2 Procedure and Materials

2.1 Preparation of materials

The constituent of the board which were used in this experiment were papyrus, cement, water and chemical additive. The papyrus used was (Typha domingensis L.) which harvested from Therthar lake near Sammara city.

The papyrus was cut into two sizes, 20 cm and 30 cm. After the particles were sorted, the two sizes of papyrus particles were dried sparately for 24 hours in an oven at a temperature of 105°c ± 2. After drying, they were kept in plastic bags to prevent them from gaining moisture from the ambient atmosphere until they were mixed with the other components of the board.

The second component of the board was portland cement type I . This type of cement was chosen since it is locally produced by Hammam Al-Alil cement plant. Also, it is widely used as a mineral binder because it is both low in price and available throughout the word. Likewise, portland cement does not burn and is not affected by insects or fungi.

The third component of the board was water. Each gram of oven - dried papyrus needed 0.85 ml water to attain fiber saturation (assuming a fiber saturation point of 85% moisture content) (10). The requirement of each gram of dry - weight cement to complete its hydration was 0.25 ml water (12). Distilled water was used to avoid reaction with dissolved chemicals possibly present in tap water.

The fourth component in the board preparation was calcium chloride. This additive is used to accelerate the setting of the cement (4,6,7,8). Calcium chloride was added as 3% relative to the total weight of cement (1). Solid calcium chloride was dissolved in the amount of water necessary for cement hydration and papyrus fiber saturation, as explained above, prior to being added to the papyrus and cement mixture.

2.2 Producing the board

After the board material were prepared, a balance was used to weigh the papyrus particles, cement and calcium chloride. The water was measured with a graduated cylinder.

The ratio (by weight) of papyrus particles and cement based on total weight, used for forming the board was 20/80. The target density of the board were 1200 kg/m^3 and 800 kg/m^3 respectively.

To produce the board, the determined oven - dry weight of papyrus particles and dry - weight of portland cement type one were mixed together in a metal container. The necessary amount of distilled water containing the desired quantity of calcium chloride was added in two portions in order to control the mixing process and to ensure that the entire mixture were moist. Mixing was accomplished by hand.

Hand - mixing was continued until the mixture was blended thoroughly. The mixture was then transferred to a locally made mold, 40 cm square by 2.5 cm deep, for casting. The mold was made of wood. After the mold was filled with the mixture, it was compressed in a hydraulic press. Pressure was applied to close the mold to 1.25 cm stops, the desired thickness of the board. Four 1.25 cm thick blocks of wood were used at the corners of the mold to provide the mechanical stops. The board was left under pressure 3 days, then taken out of the mold and stored under conditions of 21°c and 65% RH for four weeks.

After conditioning the boards, one inch was trimmed for each of its sides. The board then was cut in to specimens for four tests, static bending, tensile strength perpendicular to surface (internal bond), compression strength parallel to surface, and water absorption and thickness swelling. Two samples were cut for each test except for internal bond for which four samples were obtained. Testing was essentially according to D-1037-78 of the American Society for Testing and Materials (2).

3 Analysis of The test results

The test results are presented in tables 1 and 2. The following is a discussion of each of the tests conducted. The statistical analysis showed that the variables studied caused no significant differences in all test properties evaluated except MOE in static bending test.

Table 1. The mechanical and physical properties of papyrus-cement composite board.

Property	Range	Average
Static Bending		
Modulus of Rupture (MOR) N/cm^2	77.67-1962.06	503.20(389.49)*
Stress at proportional limit (Spl) N/cm^2	31.07-480.00	206.04(125.04)
Modulus of Elasticity (MOE) $N/cm^2 \times 10^5$	0.21-5.26**	1.26(1.14)
Internal Bond (I.B.) N/cm^2	0.39-8.53	3.64(1.91)
Compression strength Parallel to the surface		
Maximum crushing strength N/cm^2	55.13-522.56	177.94(119.75)
Water Absorption after 24-hour soaking based on		
Dry weight (%)	4.42-36.95	22.01(7.20)
Dry volume (%)	4.51-36.91	22.20(5.51)
Thickness swelling after 24-hour soaking (%)	0.34-2.69	1.58(0.70)

*Numbers in parentheses are standard deviations.
**The MOE in static bending shows statistical significance of the 5% level.

Table 2. Average test result for each combination studied

Property	Bending test MOR	MOEx10^5	Spl	I.B.	Max.Crush. Strength	Water abso. based on		Thickness Swelling
Test variables	(N/cm^2)	(N/cm^2)	(N/cm^2)	(N/cm^2)	(N/cm^2)	Wt%	Vol%	%
High density small particle	737.41	1.29	245.09	4.56	213.00	21.76	24.8	1.40
Low density small particle	295.85	0.63	149.17	4.85	151.41	20.96	18.99	1.40
High density large particle	567.61	2.29	251.02	2.65	169.29	19.60	21.23	1.91
Low density large particle	441.95	0.83	173.03	2.40	178.04	25.73	23.79	1.61

3.1 Static bending

In this test, the following measurements were determined

Modulus of Rupture (MOR) N/cm^2

Stress at Proportional limit (Spl) N/cm^2

Modulus of Elasticity (MOR) N/cm^2

The observed trends were for higher test values for MOR, Spl and MOE gave better results with high density regardless of particles size. Among the four combinations, however, MOR and MOE exhibited the best result with high density and small particle while Spl shows the best result with high density and large particle.

3.2 Tensile strength perpendicular to surface (I.B.) N/cm²
The apparent trend is that the I.B. increased the size of the particles decrease regardless of the density of the board. The highest value was observed with high density and small particles.

3.3 Compression strength parallel to surface
The following property was taken from the compression strength parallel to surface test

Maximum crushing strength (S_{max}) N/cm²

The combination of small particles and high density revealed the best value for maximum crushing strength. Reducing the density, Nevertheless, caused an apparent reduction in the crushing strength.

3.4 Water absorption and thickness swelling
Large size of papyrus particles with high density board present the lowest value of water absorption based on the original (dry) weight. Small size of papyrus particles with low density, on the other hand, exhibit the lowest value of water absorption based on original (dry) volume.

The apparent trend for thickness swelling test is that the small size of papyrus particles resulted in a considerable reduction in the amount of water absorbed regardless of the board density.

4 Conclusion

The following conclusion can be drawn based on the foregoing results and observed trends. This research would indicate that for the variable studied, the most desirable combination for a cement bonded composite board would be high density and small size of papyrus particles unless I.B. was the principle concern.

5 References

1. Al-Makssosi, Kamil S. and Shuler, G.E. (1985) Properties of wood-cement with variations in particle size and wood-cement ratio. Thesis presented at Colorado State University, Fort collins, Co. 80523 . U.S.A.
2. American Society for Testing and Materials. (1978) 'standard methods of evaluating the properties of wood-base fiber and particle panel materials. ASTM standard D-1037-78.
3. Bernard, M. Guthrie, Robort E. Torley and James K. Needham. (1982) Permawood, a wool residue and

portland cement composite used as a building material. A paper presented at the Forest products Research Society Annual Meeting. June 24, 1982. Held in the Marriott Hotel, New Orleans, LA.

4. Brunauer, S. and L.E. Copland. (1964) The chemistry of concrete. Scientific American, 210 (4), 80-92.
5. Dinwoodie, J.M. and B.H. Paxton. (1983) Wood-cement particleboard a technical assessment. Building Research Establishment Information Paper Ip-4/83. April. Scottish Laboratory, Kelvin Road, East kibride Glasgow G. 75 ORZ.
6. Dove, Leonard. P. (1945) Building materials from portland cement and vegetable fiber. Forest Research Laboratory. Oregon State University, College of Forestry, Corvallis, OR 97331, PP. 1-24.
7. Forest Products Laboratory, Forest service, USDA, (1952) Sawdust cement concrete. Report No. R 1666-15 Maintained at Madison 5, Wisconsin, in cooperation with the University of Wisconsin. U.S.A.
8. Huffaker, E.M. (1962) Use of planer mill residues in wood - fiber concrete. Forest Products Journal (7) 298 - 301.
9. Prestemon, Dean R. (1976) Preliminary evaluation of wood - cement composite. Forest Products Journal. 26 (2) 43-45 .
10. Samarai, M.A. Al-Taey, M.J. Kassir, A. F. (1986) Some chemical data and operational test Iraqi Reed and Reed products. NCCL. A paper presented at a joint symposium of a use of vegetable plants and their fibers as building materials. Oct. 7-9 , 1986. Held in Baghdad, Iraq. C 107-116.
11. Stillinger, J.R. and I.W. Wentworth. (1977) Product process and economics of producing structural wood - cement panels by the Bison - Werke system. Proc. 11th Particleboard symposium, Washington, PP. 383-410 .
12. Weatherwax, R.C. and H. Torkow.(1964) Effect of wood on setting of portland cement. Forest Products Journal 14(12) 567 -570 .

21 FIBRE-CONCRETE ROOFING TILES IN CHILE

S. ACEVEDO, M. ALVAREZ, E. NAVIA and R. MUÑOZ
University of Santiago, Chile

Abstract
One of the most important aspects of the appropriate solutions of building in Chile is the provision of a cheap durable roofing material. For some time the Intermediate Technology Program of the "University de Santiago de Chile", has been working on the development of a suitable roofing material with which to meet the needs of an important part of the Chilean population. The process described here permits the conversion of fibrous agricultural residues, common in Chile, such as wheat fibre, barley fibre, maguey fibre and curaguilla, into natural-fibre concrete roofing tile. A vibrating table manually powered was designed and constructed to produce a uniform mix. This work also included, fibre materials and their characteristics, production technology, properties of natural fibre concrete roofing tile, production methods and production cost.
Keywords: Fibre-Concrete, Tile, Vibrating, Manually, Technology, Production, Cost.

1 Introduction

The Universidad de Santiago de Chile has created the first university interdisciplinary program for the development of a technology socially appropriate to Chile.

The main objectives of the program are: to offer technological alternatives to the community so people can improve their quality of living in accordance with the enviroment and to generate small enterprises with socially appropriate technology, in order to contribute to a sustaining development of the country.

Housing is one of the main subjects and specially the development of simple, economical new building materials, which make use of the natural resources existing in the area and which do not contaminate.

In relation to housing, it is important to consider roofing materials and how limited these are in Chile clay roof-tiles, iron sheets asphalt cardboard sheets and tiles or sheets of concrete asbestos. The best quality materials are very expensive and asbestos, widely used in Chile, is an imported product, both expensive and dangerous to health. This fact encouraged us to investigate and to study the available literature, to make concrete vegetable fibre tiles. This project satisfactorily fulfills the objectives of the program.

2 Materials used in roof-tiles

The selected fibres in the research are: "curaguilla", wheat fiber, barley fiber and maguey fiber; they are all agricultural residue in the country.

The river sand used has a granulometric size - 2,5 mm (mesh 8 ASTM); and the cement is of the type Portland, made in Chile.

3 Experimental development

In order to have a more efficient compaction of the mixture into a mould, a vibrating table manually powered was designed, and constructed; the iron sheet surface is 55 cm long, 35 cm wide and 4 mm thick; the sustaining structure is made of fine steel profiles. The iron sheet is sustained by 4 fixed rubber sustainers, 45 mm high and 35 mm in diameter, by means of metal rings.

A central axis in the sustaining structure has a system of bushing and a counter weight to produce vibration; this has a small pulley of transmission to operate the system.

By means of a chemical process it was possible to obtain the plant fiber, a solution of sodium hydroxide at 10% was used for 2 hours and it was later water washed. The size of the fibers is between 0.5 and 2 cm long.

The tests carried out in order to evaluate the fiber wearing were: resistance to traction and elongating measures. The tests were made with Dinamometer Apois Type 5319.

The roof tiles were made with the following measures: 400 mm long, 200 mm wide and 6 mm thick.

The elaboration process of the roof tiles was carried out in 6 stages:

1. selection of materials
2. mixing
3. pouring the mixture into a wooden lubricated mould on the vibrating table surface.
4. manual power vibration during 20-25 seconds with a frequency corresponding to the turning of the pulley at 120 RPM.
5. slow drying process during 48 hours
6. curing under water during 7 days and storing

The roof tiles were tested according to the Chilean standards: flexural strength, impermeability and water absorption.

4 Experimental results

Table 1. Resistance to traction in dry and wet conditions (submerged in water for 24 hours)

Fiber	Conditions	Resistance (gr-wt)
	dry	770,5-1125,5
barley	wet	621,7- 966,3
	dry	771,7-1076,3
wheat	wet	642,8- 885,2
	dry	735,7-1136,3
curaguilla	wet	644,1- 971,9
	dry	834,7-1253,3
maguey	wet	636,0-1152,0

Table 2. Elongation in dry and wet conditions

Fiber	Conditions	Elongation (mm)
	dry	0.85 - 1.25
barley	wet	0.98 - 1.62
	dry	0.91 - 1.49
wheat	wet	0.98 - 1.72
	dry	0.79 - 1.31
curaguilla	wet	0.61 - 1.09
	dry	0.60 - 0.90
maguey	wet	0.65 - 1.05

Table 3. Experimental results in tests of flexion, density absorption and impermeability in roof tiles with different dosages

Prove	Age (days)	Dosages Ra/c	Rc/ar	% fiber	Tests flexion C(kg)	norm	dens. (g/ml)	abs. %	Impermeab.
	7	0.68	1:35	1	11				
A	21	0.68	1:35	1	16				
	28	0.68	1:35	1	17	r	1.72	17	r
	7	0.60	1:2.4	1	10				
C	21	0.60	1:2.4	1	14				
	28	0.60	1:2.4	1	15	r	1.68	19	r
	7	0.40	1:2	1	22				
E	21	0.40	1:2	1	27				
	28	0.40	1:2	1	30	a	1.78	15	a
	7	0.40	1:1.1	1	19				
G	21	0.40	1:1.1	1	26				
	28	0.40	1:1.1	1	34	a	1.84	13	a
	7	0.40	1:0.4	1	15				
I	21	0.40	1:0.4	1	25				
	28	0.40	1:0.4	1	34	a	1.93	15	a
	7	0.40	1:2	2 L	13				
L	21	0.40	1:2	2 L	18				
	28	0.40	1:2	2 L	21	a	1.90	18	a
	7	0.55	1:1.8	2 L	17				
V	21	0.55	1:1.8	2 L	26				
	28	0.55	1:1.8	2 L	32	a	1.86	19.5	a
	7	0.55	1:2	2 L	16				
T	21	0.55	1:2	2 L	22				
	28	0.55	1:2	2 L	30	a	1.93	20.4	a
	7	0.40	1:2	2 C	11				
M	21	0.40	1:2	2 C	19				
	28	0.40	1:2	2 C	23	a	1.87	17	a
	7	0.55	1:1.8	2 C	17				
W	21	0.55	1:1.8	2 C	24				
	28	0.55	1:1.8	2 C	31	a	1.80	14.5	a
	7	0.60	1:1.7	2 C	23				
U	21	0.60	1:1.7	2 C	27				
	28	0.60	1:1.7	2 C	35	a	1.72	13	a

L = Long fiber over 2 cm
C = short fiber less than 2 cm
R a/c= water cement relation
R c/ar= cement sand relation

P = load
a = approves norm
r = rejects norm

The chosen dosage corresponds to roof tile G

5 Cost analysis

Considering one unit.
Materials: cement US$ 0.036, sand US$ 0.008, fiber US$ 0.001,
total US$ 0.045
Labor: 2 people can make 250 roof tiles per day at US$ 0.033 per unit
Total cost = US$ 0.078.

Production yield: cement 110 roof tiles per bag of 42.5 kg
sand 994 roof tiles per cubic meter
fiber 60% per kilo of straw

Table 4. Comparative chart of characteristics of similar coverings in Chile.

Coverings	Weight (kg)	Yield/m^2	Price Unit (US$)	Cost/m^2 (US$)
Concrete vegetal fiber roof tiles	0.88	28.8	0.078	2.26
Concrete asbestos roof tiles	0.80	28.8	0.313	9.07
Clay roof tiles	3.10	28.0	0.196	5.51
Concrete asbestos roof sheets	10.30	1.18	2.963	3.50

6 Conclusions

The concrete vegetales fiber roof tiles represent an important alternative to satisfy the community needs regarding the roof covers at a low cost. The use of fibers coming from agricultural waste, without value, represent an advancement in the campaign against pollution, since these wastes are usually burned by farmers.

The roof tiles thus produced, go along with the Chilean standards and are non pollutant.

They can be manufactured by the consumers themselves and they can also originate a series of small enterprises, with the socially appropriate technology, contributing effectively to the creation of new sources of employment.

7 References

Alvarez M. and Navia E. (1989) Diseño y fabricación de tejas de fibro vegetal cemento con Tecnología Intermedia. Tesis USACH.

Hernández A. and Oddone C. (1986) Fabricación de paneles y techumbres a partir de fibras vegetales. Tesis USACH.

Swamy R.N. (Ed.) New reinforced concrete. Glasgow, London: Surrey University Press (1984) pp. 106-140.

Swamy R.N. (Ed.) Natural fibers reinforced cement and concrete. Glasgow and London: Blackie, (1988), pp. 256-285.

22 FROM RESEARCH TO DISSEMINATION OF FIBRE CONCRETE ROOFING TECHNOLOGY

T. SCHILDERMAN
Intermediate Technology Development Group, Rugby, England

Abstract
For over a decade, the Intermediate Technology Development Group (ITDG) has been intensively involved in the research, development and dissemination of fibre concrete roofing (fcr), initially in the form of sheets, later of tiles. Most of the bottlenecks in the development of the technology seem to have been overcome by now, and production equipment is available from the UK and from workshops in some third world countries. Yet the dissemination of the technology, particularly through small-scale commercial production, has not always been easy. Two recent surveys, in Kenya and Peru, highlight some of the critical factors involved: the need of support organisations, the involvement of a wider range of professionals than technical staff only from an early stage, the availability of affordable equipment and spares, a reliable and affordable cement supply, credit facilities for small producers, measures to reduce the vulnerability of small producers, and international or interregional backstopping.
Keywords: Roofing, Fibre Concrete, Natural Fibres, Dissemination, Marketing.

1 Introduction

Fibre concrete roofing (fcr) is the widely accepted terminology for a roofing material produced from a Portland cement, sand and water matrix, with the addition of a small quantity of natural fibre, such as sisal. These fibres do not act primarily as reinforcement, but are added mainly to increase the moulding ability of the concrete.

The Intermediate Technology Development Group (ITDG) is a UK-based charity with the aim to enable poor people in the third world to develop and use productive technologies which give them more control over their lives, and contribute to the long-term development of their communities. ITDG's involvement with fcr began in 1978 with the development of fcr sheets by JPM Parry & Associates (JPA), partly funded by ITDG. JPA are a UK-based private company, specialized in the development and marketing of building materials production equipment aimed at the third world. The development of fcr came in response to the need for a cheap and durable roofing material which could be manufactured in decentralized workshops in third world countries, utilizing locally available resources. Fcr sheets were originally designed to simulate asbestos-cement roofing sheets, and covered appr. 1 m^2 of roof. They were rather heavy, though, and thus required relatively large quantities of timber for the roof support structure. They also had a tendency for cracking when movement in the support structure occurred.

Consequently, fcr tiles were developed by JPA in 1983; they cover 0.08 m^2 of roof. They had the advantages of easier quality control, greater flexibility on the roof, and easier handling; in most countries, they have now entirely substituted the sheets.

By 1984, JPA had developed and was selling small-scale tile manufacturing kits, including a tile vibrator and a set of moulds as main items, and these are now increasingly sold. The current range of JPA-vibrators includes models that are manually, electrically and battery operated. Lateron, JPA also devised the semi-sheet, covering 0.25 m^2 of roof, which can be produced with a more powerful vibrator.

Fcr is gradually entering the third world roofing market; the Swiss Appropriate Technology Centre (SKAT) estimates that approximately one million m^2 of roof had been covered with fcr by the end of 1989. With markets opening up, pressure grows for the production of equipment in local workshops, which has now started in several countries. In some cases, JPA models are virtually copied, but new models have been developed as well, using quite different principles.

ITDG has been involved in the dissemination of fcr from its start, and its experience is that the development of a good technology in itself is not sufficient guarantee for a material to become widely disseminated. There are many non-technical factors involved, such as marketing. Over the past two years, ITDG has undertaken surveys of fcr production and dissemination in two of its countries of concentration: Peru and Kenya. This paper is based on these studies and derives some lessons from them.

2 The case of Peru

2.1 Development of fcr technology in Peru

Fcr sheets were introduced in Peru in 1981, during a mission of J. Parry, who helped to establish a workshop at the National Urban Transport Enterprise (ENATRU), then headed by dr. C. Michelsen Terry, who was lateron to establish the Intermediate Technology Institute, which started to disseminate fcr technology in the North of Peru. Diffusion of sheet technology was limited until 1985, when the government became interested and charged the National Housing Research Institute (ININVI) with adapting fcr to local conditions; the resulting research ultimately led to a fcr sheet production manual.

The equipment needed for fcr sheet production is very simple: a timber moulding table with a plywood surface, a 50 mm wide metal frame of 5-6 mm thickness, thick polyethylene sheets, asbestoscement sheets for moulds, a 30x70x1500 mm wooden lath used for leveling and vibrating, a 2 mm sieve for the sand, a balance, 2 trowels and a spade. The total investment in equipment is in the range of 500 US$. Besides, one would need a shed and a curing tank.

The standard mixture for making sheets is 1 sand : 1 cement : 0.02 jute fibre. Sheet sizes have not yet been standardized. The original thickness of about 6 mm has often been reduced to just under 5 mm, to economize on materials and weight. Some transversal breakage occurred with the original fairly wide sheets of 1100x1000 mm, which was a reason for changing their dimensions to 740x1300 mm, now more widespread. But local variations occur; around Trujillo, for instance, much bigger sheets are being made, of around 1200x1400 mm.

An average sheet will carry 100 kg at least in flexion, and, if well packed, can be transported over long distances without breakage. Contrary to the African experience, no durability problems have been reported, and sheets produced as early as 1982-83 are still in good shape on various roofs. This may well be due to the fact that it hardly ever rains in the coastal regions of Peru.

Late 1989, sheets sold for appr. 1.50 US$/$m^2$, and were therefore clearly a low-cost roofing material. Cement and jute fibres each made up slightly under a quarter of the production cost, and financial charges slightly over a quarter, the remainder being labour and equipment costs.

The first roofing tile kits were imported by the Swiss Development Cooperation (SDC) during late 1988, at a cost of 7000-8000 US$; one went to ININVI for research purposes, another to a private enterprise, MATECO, who is now marketing tiles in Lima. MATECO is producing standard 6 mm tiles, with a matrix of 3 sand : 1 cement; an innovation is, that he has replaced the rather expensive jute fibre with eucalyptus fibre, and reduced its percentage to 0.5, the minimum needed for proper moulding. MATECO has also adapted the semi-sheet vibrator to make 3 tiles at once, and is looking into the possibility of having vibrators and moulds produced locally.

The main market for tiles seems to be in the medium-high income group, and well for the decorative covering of roofs already made of concrete. It is therefore not astoninshing, that the demand is mostly for coloured tiles, and more for flat ones rather than ondulated. The latter are currently sold at about 2.26 US$/$m^2$; coloured, they cost 2.80-2.92 US$/$m^2$.

Both the prices of tiles and of sheets compete well with asbestos-cement sheets, sold at about 8 US$/$m^2$ for the thick grey quality, and for 5.40 US$/$m^2$ for the thin red quality; the latter are far more fragile, and probably more comparable to fcr.

In conclusion, Peru's main experience is with fcr sheets, not with tiles. This may have been due to isolation from developments elsewhere, in which tiles substituted sheets for quality reasons, or to the fact that fcr sheets in Peru did not suffer similarly from quality problems as e.g. in Africa. And than, sofar at least, sheets and tiles seem to aim at different markets, with the former catering for the low-income end, and the latter for the medium-high income end.

2.2 Dissemination of fcr technology in Peru

In Peru, a major role has been played by government agencies, and particularly by the National Training Institute for the Construction Industry (SENCICO). The ngo sector has contributed far less. With the introduction of the tiles, though, the role of the private and the ngo sector may become more pronounced in the near future.

Fcr sheets came to Peru via ENATRU, a parastatal; lateron the Housing Ministry became interested, and with that its parastatals ININVI and SENCICO. The latter has played the single most important role in disseminating fcr sheet production in Peru, and therefore deserves special attention. SENCICO is a training institution, with headquarters in Lima, 4 regional directions and 17 zonal offices. It is funded by a 0.5% tax on all construction work. Most of its programmes target the formal construction sector, and deal with the training of apprentices, skilled labourers and technicians. Its 2 programmes that are more oriented to the informal sector are of most interest for fcr dissemination. They are an employment training programme (of 7 days x 3 hours), meant to establish the un- or underemployed in the production of building elements, particularly fcr, and a support programme to communities that are involved in self-help, which provides an entrance point for various alternative materials.

Since 1987, SENCICO has tried to promote fcr sheets from most of its offices, with the best results in the North of the country, and some in the South and around Lima. SENCICO often works with other organisations involved in building, both from the public sector (e.g. PRATVIR) and ngo's (e.g. IDEAS, IDECO). Production from SENCICO workshops and related projects had reached around 25,000 sheets by late 1988. At least 1,000 informal sector producers have been trained, of which quite a few have set up workshops in peri-urban areas. Because they lack working capital, these informal workshops often work irregularly, on order, and one can only guess at the number of sheets they have produced or the number of real workplaces that have been created. These workshops have also been hard hit by the recent economic regression in Peru.

SENCICO itself is hampered by its limited funding, which does not allow for helping the informal sector in their initial investments, and which also limit staff travels to outlying workshops. An additional problem is that not all state agencies seem as convinced about fcr as SENCICO. The Banco de Materiales (Materials Bank, providing building materials credits to the poor) has repeatedly refused to include fcr sheets in its materials package, because it is not convinced of their quality.

Various ngo's have played a minor role in promoting fcr sheets, including the Intermediate Technology Institute in the beginning and Habitat Peru Siglo XXI and CEPDER more recently. They often work in relative isolation. A typical case is that of the Institute of Rural Education (IER) in Huacho, who read about the sheets in an ITDG-journal, and started production on the basis of that very limited information; that resulted in very mediocre sheets being produced, which only has been overcome with technical assistance of ITDG's team in Peru in 1989.

The international ngo's have done relatively little to support fcr dissemination in Peru sofar; ITDG was instrumental in its introduction, but could not provide a lot of follow-up from the UK; it has now established office in Peru and is again providing some support, mainly to ngo's and the informal sector. SKAT is the international focal point for fcr technology; with Swiss government support, it is now disseminating fcr tile technology to the formal sector, and hopes to establish a national centre for fcr dissemination.

2.3 Discussion of the Peru experience

Notwithstanding the problems mentioned hereafter, one has to conclude that fcr sheet production is starting to have an impact in Peru, particularly in the North of the country, thanks largely to SENCICO and various smaller organisations. Given the internal problems that each of these organisations are currently facing, especially of funding, they should in the first place join forces, in order to pool their meagre resources, and to be able to put pressure on others to join.

Peru has been relatively isolated from fcr developments elsewhere, where tiles have gained the overhand. The recent introduction of tiles to Peru may have the same effect, particularly if it gains adequate SDC funding. Sofar, however, tile dissemination takes a different route: it aims a higher income market, through production in the small formal sector. This is a pity, because a lot of useful experience obtained in sheet dissemination is not being used, and there is no reason why the informal sector would be unable to produce tiles nor SENCICO and others to provide backup.

For sheets to make the step to the formal sector, their normalisation is urgently required; the ININVI manual may provide a basis for the development of standards with respect to dimensions, production, quality control, etc... In the end, this applies to tiles as well.

Most problems in sheet production seem to have been overcome over the years. The current sheets are generally of adequate quality, can carry the weight of a person on the roof, and do not break during transport; no durability problems have been reported sofar.

But some problems are noticed with the raw materials: cement is scarce, particularly in the South, and its price fluctuates with government policy. An even bigger problem is presented by the fibre: jute is most often used, but it is supplied from the Amazon region or imported, and therefore expensive. In the North, some sisal is used, but not coir, although it is available. MATECO uses eucalyptus fibre for tilemaking, but it remains to be seen whether this fibre would be acceptable for sheets. Fibres still require further research in Peru.

Although the investment needed to start up sheet production may be well below 1,000 US$, this is still a major handicap for the informal sector. Pressure is needed on potential credit agancies, such as the Banco de Materiales, to expand their credit facilities to this sector. This will become even more urgent when production would shift towards tilemaking, requiring a much higher investment.

Most informal workshops lack management and marketing skills, and these subjects are lacking in the extension courses. They generally do not keep accounts, and it must be hard to calculate correct sales prices in times of superinflation. The workshops typically depend on small clients, buying 10-50 sheets to improve the roof on their house, and demand is not continuous. They do not actively search for clients. The current crisis in Peru, which has particularly affected the low-income sector, has hit those workshops hard for about 2 years now. It is felt, they would be less vulnerable if they could diversify production, and produce other concrete elements, such as blocks, as well. The importance of marketing and business management is also not recognized at an early enough stage by supporting agencies: they should expand their extension work beyond the technical elements, and strive to build up comprehensive support teams.

3 The case of Kenya

3.1 Development of fcr technology in Kenya

Several researchers have been working on fcr sheet technology in Kenya, from the late 70's, and sheets were succesfully produced, e.g. by dr. D. Swift at Kenyatta University College. The production of sheets was, however, never commercialized and is now altogether abandoned; durability problems may have played a role in this.

Fcr tiles came to Kenya in 1983, when JPA was asked to provide training to a group of Namibians based in Nairobi. For this purpose, a workshop was acquired at Karen, on the outskirts of the city. When the course was over, JPA maintained the workshop as a training and production site. The programme provided the company, which was only just starting to market its tile making equipment, with the start of an outlet in the third world, and since then, dozens of JPA kits have been sold in Kenya. They have set the example for tile production in the country. The process is relatively simple: a matrix is made of a 3 sand : 1 cement composition, incorporating 1 - 2% fibres, normally sisal. A measured quantity is vibrated into a flat shape on a vibrating table, covered with a plastic sheet; this is than transfered to a plastic curved mould. Moulds are self-enveloping, and can be stacked. After a day on the mould, tiles are removed to a curing tank, where they remain several weeks. Tiles are costing around 2.30 US$/m^2 of roof in Kenya, which is cheaper than any other durable roofing material, but including the roof structure, they arrive at around 7.00 US$/m^2, which is just above the price of the thinnest gci sheets.

Another important agency involved in fcr development in Kenya is the UK-based ngo Action Aid; they started producing tiles as early as 1984, initially using JPA equipment, with the dual aim of obtaining a roofing material for their schoolbuilding programme and of creating employment in rural areas. The original JPA-vibrators being electrically or battery operated, Action Aid saw the need for and therefore developed a pedal-operated machine, of which the production was subsequently taken up by a workshop of the Undugu Society. Soon after, JPA also developed a model not needing a power source. Once the capacity of local workshops to make good quality vibrating tables was established, Action Aid then went one step further, and entered into open competition with JPA. With the aim of reducing the problems and the high price related to imported equipment, they stimulated the Undugu Society to go into the production of power-operated vibrators, which are near copies of the JPA models. Without the development cost and import duties, Undugu Society is able to sell its equipment at a substantially lower price than JPA, and its sales have surpassed the latter's.

Others have been involved to a lesser degree. The Housing Research and Development Unit at the University developed a vibrator using a different principle, which was subsequently produced by a private firm, without much success. The firm of J.G. Mbau produced for some time vibrators designed by the American missionary C. Thompson, but they are also not very reliable. And ITDG is currently fieldtesting a model of its own, that has been built by the Undugu Society as well.

The most expensive part of the imported equipment are the moulds, though, and not the vibrator. A local firm, Sai Raj, now produces fibreglass moulds, which are cheaper and stronger than the imported plastic moulds. A more controversial step has been the development of concrete moulds, by C. Thompson; these 12 mm thick moulds can themselves be made on a mothermould. They are far heavier than the plastic moulds, and need a stacking frame of timber. Even than, they only cost about a fifth of the plastic moulds. But they produce slightly lower quality tiles, due to the fact that the moulds do not entirely envelop the tiles, which leads to rapid drying during the first 24 hours.

The Kenyan scene may be best characterized by its competition and division, not with respect to the choice of product, since that has come out in favour of the tiles, but with respect to the production equipment. Different types of vibrators and moulds are now available, of different quality and price; ultimately the dissemination of the product may suffer from this diversity at an early stage.

3.2 Dissemination of fcr technology in Kenya

In Kenya, the private sector and ngo's have been the main driving force behind the dissemination of fcr. Government policy is in favour of locally produced low-cost building materials, but it has taken time and a lot of pressure to get the public sector to rally behind fcr.

There is a lot of scope for private sector development in Kenya. Thus several small workshops have become involved in the production of either equipment or tiles. The private sector is in the first place concerned with increasing its sales and profit margin; they are sometimes not overly concerned with the end use of their product. These days, a lot of equipment is sold without proper instruction on tilemaking; only the JPA workshop offers training, albeit at a price, and it is not compulsive for buyers of equipment; the risk that bad tiles are produced is thus quite high. Tile producers themselves often sell tiles without proper advice on how to lay them on the roof, and some are not that concerned with the quality of their product either. In what is still the initial phase of introduction of an innovative material, this involves risk: if too much bad quality is reported, the market may no longer accept the tiles.

As it is, selling tiles is already difficult enough for many of the small producers: they offer an unknown material, they have to compete with industrial giants who offer alternative products that are sometimes not much more expensive, they have limited funds and credit problems, and they operate in a market that is seasonal. This explains to some extent why the product has been slow in taking off, particularly in rural areas. There has now been a breakthrough in the capital city, after a lot of lobbying by JPA and others, in the acceptance of tiles on a large 'low-cost' housing scheme, funded by a parastatal. The first phase of the Koma Rock (formerly Kayole) estate comprises 1700 dwelling units, currently under construction, and covered by in total 1.2 million tiles. JPA is the main contractor for the tiles, but has subcontracted several small workshops in the area. There are 3 more phases forseeen in Koma Rock, and tiles are now being considered for other large schemes as well.

The role of ngo's has been different, less commercially-oriented. Ngo's such as missions were amongst the first producers of tiles, but mainly for use on their own buildings. This has had some effect of demonstration, though.

On a different level, ngo's have provided support to the dissemination of the technology. Action Aid started producing tiles in 1984, and has since stimulated the production of equipment in Kenya itself, as explained in the previous paragraph. This ngo also supported the establishment of small tile production workshops in rural areas; it has trained several teams of school-leavers in tilemaking, and presented some with the opportunity to set up a workshop, with a guaranteed contract of supplying tiles to schools under construction, and including equipment loans. Training was purely technical and paid insufficient attention to quality control. Although these workshops had the headstart of a guaranteed market initially, they often have survival problems once they are on their own, since their tiles are sometimes not of the best, and they lack marketing and often rooflaying skills.

ITDG has tried to promote small tile production workshops, on a commercial basis. As a test case, it presented a plant to Ochembe, in Western Kenya, as early as 1984. This plant is still producing good tiles, albeit irregularly and largely on order; the workshop also produces bricks, which helps it to overcome seasonal demand. In 1986, ITDG launched a first phase of support to small producers, focusing on 10 carefully selected individuals. Technical training was provided through an arrangement with JPA, and for that purpose his workshop at Karen was enlarged and better equiped; business management advice was provided by another agency, Partnership for Productivity; both these agencies and ITDG's own technical personnel were to provide backup services. This phase was evaluated in 1988; the evaluation was very positive as to the quality of tiles produced, but less so about the marketing and business aspects. Some of the producers have not succeeded; the ones closest to Nairobi are doing best, because there is a market, particularly after the start of Koma Rock. Away from the city, the combined workshops do better, since they do not rely on the sales of tiles alone. Workshops producing only fcr tiles in rural areas have had large marketing problems, and some have gone out of production. Some informal sector enterpreneurs also were unable to obtain start-up capital from credit agencies, and had to be helped by ITDG with an equipment loan; the availability of cheaper local equipment has somewhat reduced the need for capital, but it remains a major bottleneck for the informal sector: a minimum of 5,000 US$ is required. Finally, in some areas of Western Kenya, cement scarcities have affected output.

Within the public sector, there has been early support for fcr by training and research institutions, with Kenyatta College being involved in sheet research in the early days, and the HRDU lateron in vibrators and tiles. Together with Action Aid and ITDG, the HRDU is also running a series of demonstration seminars at polytechnics throughout the country. And HRDU's voice is being heard in the ministries, which may have helped to promote the use of tiles in public housing. Under pressure from various sides, the Kenya Bureau of Standards appointed a broad committee, with representatives from the main public, private and ngo agencies involved in fcr, to draw up standards for the tiles. The resulting draft specification KS02-749 appeared in April 1989, but is not yet official. This effectively means that the use of tiles in formal construction still needs a waiver from the existing regulations. The case of Koma Rock shows that this can be achieved, but it is a lengthy process, that somehow restricts a rapid expansion of the formal market.

3.3 Discussion of the Kenya experience

Fcr tiles are penetrating the roofing materials market in Kenya; SKAT estimates that 200,000 m^2 of roof have now been covered with these tiles, which is enough of an example for further dissemination. The official recognition of the standards is the next important hurdle to overcome.

There is, however, some concern over quality; in many cases producers have had insufficient training or are insufficiently concerned with quality. Once standards are in place, a mechanism should be installed as well for controlling the producers.

Small producers face marketing problems, particularly in rural areas; their vulnerability can be reduced by producing a range of products, but this is hampered by the capital required. They are also quite weak in confrontation with large producers of alternative roofing materials; an association of producers, or a national body to support them, might overcome this problem.

Credit facilities are not easily available for the production of innovative (= risky) materials in the informal sector; several channels exist in Kenya, but they ask for too much guarantees or have very lengthy procedures.

Once production has reached the level as it has in Kenya, a country should no longer have to rely on imported equipment. With an adequate market for equipment, local companies are perfectly capable of producing it, and they deserve international backing both in the form of technical advice and of starting capital.

At the international level, more research is still required into affordable moulds: the concrete moulds designed in Kenya are affordable, but reduce tile quality; maybe an enveloping system can be developed that helps to overcome that.

Finally, it looks like problems of socio-economic nature are currently bigger than purely technical ones. The development and introduction of fcr tiles has been too much the work of isolated technical staff. If other professions would have been associated with tile dissemination at a much earlier stage, some of the problems one is facing now, might have been overcome.

4 Conclusions

There are parallels, but also differences between the cases of Peru and Kenya. The following may be some general lessons we can draw from them:

(a) The development and dissemination of an innovative technology, such as fcr is much more than a purely technical matter. Technical staff should learn to involve other professionals, particularly economists, business advisors and social scientists, at an early stage, and gradually give way to them during dissemination.

(b) It is unlikely that the introduction of fcr technology via the private sector alone is going to succeed; competition is fierce, marketing is not strong in the informal sector, and credits are hard to get.

(c) Additional support is therefore needed, by the public sector or ngo's, in: business management, marketing, troubleshooting, quality control, etc...

(d) In particular, an effort should be made to reduce the vulnerability of small producers to erratic markets and raw materials supplies: a diversification of their production surely helps, but one could also imagine central marketing/franchising and/or supply organisations serving many small producers.

(e) Credits to the informal sector also need special attention; if the formal channels do not wish to change their thinking about guarantees, other channels may have to be promoted.

(f) Once a certain level of production has been reached, equipment should be produced locally rather than imported; this comes cheaper to the producers and to the country. SKAT, as the international focal point on fcr, as well as international ngo's and aid agencies will have an important role to play in promoting the local production of equipment.

(g) There is also a role for the abovementioned international agencies in general backstopping of national organisations involved in fcr dissemination; this would include the exchange of news about worldwide developments, research into bottlenecks such as affordable moulds and alternative fibres, and maybe the development of training or extension packages.

References

SKAT (1989) **FAS Seminar No.6 - Minutes of Meeting**. SKAT, St. Gallen, Switzerland.

Wells, J. (1988) **Accelerated Dissemination of Fibre Concrete Tile Technology in Kenya - Evaluation of Phase One**. (unpublished) ITDG, Rugby, UK.

Zambrano, D. (1989) **Sistematizacion de experiencias de pequenas unidades de producccion de componentes para la vivienda popular**. (unpublished) ITDG, Lima, Peru.

23 EFFECT OF REED REINFORCEMENT ON THE BEHAVIOUR OF A TRIAL EMBANKMENT

N.M. Al MOHAMADI
Department of Civil Engineering, College of Engineering
Baghdad University, Consultant (NCCL), Iraq

Abstract
To study the suitability of the ground in marsh areas in southern Iraq to support foundations and highway embankments, an instrumented trial embankment 5.0 m high was constructed in the area. Section A of the embankment was reinforced with bundles of reeds placed across the base of the embankment to study the effect of reinforcement on embankment behavior. Section B was without reinforcement. The field and analytical study has shown that the reed bundles were effective in reducing lateral strain and hence reducing construction settlement by about 14%. The finite element analysis has shown that using reed reinforcement is equivalent to increasing the stiffness of soil by 15%.
Keywords: Trial Embankment, Marsh Areas, Soil Properties, Instrumentation, Reed Reinforcement, Finite Elements.

1 Introduction

Vast areas in southern Iraq are covered with marshes, where tall reeds (berdi) up to 3.0m high grow through about 1.0 m depth of water. The villagers build their houses from bundles of these reeds, and the villages are connected with waterways about 2.0m deep cleared of the reeds that are used by narrow boats or "canoes".

An ambitious program proposed by the Iraqi government to build roads and modern villages for the development of the area require the study of the suitability of the ground to support foundations and highway embankments. To achieve this goal a 5.0 m high trial embankment was constructed on the marsh area near qurna town where the Euphrate and the Tigris reivers confluence to form shatt Al-Arab river.

2 Soil Conditions

The soil consists of a top cohesive layer about 18.0m thick followed by a dense sand layer which extends to a great depth. The cohesive layers was considered to consist of 3 layers each about 5.0 to 6.0m thick. The upper layer was a brown clayey silt with plant roots and organic matter. The middle layer was a grey silty clay, while the lower layer was a brown to grey clayey silt. Some of the soil properties are given in table 1. Complete details of soil profiles and soil properties as determined from field and laboratory tests are given elsewhere, Al-Mohamadi, et al (1986), Al-Hamrani (1980), and Al-Timimi (1979).

Table 1. Soil properties

Soil Property	Upper Layer	Middle Layer	Lower Layer
(Average values)			
Natural Water content Wn %	35	50	44
Liquid limit L.L %	46	57	52
Plastic limit P.L %	28	24	28.6
Specific gravity Gs	2.78	2.79	2.78
Clay fraction %	50	45	56
Cohesion C' KN/m^2	17.6	7.0	---
Angle of shearing resistance	30	33	---
Undrained shear strength Cu KN/m^2	53.7	---	---
Activity	0.36	0.51	0.43
Secant modulus Kn/m^2	2870	2100	---
SPT N values, Range	9-17		
Average	12		
Cu from field Vane	59-176		
(KN/m^2), Average	89.0		

3 Trial Embankment

The trial embankment is 5.0m high with a crest 12.0m wide and 1:2.5 side slopes. The effective length is 60m composed of two sections A and B each about 30m long. Section A is reinforced with bundles of reeds and section B is without reinforcement, Fig. 1a.

4 Instrumentation

Two identical lines of instruments were placed under the middle of the two sections A and B.

Each line of instruments consists of 7 piezometers, 2 inclinometers and a horizontal settlement profile plate gage 50m long, consisting of 12 magnetic marker plates. The locations of these instruments are shown in fig. 1b.

5 Construction of the trial embankment

Bundles of reeds 15 to 20cm in diameter were laid out across the 37m width of the embankment base and spaced at 15 to 20cm centers so as to cover the 30m length of the reinforced section A of the embankment.

The 37m length was made up of 4 bundles with overlapping joints about 3m long. Other researchers have used different types of reinforcement using geotextiles, metals or other materials, Battelino (1983), Iwasaki and Watanabi (1978) and Petrik et al (1982).

Several initial readings were taken on all the instrumentations to ensure that every thing was working correctly and to establish reliable zero values prior to construction. The fill of clayey silt was spread successfully over the reed reinforcement and the marsh area of section B.

Reading of all instruments were taken frequently during embankment construction which lasted 28 days. Subsequent to the end of construction, reading of instruments was continued over a period of more than 14 months.

6 Reed Reinforcement

Samples of the reed bundles were subjected to tensile tests in both dry and saturated conditions. Both ends of the bundles were dipped in liquid epoxy resin inside steel cylinders over a length of three diameters. When the resin set, it attached the reeds to the cylinders so firmly that when the tensile force was increased to cause failure, no reeds were pulled out of the end attachments. A free

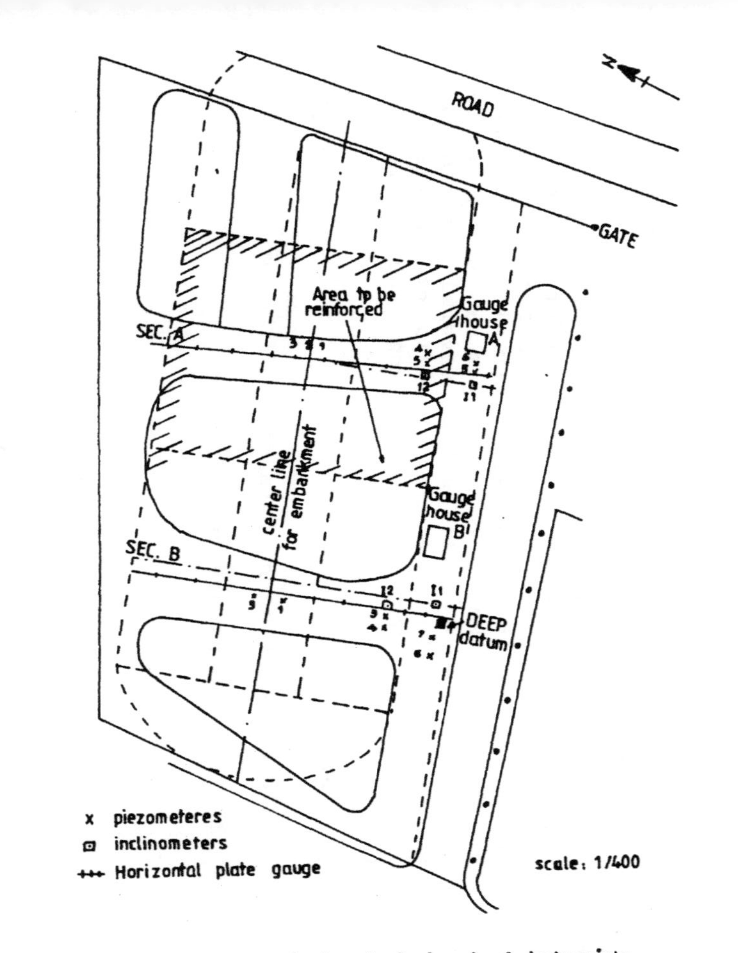

Fig 1a plan of embankment & layout of instruments

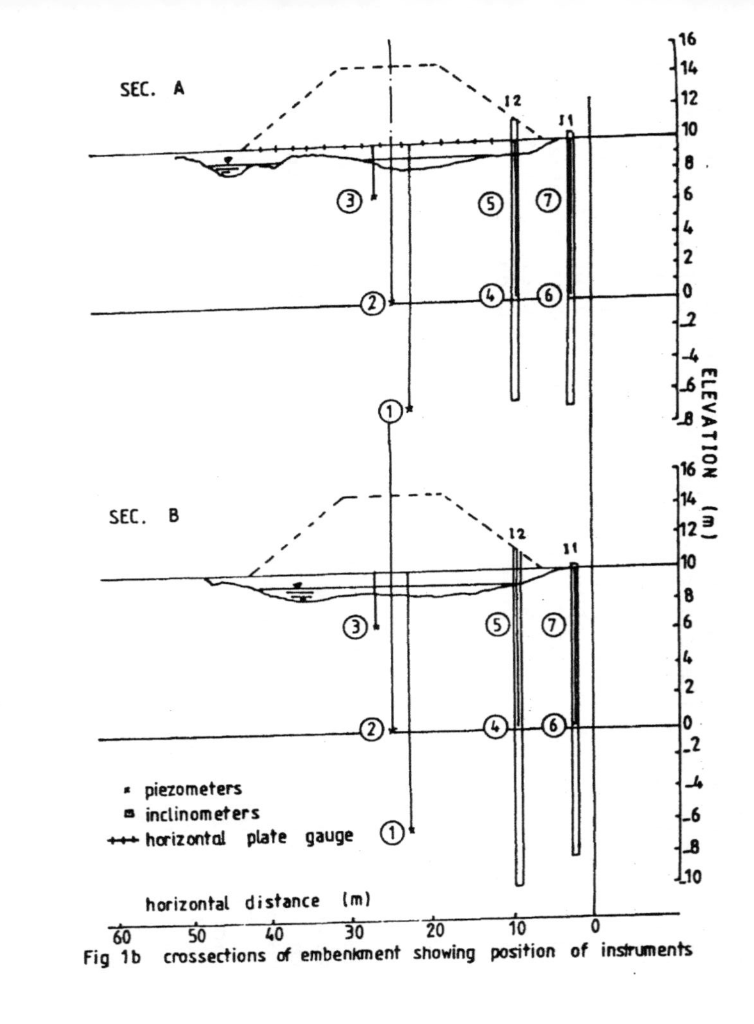

Fig 1b crossections of embenkment showing position of instruments

length of at least three diameters was left between the end cylinders.

The stress-strain relations are presented in Fig. 2. It can be seen that the dry reed bundle supported an ultimate tensile stres of 1700 KN/m^2 at an axial failure strain of 13%. The saturated reed bundle supported an ultimate tensile stress of 2160 KN/m^2 at an axial failure strain of 24%. Under both dry and saturated conditions, the reed bundles exibited linear elasticity and gave values of Young's modulus 130 MN/m^2 dry and 90 MN/m^2 when saturated. The brittle failure occured suddenly in both cases.

7 Effect of the reed reinforcement on the embankment behavior

The immediate effect of the reed reinforcement was to limit spread of the embankment and the foundation soil, thereby reducing lateral yield and hence settlement in the compressible soil during embankment construction.

The post construction settlements have been similar under both sections of the embankment, but the initial difference has been maintained. Fig. 3 shows the settlement and porewater pressure during construction under both sections. It in evident that the settlement under the reinforced section A is significantly smaller than the corresponding values under the unreinforced section B.

The porewater pressuse is also lower under section A. To study the effect of reed reinforcement on lateral displacements, the vector movements of four comparable positions under both sections have been presented in fig. 4.

It is clear that the use of reed bundles has significantly reduced the lateral spread of the embankment. As an example, the lateral displacement under the toe of the embankment (plate 4) reached 28mm for the reinforced section and 61mm for the unreinforced section.

In order to assess the equivalent improvement of soil stiffness caused by the reed reinforcement, a study has been made of the embankment by use of a model using finite elements. Trials were made using several values of stiffness modulus E for the soil, Duncan et al (1980). Poisson's ratio μ were obtained from the following relations:-

$$k_o = 1 - \sin \phi' \qquad (1)$$

ϕ' = angle of shearing resistance

ϕ' = 32° from laboratory tests

$$\mu = \frac{k_o}{1 + k_o} \qquad (2)$$

k_o = coefficient of lateral earth pressure at rest with ϕ' = 32°, μ = 0.32.

The bulk density γ = 20 KN/m³.

Vertical and horizontal movements of plates 4,5,6 and 7 given by the finite element analysis together with the measured values are presented in table 2.

Table 2. Values of deformation modulus E.

Reinforced section A.

Plate No.	Vertical movements				Horizontal movements			
	E=2.0	E=2.3	E=2.6	Meas.	E=2.0	E=2.3	E=2.6	Meas.
4	62.7	54.5	48.2	46	0.4	0.3	0.3	0
5	60.0	52.2	46.1	46	0.9	0.8	0.7	1
6	50.5	43.9	38.8	42	1.7	1.5	1.3	2
7	31.4	27.3	24.2	29	1.5	1.3	1.2	2

Unreinforced section B

Plate No.	Vertical movements			Horizontal movements		
	E=2.3	E=2.25	Meas.	E=2.3	E=2.25	Meas.
4	54.5	55.8	56	0.3	0.4	0
5	52.2	53.3	55	0.8	0.8	5
6	43.9	44.9	43	1.5	1.5	8
7	27.3	27.9	22	1.3	1.3	6

E in MN/m², movements in cm.

It can be seen that the vertical movements can be predicted very well with the finite element analysis but the horizontal movement cannot be predicted reasonably with the finite element model. The closest agreement between calculated and observed movements occur, for the reinforced section A, when E=2.6 MN/M² and for the unreinforced section B when E = 2.25 MN/m².

The effects of using the reeds under the embankment was therefore equivalent to increasing the stiffness of the soil by 15%. The secant modulus obtained from laboratory tests was 2.8 MN/m² and 2.10 MN/m² for the upper and middle layers respectively, which indicate excellent agreement with calculated values. This means that the

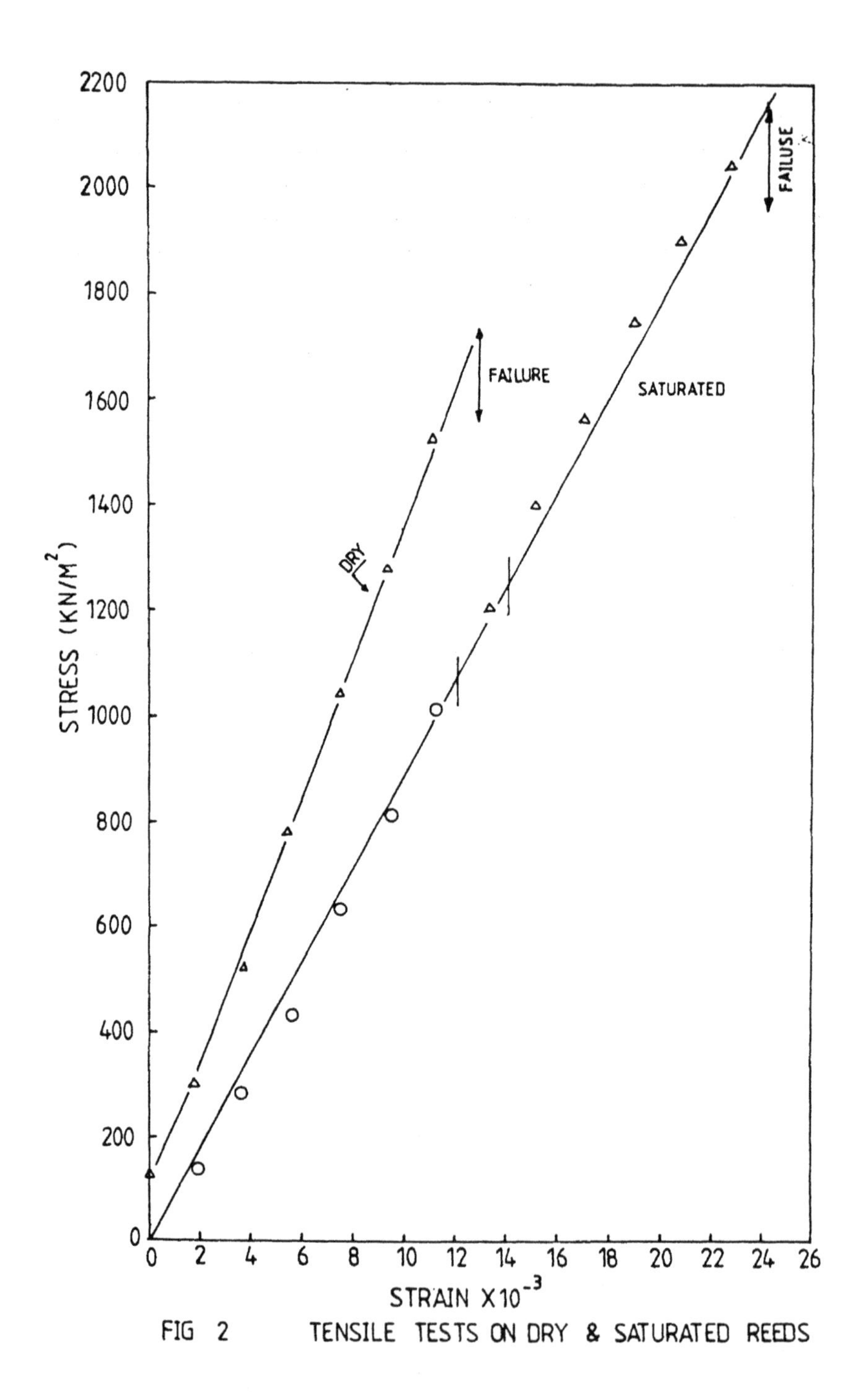

FIG 2 TENSILE TESTS ON DRY & SATURATED REEDS

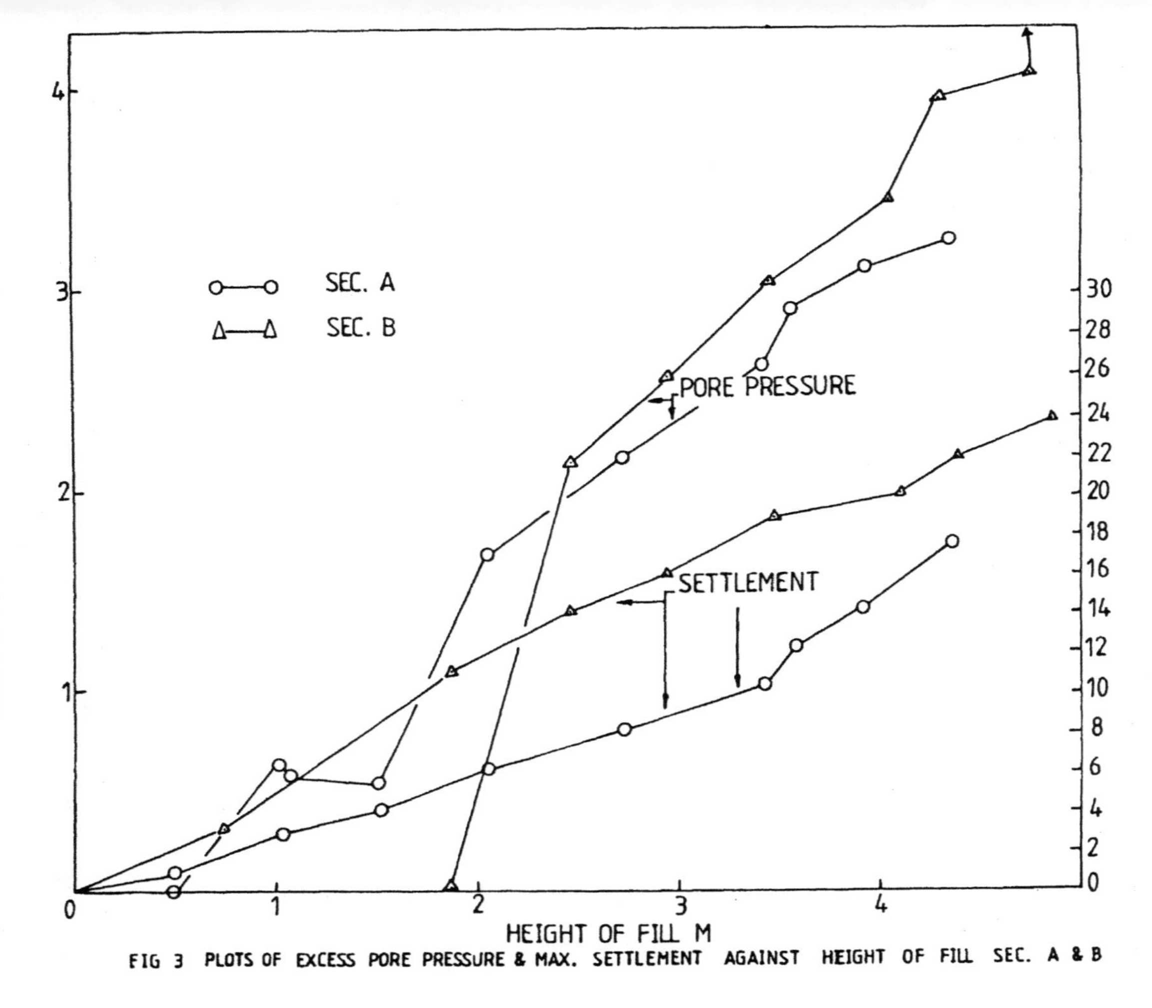

FIG 3 PLOTS OF EXCESS PORE PRESSURE & MAX. SETTLEMENT AGAINST HEIGHT OF FILL SEC. A & B

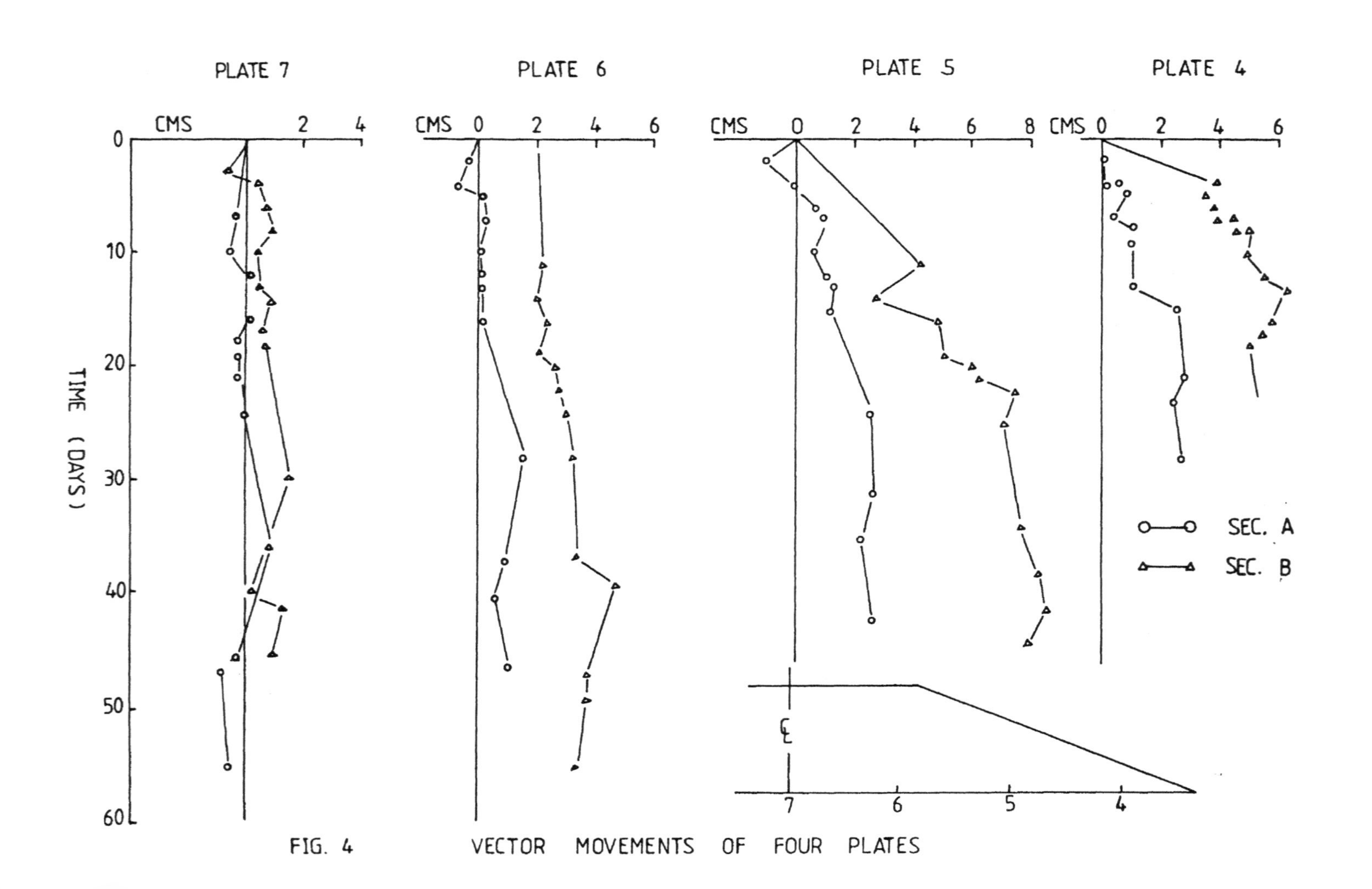

FIG. 4 VECTOR MOVEMENTS OF FOUR PLATES

tested samples were of high quality, suffering minimal disturbance.

8 Conclusions

Based on the results of this full-scale instrumented trial embankment experiment and field and laboratory tests obtained, the following conculusions may be drawn:-

1. The reed reinforcement was effective in reducing the lateral deformation of the embankment constructed on compressible soil, thus reducing the settlement by about 14%
2. The 5.0m high embankment caused moderate lateral strains that developed only a small part of the ultimate strength of the reeds. For higher embankment or softer soil lateral strain would be greater and the beneficial effect of the reeds would be more apparent.
3. The finite element model has shown that the use of the reed bundles as reinforcement under the embankment is equivalent to increasing the stiffness of soil by 15%.
4. The secant modulus obtained from laboratory tests showed very close agreement with values assumed in the finite element analysis which gave closest agreement between calculated and measured settlements.
5. The tensile strength of the reeds increases upon saturation by 27% while the Young's modulus decreases by 30.8%.

9 References

Al-Hamrani, M.M. (1980) Trial Embankment: settlement Analysis with emphasis on Stress Path method. M.Sc. Thesis, College of Eng., Baghdad Univ.

Al-Mohamadi, N.M. Al- Samarie M. and Fadel S. (1986) Behavior of Trial Embankment in Marsh Area in Southern Iraq Jr. of Building Research, Vol. 5, No.2, Iraq.

Al-Timimi, S.F. (1979) Trial Embankment: Investigation of In Situ Shear Strength. M.Sc. Thesis, College of Eng. Baghdad Univ.

Battelino D. (1983). Some experience in reinforced earth, 8th Europ. conf. SMFE, (2), Helsinki.

Duncan, J.M., Byrne, P.M., Wong, K.S. and Mabry, P. (1980). Strength, Stress - strain and Bulk Modulus Parameters for Finite Element Analysis of Stresses and Movements in Soil Masses, Report No. UCB/GT/80-01.

Iwasaki K., Watanabi S. (1978) Reinforcement of highway embankment in Japan Proc. Symp. on Earth Reinforcement, ASCE, Pittsburgh.

Petrik D.M., Baslik R., Leitner F. (1982) The behavior of Reinforced Embankment Proc. 2nd Int. Conf. Geotextiles, Las Vegas.

24 REED FIBERS AS REINFORCEMENT FOR DUNE SAND

T.O. Al-REFEAI
King Saud University, Riyadh, Saudi Arabia

Abstract
To further knowledge the behavior of fiber reinforced sand, a series of triaxial tests were conducted, in which dune sand randomly reinforced by discrete fibers. Reed and glass fibers were chosen to observe the influence of variables associated with the fiber, namely diameter, surface characteristics, aspect ratio, concentration, and stiffness, on the behavior of fiber reinforced sand. The results indicate that the presence of fiber could lead to a significiant increase in ultimate strength and stiffness of reinforced sand. Strength increase is generally proportional to fiber concentration up to some limiting content. Increasing the aspect ratio increased the ultimate strength and the stiffness of reinforced sand. At the same aspect ratio, confining stress, and volume ratio, "rougher", not stiffer fibers were more effective in increasing strength.
Keywords : Reinforced sand, Fibers, Dune Sand.

1 Introduction

Sand reinforced with discrete, randomly oriented fibers may be considered as a composite material in which short fibers of relatively high tensile strength are embedded in a matrix of lower tensile strength. The presence of the fibers will disrupt the uniform pattern of strain which would otherwise develop in unreinforced sand. This strain interference results in a more complex and less efficient deformation mechanism in the sand (Jewell, 1980), and ultimately leads to improved stiffness and shear strength.

The purposes of this study are to determine the stress-strain response of dune sand reinforced with discrete, randomly oriented fibers, and to observe the influence of variables associated with the fiber on the constitutive behavior of fiber reinforced sand.

2 Previous Work

Gray (1978) considered the role of natural fiber or the roots of woody vegetation in reinforcing soil and stabilizing slopes. The results of direct shear tests on sand reinforced with playmra (a tough natural fiber) are of particular interest. The reinforcements were placed in a random pattern perpendicular to the shear surface. The test results illustrate that fiber must be long enough or "frictional" enough to resist pullout or conversely the confining stress must be greater enough to develop sufficient traction along the fiber.

Jewell (1980) presented the results of direct shear tests of sand reinforced with fibers, which were always placed in a single line symmetrically about the central plan at different orientations. The results of the tests showed that the reinforcement must be strained sufficiently before it can have any influence, and that the orientation of the reinforcement is important. The direction of the principal tensile strain at failure in the unreinforced sand is a good guide for the optimum reinforcement orientation. Jewell also observed that the shear strength increase reached an asymptotic or limiting value with increasing reinforcement concentrations.

Gray and Ohashi (1983) performed an extensive series of direct shear tests on sand reinforced with different types of fibers (natural, synthetic, and metal wires). The fibers were placed in a circular pattern at equal spacing from each other. The results of the tests show that relatively low modulus fiber reinforcements (ideally extensible) increase peak shear strength and limit the amount of post peak reduction in shear resistance in dense sand. The amount of the strength increase was observed to be directly proportional to fiber area ratio up to a limiting value of 2 percent which was the maximum concentration used in the tests. The shear strength increases were greatest for initial fiber orientation of 60 degrees with respect to the shear surface. This orientation coincides with direction of the maximum principal tensile strain in a direct shear test. They concluded that shear strength increase for dense sand and loose reinforced sand were approximately the same, however, larger strains were required to reach peak shear resistance in loose reinforced sand.

Lee, et al., (1973) reported the results of a triaxial test on sand reinforced with firwood shavings. The results showed that the fiber increased both the strength by about a factor of 6, and the sand stiffness. Failure or reinforced sand also tended to occur at higher axial strain.

Hoare (1979) reported the results of laboratory tests on fiber reinforced sandy gravel. Compaction test results showed that the presence of the fiber increased the resistance to compaction. When a constant compactive effort was applied to a range of samples with increasing fiber content, the strength either increased hardly at all or actually decreased. This was caused by the increases in soil porosity or void ratio would tend to negate any increase in strength from fiber reinforcements.

3 Experimental Work

3.1 Properties of Dune Sand

A clean quartz dune sand was used in this study. Typical properties of the sand are listed in table 1.

Table 1. Properties Of Dune Sand

Effective particle size	.28 mm
Median particle size	.41 mm
Uniformity coefficient	1.50
Specific gravity	2.65
Max. void ratio	.78
Min. void ratio	.50
Angle of internal friction	39° (D_r = 86%)
(triaxial)	32° (D_r = 21%)

3.2 Fiber Properties

The fiber used consisted of natural and synthetic fibers. The fibers varied from 13-38 mm in length and 0.3-1.75 mm in diameter. The synthetic fibers are made of glass and are available commercially as admixes for fiber reinforced concrete. The properties of the fibers used in the testing program are presented in table 2.

Table 2. Fiber Properties

Fiber	Diameter (D) (mm)	density (gr/cc)	Tensile Stress MPa	Young's Modulus MPa
#1 Reed	1.25	.58	33.3	1470
#2 Reed	1.75	.58	33.3	1470
Glass Fiber	0.3	2.7	1250	7000

3.3 Experimental Procedures and Sample Preparation

The glass fibers used in the testing program were supplied in standard lengths of 13, 25 and 38mm. The reed fibers were cut from long fibers into smaller ones having the same lengths as the glass fibers. Although the lengths of the two types of fibers were the same, the glass fibers had much larger aspect ratios because of their smaller diameters.

Mixing is an extremely important factor in the case of discrete, randomly oriented fiber reinforcement. Blade type mixers will not work as they tend to drag and ball up the fibers. Vibratory mixer tend to float the fiber up. A special oscillatory or helical action mixer was used to avoid these problems. Even this type of mixer has limitations on the maximum concentration of fiber that can be uniformly and randomly mixed. In order to distribute the fibers as evenly and randomly as possible throughout the soil and to avoid segregation while forming the triaxial test specimen it was necessary to moisten the sand slightly. A water content of 10% was chosen for use in all the sand fiber mixes. This moisture content ensured thorough mixing and provided enough apparent cohesion to prevent segregation during forming of the samples. This moisture content had no significant effect on the stiffness and stress-deformation properties of dune sand. This sand contains no fines and is sufficiently coarse grained that capillary effects should not be important.

The amount and type of compaction affect the response of fiber reinforced soil via their influence on relative density and degree of fiber entanglement or distortion. The lower the porosity, the greater will be the strength of the mixture. The presence of fiber tends to increase porosity (decrease relative density) for a given type and amount of compactive effort (Hoare, 1979). The type of compaction also plays a role (Hoare, 1979). Vibratory compaction tend to "float" the fibers where as kneading methods (e.g. ramming) distort and entangle the fibers and bring them into closer and more intimate contact with the sand.

Prior to testing, it was necessary to choose a suitable void ratio at which to test fiber reinforced sand specimens under triaxial loading. Trial sand-fiber mixes with different concentrations were compacted in a triaxial test mold, by tamping successive layers of moist sand-fiber mix with a tamper consisting of a circular glass disk (34mm in diameter) attached to the end of a steel rod. It was necessary, when using this tamping procedure, to increase the number of sand-fiber mix layers to maintain the same void ratio with increasing fiber concentration and fiber length. A void ratio of about 0.62 was finally selected because it was easily and efficiently achieved for most of the fiber concentrations of interest.

Selected proportions of dry sand were mixed with the fiber. Water was then added (10% by weight). The sand-fiber-water was then mixed until the fibers were evenly distributed and randomly oriented throughout the sand. This mix was next placed into small containers with tight covers to protect the mix from drying prior to forming the triaxial specimens. The samples were formed inside a triaxial mold lined with a 0.35mm thick rubber membrane which is sealed to the base of the triaxial cell. After forming the sample in the manner previously explained a loading cap was placed on top and the membrane was sealed to the top cap.

4 Test Results and Discussion

4.1 Effect of Confining Pressure on the Strength of Reinforced Sand

The few available published investigations of the properties of randomly oriented discrete fiber reinforced soil (Lee, et. al.,1973; Andersland and Khattak, 1979; and Hoare, 1979) have dealt principally with stress-deformation behavior at one or two values of confining stress in triaxial tests. A serious lack of information exists on the behavior of fiber rainforced sand over a range of confining stress. This dependency on confining stress is significant because the response to confining stress is basically what differentiates the behavior of fiber reinforced, granular soil versus other fiber reinforced material (e.g., concrete, mortar, asphalt, and plastic).

Fibers that are short, that have low specific surface areas (or low aspect ratios = length/diameter), or that are very stiff or smooth required much higher confining stress in order to mobilize their tensile resistance and interlocking effectiveness.

Figure 1 shows the relations between the confining stress and the principal stress at failure. For sand reinforced with 13mm, #2 reed fibers. This case represents the worst conditions, i.e. the shortest fibers and the largest diameter (smallest aspect ratio).

Figure 2 shows the same relationships for sand reinforced with glass fiber. The break in the curves correspond to the critical confining stress in question, below this critical confining stress the fibers tended to slip as opposed to stretching. It is interesting to note that in spite of the fact that the reed fiber has the smallest aspect ratio (L/D=7) of any of the fiber tested, it nevertheless required a much smaller confining stress (1.0 kg/cm^2) compared to any of the glass fibers with much higher aspect ratios (L/D=84-125). This behavior can be explained by the fact that the reed fiber with its rough, slightly irregular surface has better interface friction characteristics than the smooth and slippery glass fibers.

A confining stress equal to or exceeding 2 kg/cm^2 was used in all the subsequent tests in order to mobilize fiber tensile resistance as opposed to merely pullout resistance.

4.2 The Effect of Fiber Reinforcements on Shear Strength of Reinforced Sand

The need for information on the behavior of fiber reinforced sand and the desire to clearly establish the influence of variables associated with the fiber, namely diameter, surface characteristics, aspect ratio, concentration, and stiffness were the basis for performing triaxial compression tests on fiber reinforced sand.

Due to space limition only the results for sand reinforced with reed fiber with L/D=7 and glass fiber with L/D=84 are shown.

Figure 3 shows the stress-strain relationship for sand reinforced with #1 reed fiber. Figure 3 shows that increasing the fiber concentration (W_f) increases both the stiffness and the strength of the reinforced sand. The increase in the strength of the reinforced sand is very pronounced as the fiber concentration increases. At low aspect ratio (L/D=7) although the presence of fiber increased the stiffness and strength of reinforced sand no appreciable advantage is gained by doubling the fiber concentration from .5 to 1% by weight. For fiber with high aspect ratio (L/D=20-30) the increase in the stiffness or the strength of the reinforced sand is very pronounced as the aspect ratio increases. The peak stress in the sample reinforced by 1% by weight reed fibers having an aspect ratio of 30 is almost twice that of the sample with fibers with an aspect ratio of 15 at the same weight concentration.

The trend towards greater stiffness and strength are similar in the case of glass fiber as shown in figure 4, which indicates that although increasing the fiber concentration increases the strength it slightly reduces the stiffness at low strain. However, in no case is the stiffness of a reinforced sample less than of the sand alone. Andersland and Khattak (1979) reported similar results.

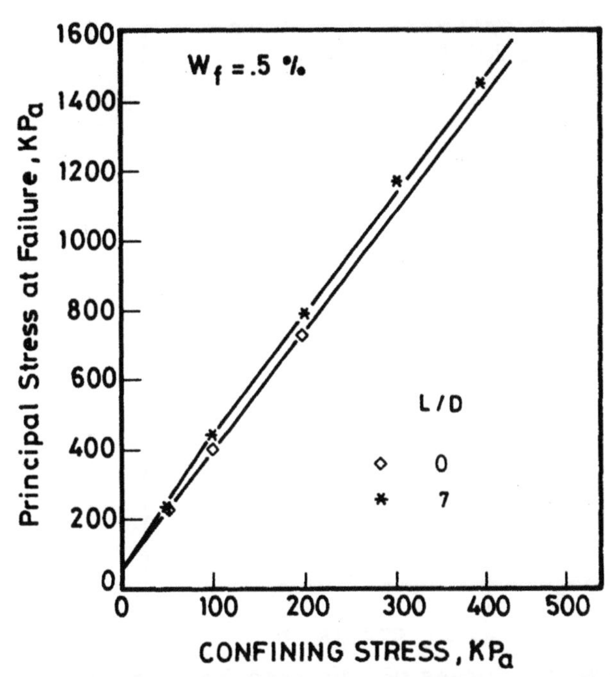

Fig.1. Confining Stress Vs. Principal Stress at Failure #2 R.F.

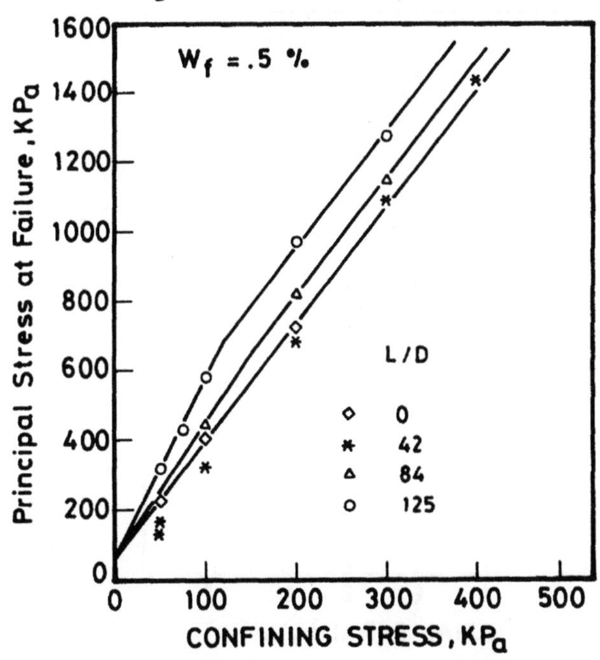

Fig. 2. Confining Stress Vs. Principal Stress at Failure - Glass Fiber

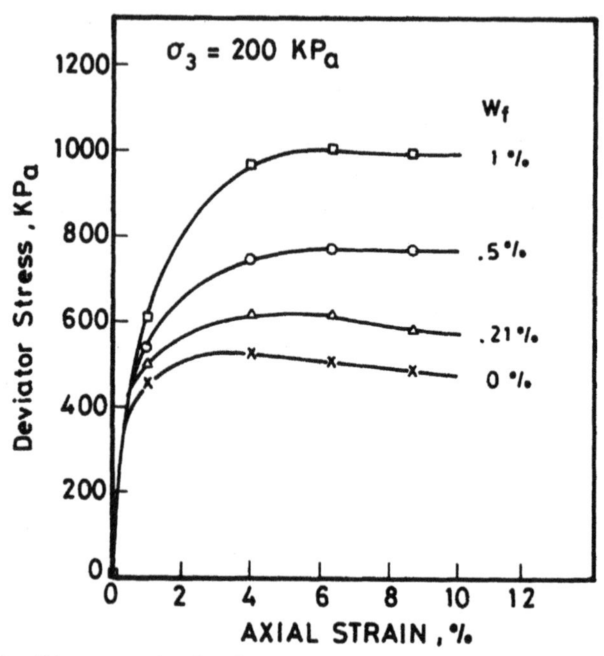

Fig. 3. Stress_strain Curves of Reinforce Sand_#1 Reed Fiber

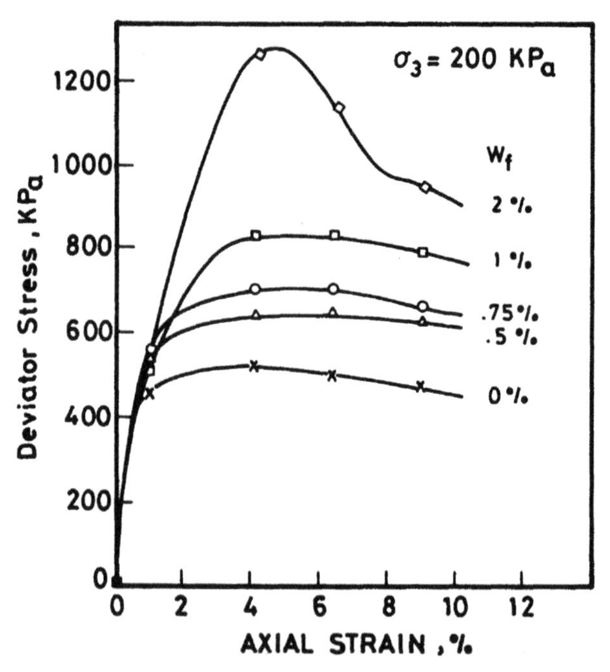

Fig. 4. Stress - Strain Curves of Reinforce Sand - Glass Fiber

Figure 4 indicates that increasing the fiber concentration results in a great increase in the peak strength of the reinforced sand; a possible reason for this increase should be mentioned. Considerable difficulty was encountered in achieving the desired density of fiber reinforced sand (e = .62) with more that 1 by weight fibers. The high peak strengths of the sand samples with high fiber contents may be associated in part with fiber entanglement and prestress tension in the fiber due to the increased compactive effort required to achieve the desired density. It is also interesting to note that considerably more post peak loss of strength occurred in samples with high fiber concentration ($\geq$ 1%). This lose of strength can be explained by fiber breakage that was observed in samples with high fiber contents.

4.3 Fiber Concentration and Aspect Ratio Effects

Figure 5 and 6 show that the strength increase is approximately proportional to fiber concentration for a given aspect ratio; and the magnitude of this strength increment generally increased with increasing the aspect ratio. With the short glass fiber (L=13mm) there was no change due to increasing the fiber concentration.
Strength appears to increase with fiber concentration up to some limiting fiber content. For glass fiber with L/D = 84, for fiber concentration greater than 4% the increase in the strength appears to reach a maximum value. With high fiber concentration (> 4%) the relative amount of sand in the reinforced specimen decreases, and number of frictional contacts between the fibers and soil is reduced, thus resulting in decreased strength. An extreme case is when the soil content is nil (fiber concentration of 100% and the fibers alone must carry the load. The efficiency of reed fibers vs. that of glass fibers is shown in Figure 7.

4.4 The Effect of Fiber Reinforcement on the Stiffness of Reinforced Sand

Figures 8 and 9 show the variations of the ratio of secant modulus of fiber reinforced to unreinforced sand at different fiber concentration and aspect ratio. These figures show that increasing the fiber aspect ratio or concentration for both reed and glass fibers increases the stiffness at any strain level.

5 Conclusion

The following conclusions can be drawn from the results of the study:
Discrete, randomly oriented fiber reinforcements increased both the ultimate strength and the stiffness of reinforced sand.

Increasing the aspect ratio increased the ultimate strength and the stiffness of reinforced sand.
At the same aspect ratio, confining stress, and volume fraction, "rougher," not stiffer fibers were more effective in increasing strength.

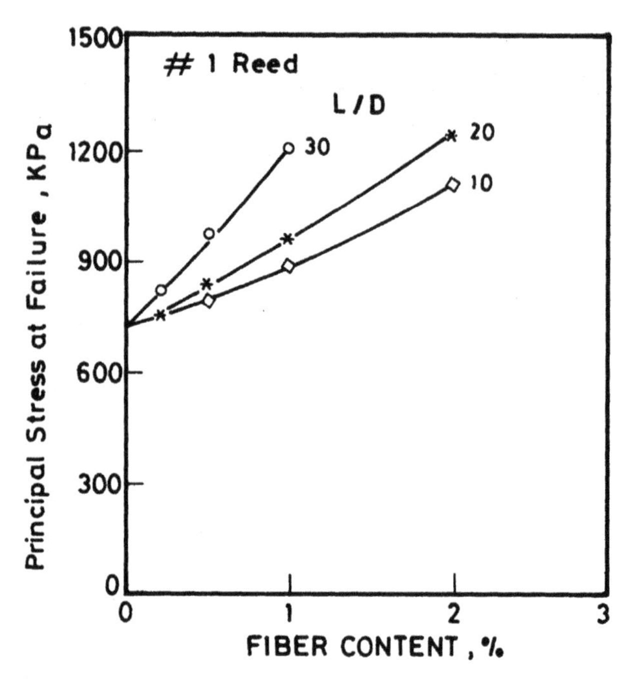

Fig. 5 . Principal Stress Vs. Reed Fiber Concentration

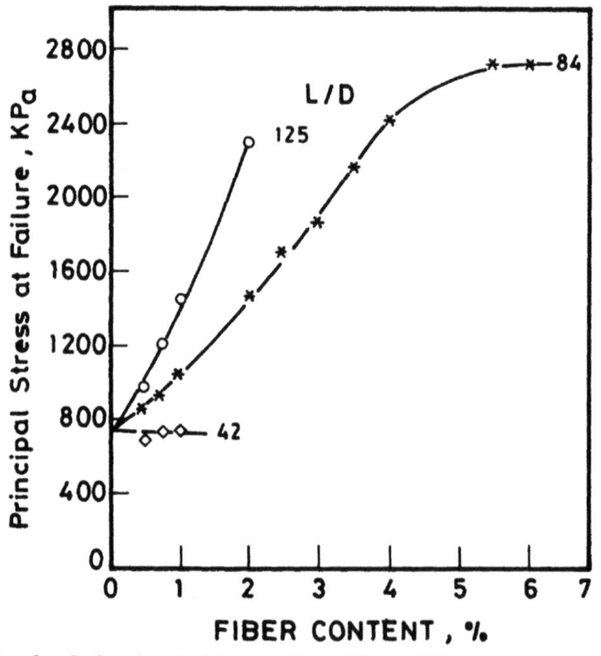

Fig. 6 . Principal Stress Vs. Glass Fiber Concentration

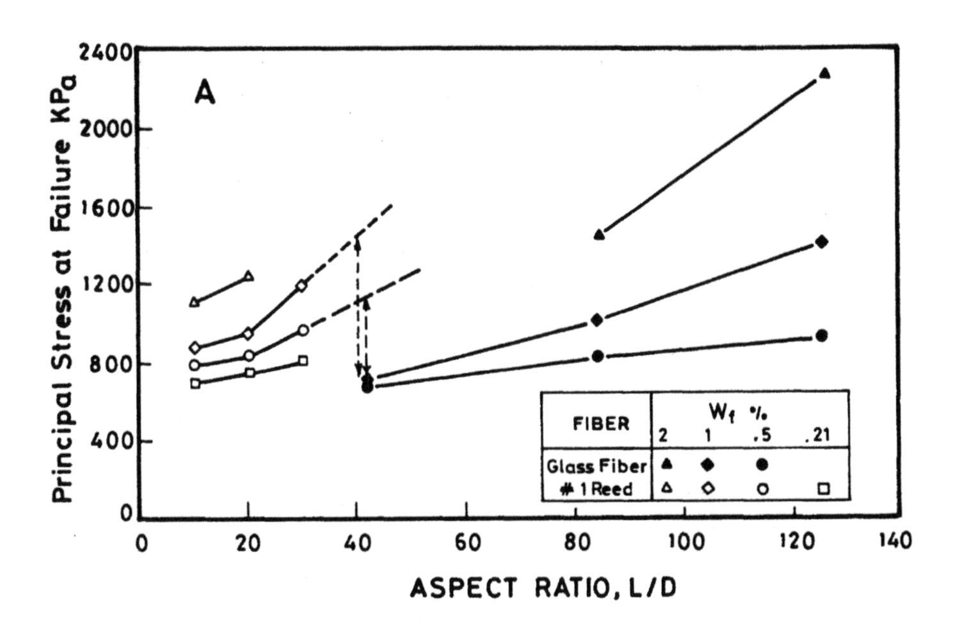

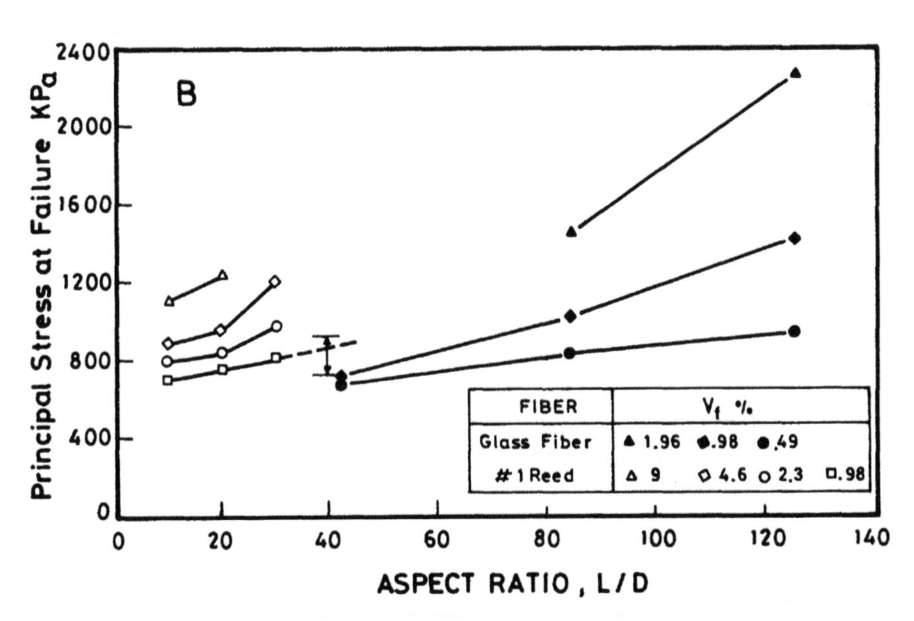

Fig. 7. Principal Stress Vs. Aspect Ratio; (A) Weight Concentrations; (B) Volume Concentrations.

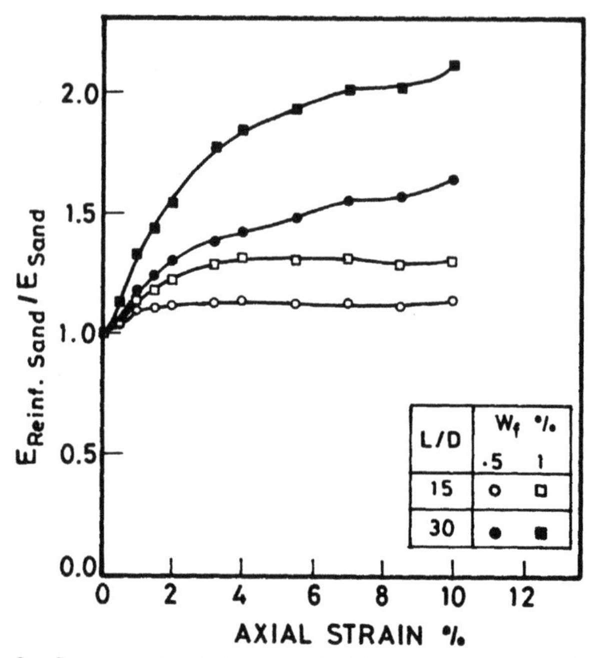

Fig. 8 . Secant Modulus Ratio Vs. Axial Strain-Reed Fiber

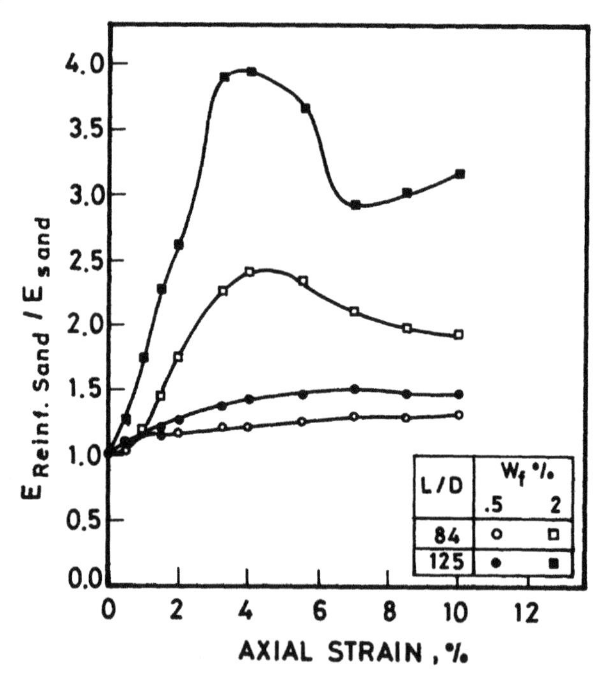

Fig. 9 . Secant Modulus Ratio Vs. Axial Strain-Glass Fiber

6 References

Andersland, O. B., and Khattak, A. S. (1979). "Shear Strength of Kaolinite/fiber Soil Mixtures," Proceedings, Intl. Conf. on Soil Reinforcement: Reinforced Earth and Other Techniques, Paris, Vol.I, pp. 11-16.

Gray, D. H. (1978). "Role of Woody Vegetation in Reinforcing Soils and Stabilization Slopes," Proceedings, Symposium on Soil Reinforcing and Stabilizing Techniques in Engineering Practice, Sydney, Australia, pp. 253-306.

Gray, D. H. and Ohashi, H. (1983). "Mechanics of Fiber Reinforcement in Sand," Journal of the Geotechnical Engineering Division (ASCE), Vol. 109, No. 3, pp. 335-353.

Hoare, D. J. (1979). "Laboratory Study of Granular Soils Reinforced with Randomly Oriented Discrete Fibers,"Proceedings, International Conference on Soil Reinforcement, Paris, Vol. I, pp.47-52.

Jewell, R. A. (1980). "Some Factors which Influence the Shear Strength of Reinforced Sand," Cambridge University Eng. Dept., Cambridge, England, CUED/D-Soils/TR85, 37 pp.

Lee, K. L., Adams, B. D., and Vagneron, J. J. (1973), "Reinforced Earth Retaining Walls," Journal of SMF ASCE, Vol. 99, SM10, pp. 745-764.

PART FIVE

BUILDING COMPONENTS MADE WITH WOOD

25 PROPERTIES AND DURABILITY OF RAPIDLY CURING CEMENT-BONDED PARTICLE BOARDS, MANUFACTURED BY ADDITION OF A CARBONATE

M.H. SIMATUPANG
Institute of Wood Chemistry and Chemical Technology of Wood, Federal Research Center for Forestry and Forest Products, Hamburg, Federal Republic of Germany

Abstract
The manufacturing process and the properties of cement-bonded particleboards are shortly described. It is feasible to reduce the press time from 6-8 h to 10-15 minutes, if potassium, sodium or ammonium carbonate and waterglass are added as accelerator. Properties and durability of experimental particleboards made according to the new process are presented and compared with conventionally made cement-bonded boards. Feasibility and limitations of the novel process are discussed.
Keywords: Cement-bonded particleboards, Ammonium carbonate, Sodium carbonate, Potassium carbonate, Waterglass, Durability.

1 Introduction

Presently, cement-bonded particleboards are manufactured in 12 countries. The total capacity is about 3,500 m^3/d. The board comprises about 20 wt% of wood particles and 80 wt% of hardened portland cement. The modulus of rupture of the boards is about 8 to 10 MPa and the density is between 1,100 and 1,250 kg/m^3 (Simatupang 1989). The manufacture of cement-bonded particleboards is based on the patent of Elmendorf (1966). A scheme of the industrial process is shown in Fig. 1. Suitable wood is converted to appropiate wood particles, and mixed with water containing additives and cement. The furnish is a spreadable mixture and is relatively dry, compared to cement mixtures, generally applied in concrete technology. The furnish is continuously spread on metall cauls. Subsequently it is prepressed and the cauls are stacked. On the bottom and on top of each stack, two heavy iron frames are placed. With a hydraulic press the stack is pressed to the desired height, and also to the desired board thickness. Due to the long setting time of portland cement, the stack has to be kept under pressure until the green boards gain enough strength to permit stripping without damage. This is performed with heavy steel wedges and edges. In most plants the pressed stacks are put into a heated compartment, with a temperature of about 60 to 80°C, during 6 to 8 h, in order to accelerate the setting of the cement. After stripping, the green boards harden within 10 to 14 days to attain the final strength properties. The cauls are returned to the manufacturing process after cleaning, and spraying with an emulsion to avoid sticking of the boards. The stripped boards are subsequently sized, eventually dried and sanded.

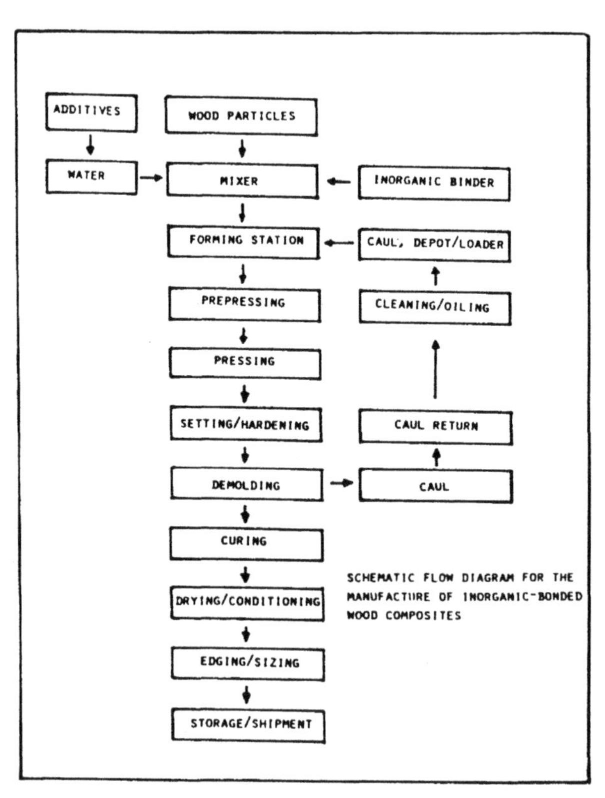

SCHEMATIC FLOW DIAGRAM FOR THE MANUFACTURE OF INORGANIC-BONDED WOOD COMPOSITES

The long setting time of the cement binder and the incompatibility of some woods with portland cement are the main hindrances for the installation of more plants. Presently, most of the plants utilize conifers, especially spruce or pine (Simatupang et al. 1987). In Malaysia seasoned rubber wood is used (Schwarz 1989). In Japan larch wood can be used after spraying with parafin or phosphoric acid ester (Yamagishi et al. 1983), but the main raw materials are lauan or pine.

To reduce the setting time of cement during the manufacture of cement-bonded particleboards the following processes are known: 1. Utilizing a mixture of portland cement, alumina cement and plaster or gypsum anhydrite (Sekusui Chem. Co. 1981; Odler 1986; Paulisan 1989). 2. Injection of carbon dioxide to the furnish during pressing. Calcium carbonate is formed, which bonds the wood particles into a stable composite. The ultimate strength properties are attained after two weeks of curing (Konkola 1989; Schmidt 1989).

The new technologies to accelerate the cement setting, avoiding the long setting time, require a special cement e.g. alumina cement, or a special press, which enables the injection of carbon dioxide during pressing. In the following examination a method is described, which uses carbonates as additives. Appropiate carbonates decompose on heating evolving carbon dioxide. The gas reacts with calcium hydroxide and forms calcium carbonate.

2 Material and Methods

2.1 Determination of the hydratation time

10 g of cement were mixed in a reaction tube with 6 ml of distilled water. The chemical analysis of the cements used was already reported (Lange, Lieber, Simatupang, 1986). Additives were first dissolved and cooled before they were mixed with the inorganic binder. The filled reaction tube was shaken with a reaction tube shaker, and placed into a brass pressure vessel. In the cover of this vessel was a thin tube with a thermocouple to measure the temperature in the cement mixture. The pressure vessel was placed in a thermostate. A two point recorder registered the temperature of the cement mixture and the water in the thermostate. Due to the exothermic reaction, the temperature of the cement can exeed that of the water bath. After some time the temperature drops to the temperature of the water bath. When the temperature exeeds the water bath temperature the setting is considered to start, when it has dropped to the water bath temperature again, the setting is over. The hydration time is the time to reach the maximum temperature of the setting cement.

2.2 Manufacture of cement-bonded boards

Freshly cut wood of spruce and larch were debarked and chipped in an industrial laboratory. The chips were ringflaked to wood particles with a nominal thickness of 0.2 to 0.3 mm and a length of 20 to 30 mm. They were fractionated by appropiate sieves. Particles smaller than 2 mm were used for the outer layers of three layer particleboards. The wood particles for the middle layer were 4 to 6 mm wide. All wood particles were air-dried in a room (temperature about 30°C) to a moisture content of about 7 %.

To manufacture three layer particleboards, wood particles for the outer and middle layers were separately mixed with water, or water containing additives in a pug mixer. The cement was added and the mixture further mixed. The cement/wood (oven dry) ratio in the surface layer and middle layer was 3.0 and 2.6, respectively. The available water/cement ratio (Simatupang 1979) was 0.35. The furnish was manually felted in a woody frame (40 x 40 cm) on a stainless steel caul, and slightly prepressed before it was placed in the hydraulic press. Pressing was performed in an electrically heated hydraulic press. The thickness of the board (12 mm) was determined by steel distance bars. Conventionally made particleboards contained 3 wt% (based on cement) of calcium chloride as accelerator. They were pressed at 40°C during 8 h. Rapidly setting boards were pressed at 80 or 85°C during 15 to 18 minutes. At least 3 boards for each variation were manufactured. The green boards were subsequently cured for 14 days in a closed container and cut into samples of 23 x 5 cm. Eight samples could be obtained from each board. These samples were conditioned for 14 days in a room of 20°C and 60 % RH and tested according to DIN 52362. Undamaged parts of the tested samples were cut to pieces of 5 x 5 cm and used for the determination of the internal bond according to DIN 52365.

To test the water resistance of the manufactured boards, 10 samples of 23 x 5 cm were treated according to the French Standard B 51263, commonly designated as V313 test. According to this test the samples were soaked in water during 3 x 24 h; put in a freezer at about -18°C for 24 h, and dried in an electrically heated oven with air circulation at 70°C during 3 x 24 h. Three test cycles were performed. The samples were cooled in a desiccator, weighted, and their bending strengths were determined. Undamaged parts of the samples were used for the resistance test, 6 boards for each variation were manufactured. They were cut to samples of 23 x 5 cm and mixed. The bending and internal bond strength of untreated samples were determined on 10 samples 28 days after manufacturing. Each of 10 samples were stored in a room (temp. 20°C anf RH 60 %) during 64 weeks, and exposed towards South in the open on a slightly inclined frame during 32 or 64 weeks. After each test the samples were conditioned at 20°C and 65 % RH until constant weight. Then the bending and internal bond strengths were determined.

3 Results and Discussion

3.1 Influence of carbonates and waterglass

Under the influence of ammonium carbonate, potassium carbonate or sodium carbonate the setting time of cement can be reduced to about 10 to 15 minutes. Some shotcretes contain such carbonates. In the manufacture of cement-bonded particleboards the furnish for the felting of boards are generally stored for a short time in a bin, in order to equalize the uneven supply from the discontinuous mixer. The open time for cement mixed with solid ammonium carbonate, potassium carbonate or sodium carbonate is only 2 to 3 minutes. In such mixtures the solid carbonates dissolve in the mix water. The temperature increases because of the heat of solution enhancing the decomposition of the carbonate. The emitted carbon dioxide reacts with calcium hydroxide and the cement sets before the board is placed into the press. Such material is unsuitable to manufacture stable cement-bonded particleboards. Due to the spring back forces of wood particles the green cement-bonded particleboards should have a bending strength of about 4 to 5 MPa to permit demolding without damage. It was found that such strength properties could be attained by the addition of at least of 7.5 to 10 wt% (based on cement) of carbonates. However, if the carbonates are added as a solution, the open time can be prolonged. Also the addition of waterglass, especially as a solution, will prolong the setting time of cement. With increasing temperature the difference between the begin and the end of the setting time will decrease. The various cements show different effects (Table 1). It is feasible to use a mixture of carbonates and waterglass as an additive to accelerate the setting of cement, and also to prolong the open time of such mixtures before pressing. The open time is comparable with that of cement without additonal potassium carbonate. If heat is applied, the mixture of potassium carbonate and waterglass will set within 10 to 15 minutes and bond the wood particles to a stable board.

3.2 Properties of the cement-bonded particleboards

It is feasible to manufacture particleboards setting within 10 to 15 minutes at a press temperature of 80°C. The preliminary experiments show the following bending strenght and density of cement-bonded particleboards using various carbonates (Table 2).

The addition of ammonium carbonate caused the emission of ammonia, which is injurious to health. Only a limited number of such boards were manufactured. The nominal density is 1.200, and the observed value is only 1.060.

This is due to the spring back of the wood particles, which is an indication of the unsatisfactory bonding properties of the inorganic binder. Sodium carbonate caused a faster setting of cement. The setting time of cement mixed with a solution of this carbonate is less than 3 minutes. It is also important, that the carbonates, expecially ammonium carbonate and sodium carbonate, are free of ammonium hydrogen carbonate and sodium hydrogen carbonate. The presence of small amounts of these acid salts will enhance the setting of cement.

Table 1. Influence of temperature and waterglass on the setting of various cements*)

Cement	Temp.	Setting time			
		without		+1.5wt% watergl.	
		start	end	start	end
	°C	min	min	min	min
PZ 35 F	25	25	39	60	93
	40	4	37	50	85
	60	3	25	10	20
	80	2	23	5	10
Highfurnace slag	25	55	130	65	205
	40	20	90	40	180
	60	5	50	10	20
	80	5	20	10	17
Trass	25	12	28	13	117
	40	10	20	10	100
	60	8	15	10	25
	80	5	12	7	14

*) Additive: 13 wt% K_2CO_3

Table 2. Properties of rapidly hardening cement-bonded particleboards*)

Additive wt % (based on cement)	Bending strength MPa	Density kg/m²
13 % $(NH_4)_2CO_3$ + 1.5 % waterglass	6.0	1,060
13 % Na_2CO_3 + 1.5 % waterglass	6.0	1,150
13 % K_2CO_3 + 1.5 % waterglass	10.0	1,252

*) Cement: PZ 35F. Wood particles: Spruce.
Press temperature: 80°C. Press time: 15 min.

When hydrogen carbonate was present in sodium carbonate it proved to be impossible to make cement-bonded particleboards, due to a presetting of cement. The slower reaction of potassium carbonate enable a sufficient open time. In most of the experiments this carbonate was used as additive.

To attain the required strength properties (bending strength 8 to 10 MPa) the rapidly hardening boards have to be stored at least 14 d after manufacturing. In this examination all boards were tested after 28 days. Larch wood, which is incompatible with cement, can be used to manufacture cement-bonded particleboards, if potassium carbonate and waterglass solution are used as additives. The properties of such boards are not as good as those of spruce (Table 3), certainly due to the high concentration of soluble carbohydrates in the wood.

3.3 Water resistance and durability

The results of the V313 test is depicted in Table 3.

Table 3. Water resistance of rapidly setting cement-bonded particleboards*)

Wood	Cement	Untreated			After testing		
		Bend. strength	Inter-nal Bond S	Densi-ty	Bend. strength	Inter-nal Bond S	Densi-ty
		MPa	MPa	kg/m^3	MPa	MPa	kg/m^3
Spruce	PZ 35F	9.98	0.70	1,252	11.57	0.47	1,165
	Trass	12.05	0.79	1,233	13.23	0.57	1,156
	Highfurnace slag	11.08	0.70	1,228	12.60	0.49	1,145
Larch	PZ 35F	6.91	0.32	1,195	4.96	0.16	1,085
	Trass	7.51	0.24	1,186	5.60	0.15	1,091
	Highfurnace slag	5.13	0.20	1,142	4.51	0.11	1,035

*) Press temperature: 85°C. Press time: 18 min

After testing, the bending strength of all samples made from spruce increased due to further hydration of the inorganic binder, However, the internal bond strength decreased, although still exeeding 40 % of the original value. According to the French Standard the boards are considered to be water resistant. Cement-bonded particleboards made from larch have always lower strength properties than those made of spruce. After testing all boards showed lower densities, due to swelling, but also because soluble salts of potassium and sodium are leached out.

During the first and the second water treatment the water becomes light brown. The results of the durability test is shown in Table 4.

Table 4. Comparative durability of conventionally and rapidly setting cement-bonded particleboards from spruce.

Cement	Bending Strength (BS) and Density (D)							
	28 days		32 weeks		64 w (outd.)		64 w (ind.)	
	BS MPa	D kg/m^3	BS MP	D kg/m^3	BS MPa	D kg/m^3	BS MPa	D kg/m^3
Press time 8 h:*)								
PZ 35F	14.38	1,117	17.92	1,189	22.19	1,207	16.85	1,141
Trass	16.30	1,128	18.35	1,184	19.62	1,184	11.16	1,109
Highfurnace slag	11.55	1,049	13.52	1,122	15.70	1,159	10.79	1,079
Press time 18 min:**)								
PZ 35F	11.84	1,089	10.18	1,157	12.98	1,154	12.01	1,155
Trass	11.56	1,092	8.58	1,069	9.25	1,054	17.14	1,150
highfurnace slag	11.03	1,039	8.67	1,076	10.04	1,079	11.45	1,107

*) Accelerator: 3 wt% $CaCl_2$. Press time: 8 h at 40°C.
**) Additive: 13 wt% K_2CO_3 + 1.5 wt% waterglass. Press time: 28 min at 85°C.
outd. = outdoor; ind. = indoor; w = weeks

Conventionally made boards were compared with rapidly setting ones. The duration of the test is only 64 weeks, which is rather short. However, the results give an indication of the behaviour of rapidly setting cement-bonded particleboards for outdoor purposes.

Conventionally made cement-bonded particleboards have higher bending strength properties and are also more durable. However, the properties of the rapidly setting boards fulfill the requirements of existing standards for such boards. After 64 weeks of weathering no substancial loss of bending strength can be observed.

4 Conclusion

Rapidly setting cement-bonded particleboards can be manufactured within 15 minutes at a press temperature of 80-85°C. The properties of such boards are not as good as conventionally made ones. However, they fulfill most of existing standards. Due to the short pressing time it should be feasible to manufacture the boards in slightly modified wood particleboard plants. The rather high price and the high amounts of carbonates required for the process may hinder the application of this technology.

5 References

Elmendorf, A. (1966) Method of making an non-porous board composed of strands of wood and portland cement. **US Patent** No. 3,271,492.

Konkola, J. (1989) Universal building material - entirely new technology, in **Fiber and Particleboards bonded with Inorganic Binders** (ed. A.A. Moslemi), Forest Products Research Society, 2801 Marshall Court, Madison, WI 53705.

Lange, H., Lieber, T., Simatupang, M.H. (1986) Investigations on the relationship between strength properties and dimensional stability of cement-bonded particleboards. **Holz Roh- Werkstoff**, 44, 127-132.

Odler, I. (1986) Neuartige umweltfreundliche Werkstoffe und Technologien. **Hannover Fair 1986**. Inst. für nichtmetallische Werkstoffe, Zehntnerstrasse 2A, D-3920 Clausthal-Zellerfeld.

Paulisan, P. (1989) Verfahren zur Herstellung plattenförmiger Verbundstoffe, **E.P.** 0340 620 A2.

Schmidt (1989) Requirements and demands for further processing of cement-bonded boards, in **Fiber and Particleboards Bonded with Inorganic Binders** (ed. A.A. Moslemi), Forest Products Research Society, 2801 Marshall Court, Madison, WI 53705.

Schwarz, H.G. (1989) Cement-bonded boards in Malaysia in **Fiber and Particleboards Bonded with Inorganic Binders** (ed. A.A. Moslemi), Forest Products Research Society, 2801 Marshall Court, Madison, WI 53705.

Sekisui Chemical Co. (1981) Wood chip cement products. **Jap. Patent** 56/140059. CA. 96(10): 73740h.

Simatupang, M.H. (1979) Water requirement during the manufacture of cement-bonded particleboards. **Holz Roh- Werkstoff**, 37, 379-382.

Simatupang, M.H., Lange, H., Neubauer, A. (1987) Influence of seasoning of poplar, birch, oak and larch and the addition of condensed silica fume on the bending strength of cement-bonded particleboards. **Holz Roh- Werkstoff**, 45, 131-136.

Simatupang, M.H. (1989) Mineral-bonded wood composites, in **Concise Encyclopedia of Wood and Wood-Based Materials** (ed. A.P. Schniewind), Pergamon, pp. 196-201.

Yamagishi, K.Y., Sano, Y. Sakakibara, A. (1983) Wood cement boards. **Rinsan Shikenjo Geppo** (Hokkaido) 375, 11-17. CA 99: 55274 (1984).

26 THE LONGITUDINAL MODULUS OF ELASTICITY OF WOOD

E. CHAHUD
Structures Department, School of Engineering,
Federal University of Minas Gerais, Brazil

Abstract
The aim of this work is generate informations for the revision of NBR 7190 (Brazilian Standard for Design and Building of Wooden Structures).This paper presents and analyses results of tests for the compression and tension parallel to the fibres and for bending, with the objective of determining the respectives moduli of the elasticity (Ec,Et and Ef). The statistical equivalence between the moduli of the seven wood species studied becomes evident from tests results.
Keywords: Wood Properties, Elastic Constants, Modulus of Elasticity.

1 Introduction

The wood longitudinal modulus of elasticity (E) is a necessary paramete for the valuation of the limit states of serviceability of members of the many types of structures built with the former material.Thereafter it is important to know the value of E, and discuses the experimental methods for its determination.

Nowadays the Brazilian Method for Physical and Mechanical Testing of Woods (NBR 6230), from the Brazilian Standards Association (ABNT) recomends two types of tests for the determination of the wood longitudinal modulus of elasticity: one the compression parallel to the fibers and the other of the bending with the load applied on the middle of the span, and ratio between span and height of the test specimen (l/h) set equal to 14.

The making of those tests permits to obtain two values: the modulus of elasticity in compression parallel to the fibres (Ec) and the modulus of elasticity in bending (Ef). ABNT also recommends that for structural design only Ec must be used, and Ef must be used only for wood species comparison.

On the other hand the method "Woods: determination of its characteristics", endorsed by the Wood and Wooden Structures of Laboratory (LaMEM) of the EESC-USP and by the Brazilian Institute of Wood and Wooden Structures (IBRAMEM) recommends the determination of the longitudinal moduli of elasticity in compression parallel to the fibres (Ec), in tension parallel to the fibers (Et) and in bending (Ef) with the ratio l/h set equal to 21.In this situation,according to Rocco Lahr the contribuition of deformations concerning from shear in the vertical displacement can be considered negligible.

As a consequence, for l/h greater than or equal to 21, Ef can be found by the equation:

$$Ef= (Pl^3)/(48vI) \qquad (1)$$

In equation (1),P is the load applied until the proportionality limit, at most, and causes the vertical displacement v in the specimen with the moment of inertia I.

Rocco Lahr, based in results of tests for compression parallel to the fibres and tests for bending with l/h set equal to 21, showed the statistical equivalence between Ec and Ef, with a confidence level of 95%. This is an interesting conclusion that sanctions the concept of the wood's longitudinal modulus of elasticity.

Therefore the possibility of comparisons between results of Ec and Et with values of the modulus of elasticity obtained from tests for tension parallel to the fibres (Et) according to the methodology proposed in the method of testing adopted by LaMEM and by IBRAMEM was considerated.

2 AIM OF THE WORK

Taken into account what was said in the preceding item, this work was conceived with the aim of comparing the wood longitudinal moduli of elasticity obtained from tests for compression and tension parallel to the fibers and for bending with l/h equal to 21.

3 MAKING OF THE TESTS

The experimental part of this work was the testing for the determination of Ec,Et and Ef for the following wood species:

*Castanheira (Bertholletia excelsa)
*Cumaru (Coumaruna alata)
*Garapa (Apuleia leiocarpa)
*Jatobá (Hymenaea stilbocarpa)
*Maçaranduba (Manilkara sp)
*Peroba Rosa (Aspidosperma polyneuron)
*Pinus hondurensis (Pinus hondurensis)

These species were chosen because they showed properties deemed compatible for structural aplications. The species Pinus hondurensis was included because of the good out look of its use in the making of panels, formwork and shoring, and in prefabrication of trusses to cover small spans.

3.1 The making of tests specimens

All the test specimens were taken from sawed lumber (rafters and beams) bought in different occasions, during a three years span, in sawmills of the city of São Carlos, SP, Brasil. From each member, five test specimens were taken, one being used in the bending test, two being used in the tests for compression parallel to the fibers and the other two in the tests for tension parallel to the fibers.

3.2 Bending tests

The bending tests were made on test specimens with a cross section of 6cm x 8cm, a length of 170 cm and a free span of 160 cm. A concentrated load was applied in the middle of the span.

The application of this load P was made in a continuus manner from zero until a value that causes stresses nearing the proporcionality limit on the material of the test specimen.

For constant increments of the load, the respectives vertical displacements in the middle of beam were tabulated, the measures made by a comparator clock with a precision of 0,01 mm set on the mean height of the section. When the load was taken out, the test specimen recovered its initial configuration. After that, the load was applied again, and the vertical displacements measured by the comparator clock were tabulated. For each testing, this procedure was made three times, and mean value of the vertical displacements measured was taken as the vertical displacement for each increment of the load.

Applying mean value of the vertical displacement in equation (1), the value of Ef was obtained for the test specimen under load.

Figure 1 shows the sketch of the testing apparatus for bending.

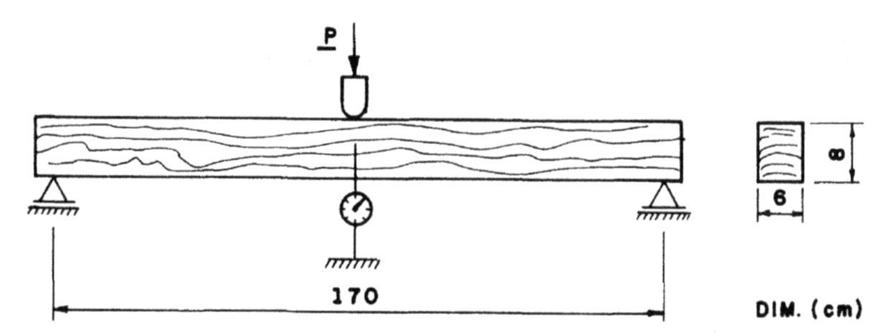

Fig.1 Sketch of bending test

3.3 Test for compression parallel to the fibres

Two test specimens were taken from each member, with dimensions of 5 cm x 5 cm x 20 cm, and used in tests for compression parallel to the fibres.

The deformations were measured by two comparator clocks with a precision of 0,01 mm, attached to opposite sides of the test specimens. The distance between the points of affixation was 10 cm, and this distance was used in the computation of the specific deformations. The values of shortening for constant increments of loading, measured by the two comparator clock, were tabulated during the tests. This procedure was used for stresses varying from zero to values nearing the prporcionality limit of the material.

The specific deformation for every level of loading was obtained dividing the mean value of the deformation by the distance between the points of affixation of the comparator clocks, set equal to 10 cm.

The value of Ec was obtained from the linear regression between the pairs of values of the stress applied and the tabulated specific deformations.

Figure 2 shows a skech of the test specimen used in the test for compression parallel to the fibres .

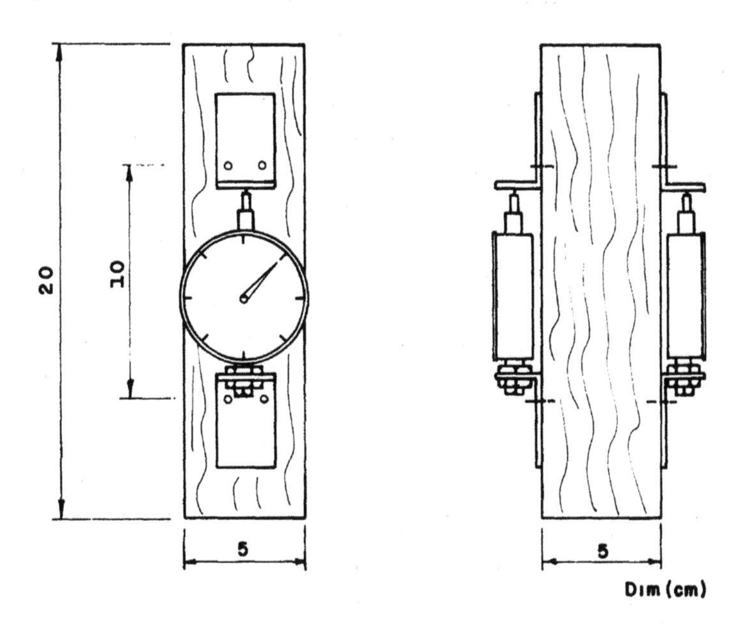

Fig. 2 Test specimens used in the test for compression parallel to the fibres.

3.4 Test for tension parallel to the fibres.

Two test specimens, whose dimensions are shown in figure 3, were taken from each member. The procedures and methodology for testing are similar to those shown in item 3.3, and in this manner the values of the modulus of elasticity in tension parallel to the fibres were obtained.

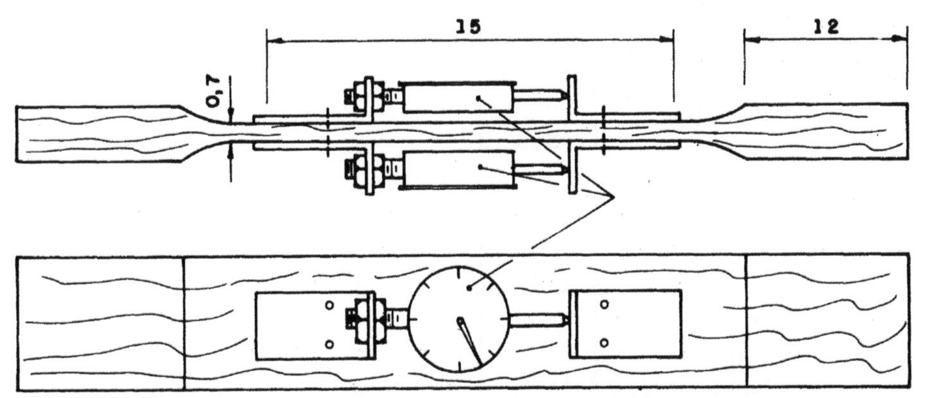

Fig.3 Test specimen in test for tension parallel the fibres

4 Presentation of results

4.1 Jatobá

Specimen	Ef (MPa)	Ec (MPa)	Et (MPa)
01	12471	10841	11247
02	8801	9194	9989
03	8738	10147	9094
04	11881	12433	12521
05	15286	15467	15826
06	12188	14334	14981
07	18678	18024	23109
08	11515	11635	11247
09	18722	19689	18073
10	10633	10574	13960

4.2 Cumaru

Specimen	Ef (MPa)	Ec (MPa)	Et (MPa)
01	24269	24950	21523
02	22944	22260	16991
03	22738	22841	23166
04	21627	24479	24784

4.3 Garapa

Specimen	Ef (MPa)	Ec (MPa)	Et (MPa)
01	11928	13684	12735
02	14810	16639	12414
03	14365	15457	18867
04	15642	16048	13565
05	16598	14823	15565
06	11563	14818	12700
07	11072	13899	15968

4.4 Peroba Rosa

Specimen	Ef (MPa)	Ec (MPa)	Et (MPa)
01	8777	10330	11364
02	9868	9194	11176
03	11144	11718	11392
04	13830	10700	15555
05	12747	10640	12592

4.5 Pinus hondurensis

Specimen	Ef (MPa)	Ec (MPa)	Et (MPa)
01	7272	9320	6750
02	6191	6943	5237
03	6469	7981	5120
04	7009	9003	6445
05	7083	7699	12315
06	8492	8328	7742
07	8048	7211	6882
08	9194	8904	10031
09	6903	7007	7894
10	7711	7338	8042
11	6901	6631	6290
12	8006	6480	6816

4.6 Maçaranduba

Specimen	Ef (MPa)	Ec (MPa)	Et (MPa)
01	19108	21634	22391
02	18812	26677	22793
03	19434	19521	21064
04	18230	18638	17931
05	22092	20481	21238
06	14783	16098	13941
07	16573	14731	17295
08	19822	18947	16338
09	17031	17575	18283
10	20255	18084	19308
11	21740	20223	19661
12	20882	21038	20298

4.7 Castanheira

Specimec	Ef (MPa)	Ec (MPa)	Et (MPa)
01	12213	9807	11350
02	9446	10592	11974
03	12913	11628	11170
04	7429	8588	8916
05	8272	8591	8913
06	13725	14233	12875
07	14877	13856	12613
08	14081	14679	12860
09	10717	11164	11393
10	9645	11159	10777
11	9680	10267	11269

Specimen	Ef (MPa)	Ec (MPa)	Et (MPa)
12	8963	10843	10330
13	9091	8702	10375
14	7826	9643	10241
15	8258	9964	10505
16	6673	8361	8043
17	9395	10799	10392
18	9651	10786	11194
19	13913	11403	11263
20	15507	11451	11363

5 Analysis of Results and Conclusion

The comparison of results was made by 2 statistical tests, envolving two of the moduli (Ef,Ec anf Et) at each time. The first test, variancy analysis, verify if the means are statistically equivalents, at a level of confidence taken equal to 95%.

The statistical analysis made for individual species, and for more than one species, makes possible to conclude that there is a statistical equivalence, at a level of 95%, between the wood longitudinal moduli of elasticity found in the tests for compression and tension parallel to the fibres and for bending. This shows the possbility of the use, in design , of the values of Ec,Et or Ef, without significant incorrections.

6 REFERENCES

ASSOCIAÇÃO BRASILEIRA DE NORMAS TÉCNICAS- Ensaios Físicos e Mecânicos de Meiras - NBR 6230. Rio de Janeiro, ABNT, 1980

LABORATÓRIO DE MADEIRAS E DE ESTRUTURAS DE MADEIRA - Madeira: determinação de suas características. São Carlos, LaMEM-EESC-USP, 1988.

ROCCO LHAR, F.A. - Sobre a determinação de propriedades de elasticidade da madeira. São Carlos, LaMEM-EESC-USP, 1983. Tese.

CHAHUD,E. - Módulo de elasticidade longitudinal da madeira e proposta de cálculo para peças fletidas. São Carlos, LaMEM-EESC-USP, 1989. Tese.

FUSCO, P.B. - Estruturas de concreto: fundamentos estatísticos de segurança das estruturas. São Paulo, McGraw-Hill do Brasil, EDUSP, 1977.

SNEDECOR, G.W. - Statistical methods. 5.ed. Ames, Iowa State University, 1956.

27 MULTI-STOREY LIGHTWEIGHT PANEL BUILDINGS

H. GALLEGOS
Catholic University, Lima, Peru

Abstract
The main use of natural fibers in future urban construction, if they are successful as building materials, will be in the form of lightweight panels. This paper -a conceptual document and the starting point of an integral study- analyzes the structural feasibility of multi-story cellular buildings -or panelized building systems- contructed of these lightweight panels. It reviews the structural behavior of panel buildings, provides the basic information on the tests needed to obtain the relevant mechanical properties and, provides also, the details of the preliminary analysis of a prototype panel structure.
Keywords: Fibers, Panels, Structures, Timber

1 Introduction

In developing countries the urban house is mainly the end-product of a non-engineered self-help process that lasts for many years and that takes place in an environment of poverty. This process produces low density, physically extended and seismically unsafe cities, in which the total urban cost per person is far from minimum, and where it is becoming increasingly difficult and unnecessarily costly to provide adequate infrastructure and services. As a result, the present housing policy in some developing countries includes the objectives of reducing self-help construction and providing urban housing in low-rise, three to five story, buildings. This policy will, in addition, aid in urban renewal by attenuating the social problems of the slum and helping to protect, in some cases, a valuable architectural heritage. Gallegos (1979).

The most efficient structure for these multi-story buildings is the cellular (or surface) structure -which, in some aspects, resembles the monocoque structure used in the automobile and aircraft industries- because of the following: a) the wall acts simultaneously as structure, enclosure and partition, thereby satisfying one of the basic philosophical conditions for structural efficiency; b) the floor space occupied by the vertical elements is minimum; c) the volume of building materials per square meter of floor space or cubic meter of building volume, is minimized, and d) as the weight is minimized, weight dependent loads -such as earthquake loads- are reduced.

Thus far this type of multi-story structure has been built mainly of reinforced concrete or reinforced masonry walls and reinforced concrete floors and roof. Nevertheless, it is feasible to build these structures using lightweight panels and lightweight diaphragms in order to provide a panelized building system (PBS). The adoption of these elements will provide additional advantages: a) promote the use of locally available vegetable fibers; b) further the use of panels for partitions and furniture, as well as for loadbearing purposes, and c) optimize, even more, the weight reduction potential of cellular buildings.

Two basic types of panels, shown in Fig. 1, are essential -Gallegos (1986)- to the solution of the PBS:

Panel A. This panel consists of a strong timber framed core. The dimensions of the timber frame are the minimum necessary to allow for proper connection between panels and between panels and diaphragms. This type of panel can be plywood or particle board or fiber board sheathed when necessary for lateral strength. The core can be solid or consist of two boards with an air-space between them.

Panel B. This panel consists of a strong sheathed frame and a weak core or no core at all. This is the preferred type of panel because it can be made lighter than panel A.

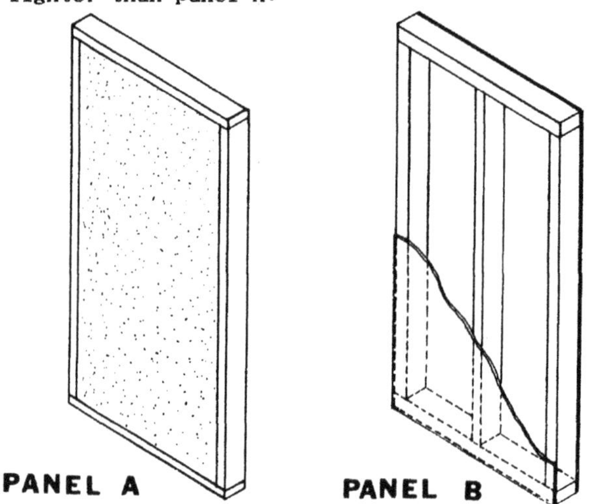

Fig. 1. Types of panels for PBS

The diaphram, on the other hand, is essentially a trussed timber structure -designed for maximum in-plane rigidity- with the necessary sheathings to cover practical floor and/or roof and ceiling needs. Smith (1986).

The behavior of the PBS -a diaphragmed type of structure- is not well understood, particularly in connection with lateral -wind or seismic- actions, and is confused with the behavior of panel systems. This misconception is liable to lead to unsafe or uneconomic solutions.

This paper provides an analysis of the basic concepts of panel structures behavior and identifies the mechanical characteristics that have to be known for design purposes and the methods to be used for their determination.

2 Glossary

2.1 Cellular (or surface) structure
Structure consisting of walls and rigid diaphragms in which the space limitation is also the structure.

2.2 Diaphragm
Horizontal structural member which has the capacity for integrating the vertical structural elements of a building to form a three-dimensional resisting structural system.

2.3 Fiberboard
Sheet material manufactured from vegetable fibers; its basic strength is derived from the felting together and inherent adhesion of the individual fibers.

2.4 Loadbearing wall
A wall with the capacity for supporting superimposed coplanar loads.

2.5 Non-loadbearing wall
A wall that supports no coplanar loads other than its own weight. For example: partitions and curtain walls.

2.6 Panel
A thin prefabricated wall.

2.7 Panelized building system (PBS)
Cellular structure in which the loadbearing panels and prefabricated or site-assembled diaphragms resist and distribute all loads, provide the three-dimensional strength and rigidity of the building, and eliminate the use of a structural skeleton. Fazio (1971).

2.8 Panel system
A wall system composed of non-loadbearing panels attached to the building frame and to each other.

2.9 Particle board
Sheet material made from wood chips or shavings bonded together under heat and pressure using a synthetic resin adhesive.

2.10 Plywood
Composite sheet material made by bonding thin wood veneers with adhesives and with alternate layers laid al 90 degrees to each other.

2.11 Sheathing
Sheet material -normally 8 to 12 mm thick- that is glued or nailed to the frame in panel construction.

3 Panel structures

A panel by itself is not stable. If it is not propped up by an adjoining structural member or attached rigidly to a foundation, it will topple over perpendicular to its plane (Figs. 2a and 2b). The collapse can be avoided if stability is providad by another panel more or less perpendicular to it: the two panels support each other since the supporting panel also requires stability perpendicular to its plane (Figs. 2c, 2d and 2e). In this case it is essential for the two panels to share a common edge. The structural design of a system like this one consists basically of verifying the stability and competence for providing stability of each and every panel.

When a diaphragm is introduced (Fig. 2f) panels become loadbearing and the previous system undergoes two essential structural modifications: a) the need for a common edge is eliminated (the panels need not touch each other to ensure stability), and b) the structural design must verify not only the stability of the panels, but also their coplanar load carrying capacity.

A PBS is simply a development of this diaphragmed system, to form a typical multi-story building (Fig. 3). These structures must comply with two additional conditions: a) the panel-foundation connection must be able to transmit coplanar bending moments and shear forces, and b) panels must run continuously from the foundation to the top of the building, which means that all vertical coplanar connections -which are usually needed on each floor- must also resist flexure and shear.

The PBS is a diaphragmed structure and because of that, consideration must be given to the stiffness of the diaphragm and to the strength of the connections between the panel and the diaphragm (Fig. 4). As the diaphragm must have the necessary stiffness or rigidity to integrate the structural system this property must be verified. When diaphragms are diagonally braced, they can be assumed to be rigid enough for this purpose -in other words, able to distribute lateral loads according to the relative rigidity of the vertical elements and their position in terms of the horizontal center of rotation-, and the calculations can assume them to be infinitely rigid. The diaphragm-panel connection, on the other hand, must be capable of transmitting shear forces in the intersection of the diaphragm and the panel and moments on the vertical plane. Moment-resisting connections are not necessary in the horizontal plane.

In a PBS, lateral loads are distributed by the diaphragm as coplanar horizontal loads, and the load in each panel is determined by the flexibility of the diaphragm relative to the flexibility of the lateral load resisting system. Gravity loads are allocated to the panels in proportion to tributary areas, and act as coplanar compressive forces. Panels are then able to resist these loads and to transmit them to the ground by acting as loadbearing walls -somewhat like foundation-anchored cantilevers-, which are subjected to a combination of these coplanar loads: a) vertical compressive, and b) lateral shear.

Coplanar compressive forces are the result of self-weight, diaphragm loads and live loads. Large coplanar compressive forces supress the development of flexural and diagonal tensile stresses in the panel and usually provide adequate resistance to panel sliding on the foundation.

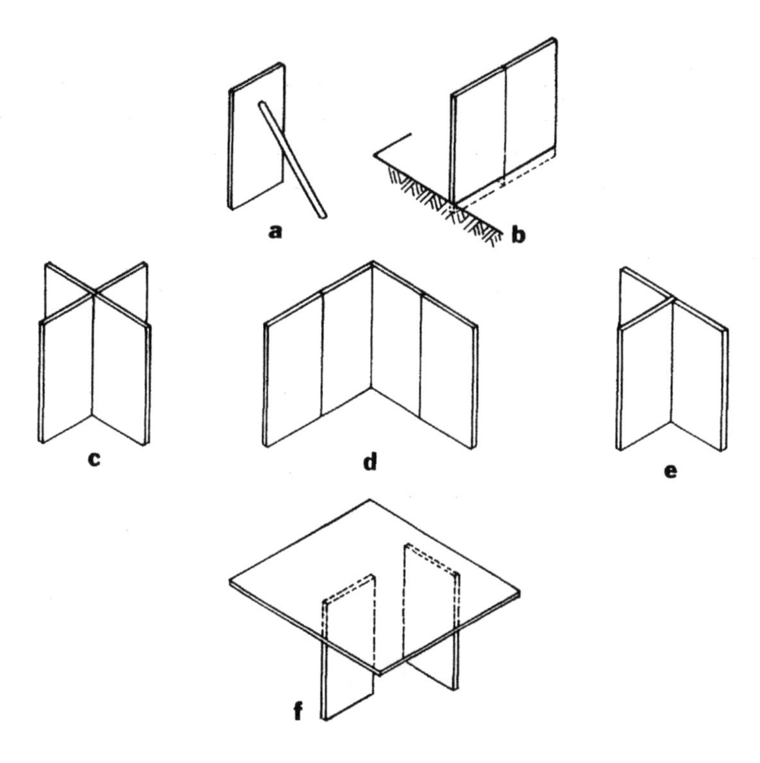

Fig. 2. Panel stability

Lateral load design in this case must deal mainly with the compressive strength of the panel.

Where coplanar compressive forces are relatively small -which is normally the case for lightweight cellular structures-, the tensile strength of the panel and the strength of connections become critical, together with the coplanar compressive strength of the panel.

It is usually uneconomic to design structures to remain elastic during severe earthquake motion. Provided that the structure is designed, and detailed, to be able to accept displacements greater than those attained at its design load, it can be safely designed to resist horizontal forces substantially inferior to those predicted for an elastically responding structure. This "ductile behavior" can be predicted and provided in steel and steel-reinforced structures, but has not been fully investigated and developed for panel structures. There is evidence: a) that sheathed panels can develop considerable ductility through the slip action of the nails connecting the sheathing to the perimetrical timber frame, Dean et al. (1986), and b) that glued panels fail in a brittle maner, Dowrick et al. (1986).

Finally, it is noted that the good performance of panels under loading -specially lateral loading- depends strongly on correct detailing. In this respect careful attention, specially to nailing parameters such as edge distance and spacing, is essential.

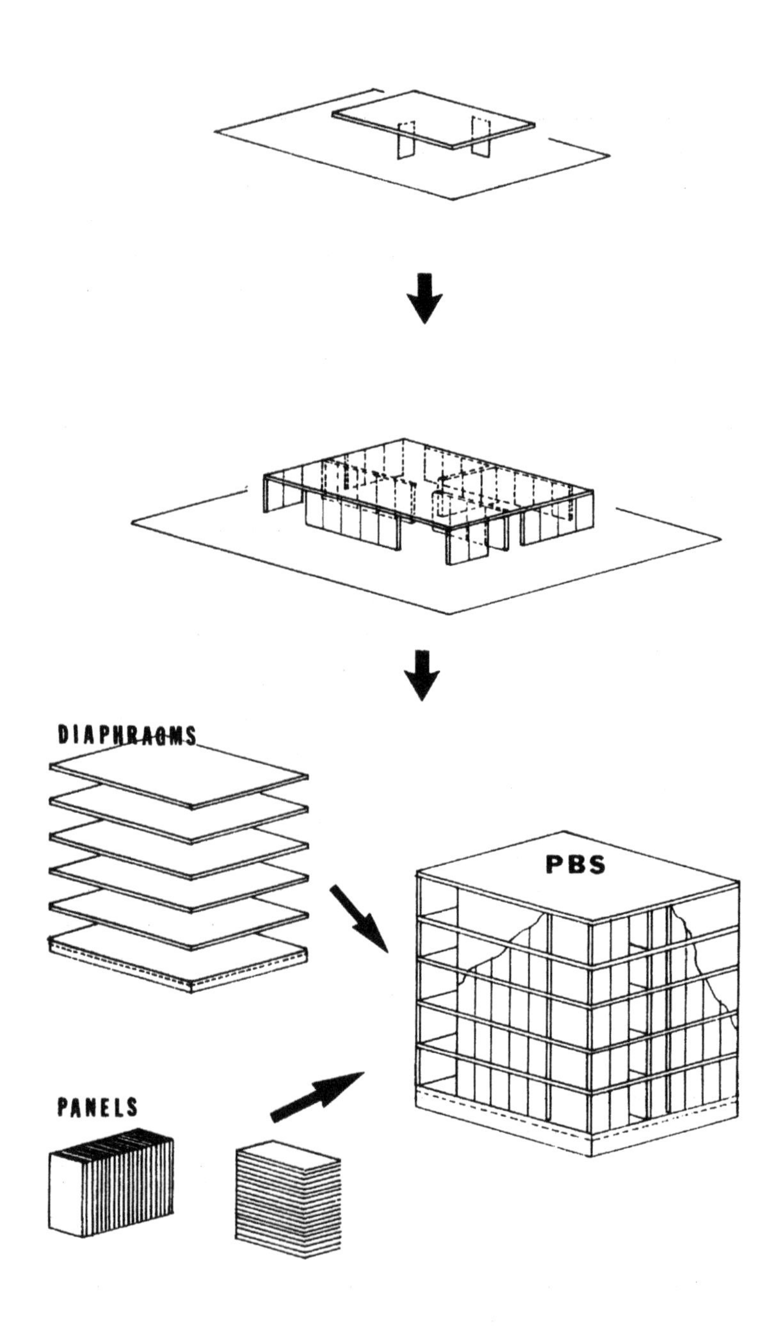

Fig. 3. Development of the PBS structural system

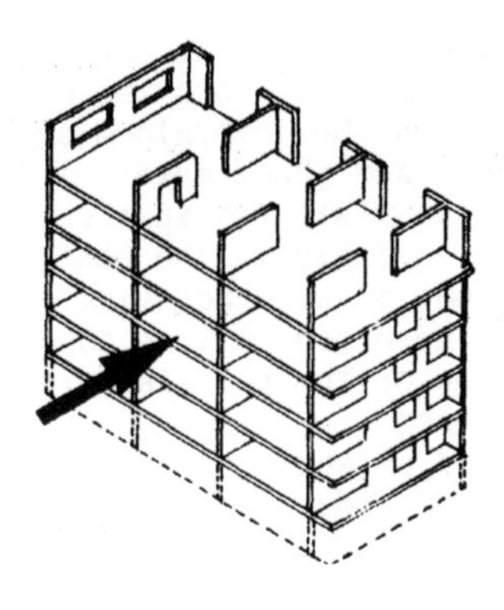

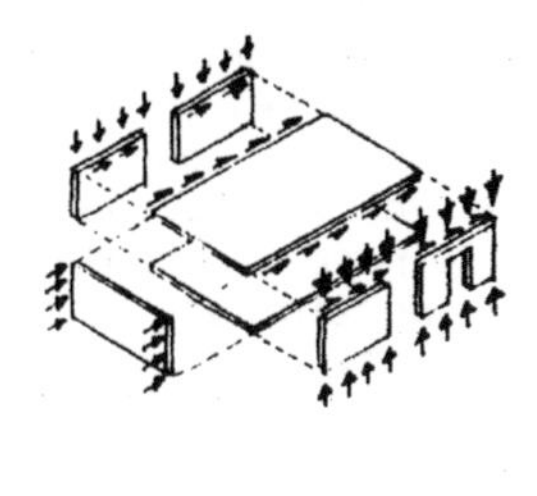

Fig. 4. Lateral load path in cellular structures

4 Panel tests

The mechanical properties of panels can only be estimated from the properties of their individual components. Tne panel is actually a complex structure whose properties depend not only on those of its components, but also on the ways in which they are bonded or nailed, and assembled. Accordingly, the only way to determine their properties is through direct panel testing, in which the specific mechanical properties must be isolated and the methods and procedures to be used must be carefully designed. As is normally the case with product development and expensive testing, the procedure is one of analysis-design-test-analysis-design. A basic condition of tests must be to reduce unforeseen results to a minimum.

The relevant mechanical properties to be investigated can be derived from the structural role that the individual panel plays in the completed structure. In addition to isolated panel testing, the connections, connected panels, panel-foundation conncetions and panel-diaphragm connections must also be tested.

The results obtained must be evaluated within an statistical context in order to determine charcteristic values for the differente relevant mechanical properties.

The basic properties that need to be determined for the panels are the following:

a) Panel coplanar compressive strength (Fig. 5a).

b) Panel out-of-plane bending strength (Fig. 5b). For PBS structures, panels can be assumed to be exclusively top and bottom supported for this test.

c) Single panel coplanar lateral load strength (Fig. 5c). This test is essential. Failure can occur because of any one of the following: tensile flexure in the panel, compressive flexure in the panel, diagonal tension in the panel, diagonal compression in the panel, sliding between the foundation and the panel and, finally, tensile failure of the panel-foundation connection. Because of the excessive cost of the tests, and the complexity of the analysis of the results obtained when no control is exercised over the strength parameters, and when any type of failure can occur, the best procedure is to predetermine -by means of a preliminary strength analysis and a correct conceptual design- the type of failure desired. Then the different dimensions and sections of the panel can be determined and tests surprises practically eliminated. As timber and sheet material fail in a brittle manner in tension, and compression failures must be avoided because of buckling -which is also a brittle failure-, the best type of failure is that which is associated with some plastic behavior and with maximum inelastic lateral deformations. Therefore, either the tensile failure of the panel-foundation connection or the slip failure of the sheath-frame connection are normally preferred.

d) Single panel coplanar lateral load strength when the panel is subjected to coplanar compressive loads (Fig. 5d). This test should be conducted only in the final stages of panel design as a verification test -like some sort of theater full-dress rehearsal- for, although, the information it provides on real panel behavior is valuable the test set-up is complex and costly.

e) Single panel pseudo-static lateral load (Fig. 5e). This test is needed when hysteretic energy absorption and dissipation and ductile behavior have to known. It is associated exclusively with seismic loads, especially when inelastic structural deformations are expected.

f) Connection tests (Figs. 5f, 5g and 5h). These tests determine tensile and shear strength in the different critical portions of the connections. As normally they are designed for lower strengths and greater ductilities than those corresponding to the panel, and as they are responsible for adequate inelastic behavior, these tests are as essential as are those tests of coplanar loaded panels.

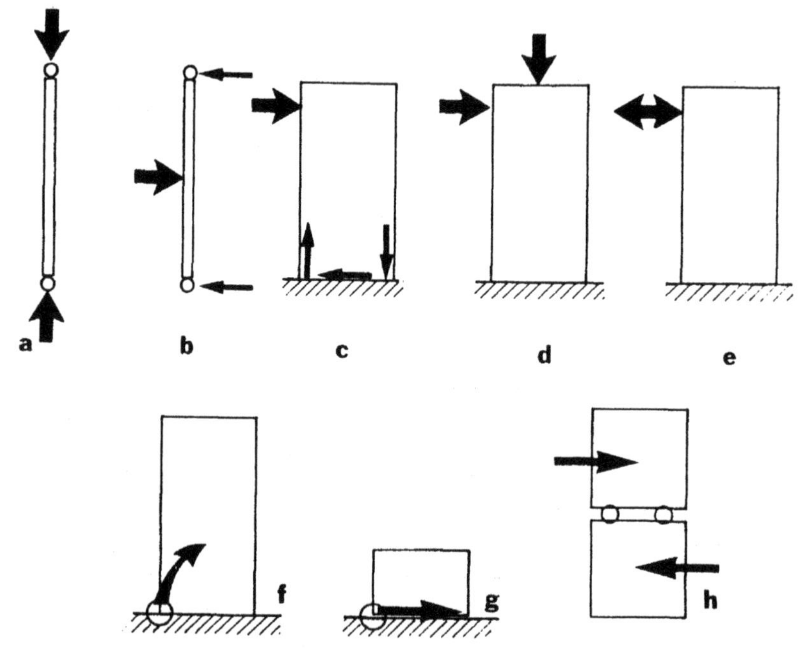

Fig. 5. Panel tests

5 Preliminary analysis of a prototype PBS atructure

The cross section of a 12 cm thick, 120 cm wide and 240 cm high (nominal dimensions), type B panel, weighing 82 kg/unit, is shown in Fig. 6. This prototype panel has been used as the basic structural component for the four-story PBS, whose floor plan is shown in Fig. 7. The PBS structure is assumed to posses infinitely rigid diaphragms (floors and roof), and, it is also assumed, that governing lateral loads are seismic loads. These loads have been determined -as equivalent static loads- from the acceleration spectrum for the coastal seismic zone of Peru, for an estimated fundamental structural period of 0.8 seconds, a fundamental soil period of 0.3 seconds and a ductility factor of 2. It has been further assumed, that for these seismic conditions the "weak link" of the structural resisting chain will be the steel rods -continuous from the foundation to the top plate- that provide all panel connections and vertical continuity; this "weak links" are ductile and able to prvide a displacement ductility factor of more than 4. Finally, it has been assumed that all the other links of the structural chain -esentially panel flexural and shear strengths- will be brittle but much stronger than needed; this condition of the "strong links" will demand for each and any of them combined load and capacity reduction factors of at least 5. The maximum loads of the critically loaded panel are shown in Fig. 8.

Panel properties are as follows: a) compression area -provided only by the vertical elements of the frame- 250 cm^2; b) shear area -provided only by the sheathing- 240 cm^2, and c) section modulus 12,000 cm^3.

The combined load and capacity factors obtained, using white fir for the frame and medium density (600 kg/m^3) particle board for the sheathing, are the following: a) panel compression buckling 7; b) panel local compression crushing 12, and c) panel shear 6.

The moment at the base of the panel is $M = 4.8$ KN x 7.5 m = 36 KNm, and the maximum force in the steel anchoring rod is $T = M$ / resisting arm, that is $T = 36 / (1.20 - 2 \times 0.13) = 38$ KN. This strength can be provided by a 12 mm diameter grade 420 MPa steel bar, which yieldsat a characteristic load of 41 KN. It must be noted that it is essential that no excessive over-strength be provided in the "weak link" of a ductile structure, as it must remain the weakest, but ductile, part of the chain.

From an economic point of view it is interesting to point out the basic structural materials needed per square meter of total floor area of the PBS: a) timber 0.12 m^3; b) particle board 0.08 m^3; c) concrete foundation 0.05 m^3, and d) steel 4.5 kg. These quantities are associated in Peru with a very competitive structure.

6 Conclusions

a) Three to five story buildings are needed for urban housing in developing countries.

b) The most efficient structure for these buildings is the cellular structure.

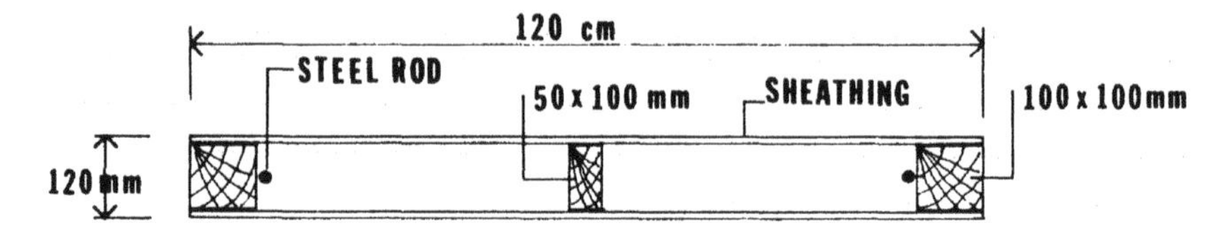

Fig. 6. Cross section of prototype panel

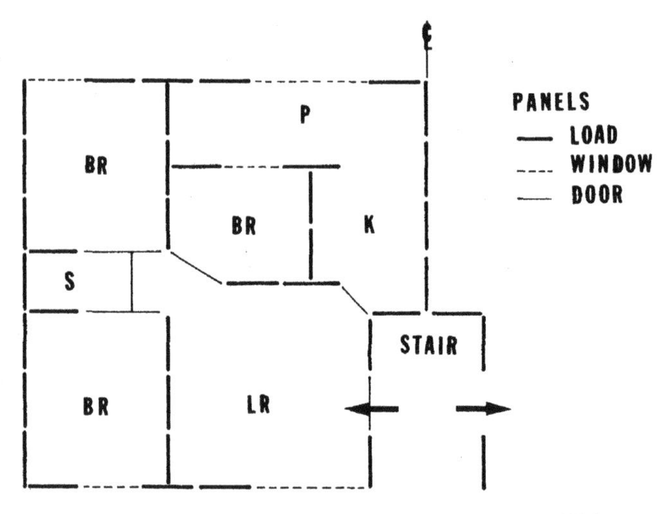

Fig. 7. Floor plan of 4 story prototype PBS

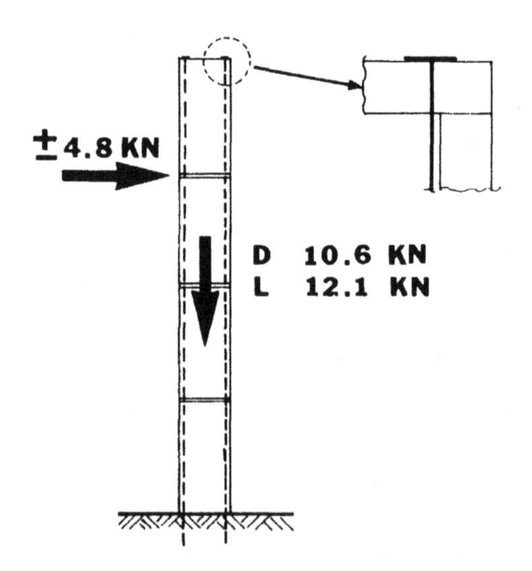

Fig. 8. Maximum coplanar loads on critical panel

c) Even in high-risk seismic areas, the PBS -a lightweight panelized building-can be structurally competent.

d) The PBS -as determined by a preliminary analysis- can also be cost-competitive.

7 References

American Concrete Institute, ACI (1966). **Symposium on Precast Concrete Panels**. Publication SP-11. Michigan, USA.

American Society for Testing and Materials, ASTM (1989). **Wood-base Fiber and Particle Panel Materials**. Designation D-1554. USA.

American Society of Civil Engineers, ASCE (1975). **Wood Structures**. New York, USA.

Angerer, F. (1961). **Surface Structures in Building**. Reinhold Publishing Corporation. USA.

Dean, J.A., Stewart, W.G. and Carr, A.J. (1986).The seismic behavior of plywood sheathed shearwalls. **Bulletin of the New Zealand National Society for Earthquake Engineering**. Vol. 19. Nº 1.

Dowrick, D.J. ans Smith, P.C. (1986). Timber sheathed walls for wind and earthquake resistance. **Bulletin of the New Zealand National Society for Earthquake Engineering**. Vol. 19. Nº 2.

Fazio, P.P. (1971). Study of modular panelized building. ASCE National Structural Meeting, Maryland, USA.

Gallegos, H. (1989). Seismic resistant low-cost buildings in Peru. Proceedings of the IAHS World Congress on Housing, Oporto, Portugal.

Gallegos, H. (1986). Use of vegetable fibers as building materials in Peru. Proceeding of the RILEM/NCCL/CIB Joint Symposium on the use of vegetable plants and their fibers as building materials, Bagdad, Iraq.

Smith, P.C., Dowrick, D.J. and Dean, J.A. (1986). Horizontal timber diaphragms for wind and earthquake resistance. **Bulletin of the New Zealand National Society for Earthquake Engineering**. Vol. 19. Nº 2.

Smith, R.C. (1963). **Principles and Practice of Light Construction**. Prentice-Hall Inc. New Jersey, USA.

28 THE SCALE INFLUENCE ON LAMINATED GLUED TIMBER BEAMS STRENGTH (GLULAM)

E.V.M. CARRASCO
Structures Department, School of Engineering,
Federal University of Minas Gerais, Brazil

Abstract
The study presented in this paper is a contribution for the scale factor determination on laminated glued timber beams. Such factor is part of the ponderation coefficients for determining the design strength using a dimensioning method based on limit analysis. The aim of this work is to determine the scale factor experimentally. The paper consists of the description of the experimental program, the presentation of the elastic line graphics obtained experimentally and the resulting scale factors or ponderation coefficients ($\gamma m2$). From the obtained results one can conclude that by reducing the scale, the elastic constants and the rupture modulus increase. As a consequence such results obtained using a reduced scale model must be ajusted through the scale factor.
Keywords: Laminated Timber, Glulam, Wood Products.

1 Introduction

Nowadays, the glues laminated timber (glulam) is known as one of the best engineering materials. This fact is due to its high strength, adaptability and durability.

The glued laminated timber has extended the utilization of wood as a structural material, because the glulam can be obtained in almost all sizes and forms. However, its practical utilization must be based on a comprehensive knowledge of its mechanical properties. In the last decades, various related researches were developed: FREAS and SELBO (1954), CHUNG (1964), MOODY (1978) and ABBOTT (1987).

Among the methods proposed to the timber structural elements design the semi-probabilistic and the limit state are the currently most adopted ones. The usage of these methods presupposes the calculus strength determination, which depends on the ponderation factors. Although various international codes and researchers have prescribed ponderation factors to timber structures,the researches concerning the ponderation factors to the glulam timber are not conclusive.

This paper intends to be a contribution to the ponderation factor ($\gamma m2$) evaluation. This factor depends on the size of the specimen in which the mechanical properties were obtained. In order to analyse this problem, a experimental program was conducted. This program consists of evaluating the mechanical properties of glulam beams in various sizes and further correlating the results.

2 Previous works

Only few works were realized concerning the evaluation of the size effect on the mechanical properties of glulam beams. The first approach has dealt with the evaluation of the height effect on stength of timber beams. Although this factor does not precise the size effect on the strength, it is very important in design practice.

According to several references, height and cross-sectional shape effects on timber are well known. This led to the obtention of empirical factors for the evaluation of these effects, which were used in the bending equations. Before the advent of glulam timber, the height of timber beams was limited, then little interest was devoted to this problem. With the progressive utilization of glulam timber, this limit was overcome, and now we must pay attention on the height effect in the design of glulam beams.

Since 1924, various researchers have been interested on the study of the height influence on the mechanical properties of bending elements. Figure 1 presents a comparison among the results of the works of Newllin and Trayer, Dawley and Youngquist, and Bohannan.

In 1924, Newllin and Trayer (apud MASCIA, 1985) derived the equation 1 (figure 1), which represents the empirical relationship between the strength ratio (height factor) Fh and the ratio of the beam d (in inches). These results were obtained in tests of timber beams with 5 cm. of width, 70 cm. of span, and heigth ranging from 5 cm. to 30.48 cm.

In 1947, Dawley and Youngquist (apud BOHANNAN, 1968) derived the equation 2, which extends the evaluation of Fh to height equal to 40.64 cm. These results were published by FREAS and SELBO (1954). Equation 2 was derived from experimental tests of timber beams with a concentrated load in the middle of the span.The small number of specimens and the maximun height (40.64 cm.) of beams studied by Dawley and Youngquist limit the application of these results to the design of glulam beams.

Trying to overcome these problems, BOHANNAN (1966, 1968) derived equation 3, which is based on the strength of materials statistical theory, suggested by Weibull in 1923 (apud BOHANNAN, 1968). In order to assure the validity of the expression derived, Bohannan compared his theoretical results with those obtained experimentally for more than 2000 small glulam beams (maximum height 40.64 cm.) and also large beams. By these results, Bohannan observed a good agreement between theoretical and experimental results, not only to small, but also to large glulam beams.

Equation 3 derived by Bohannan, generally is presented for a standard height 30.58 cm. (12"), ratio height/span 21 and uniformly load, then it assumes:

$$Fh = (12/d)^{1/9} \tag{4}$$

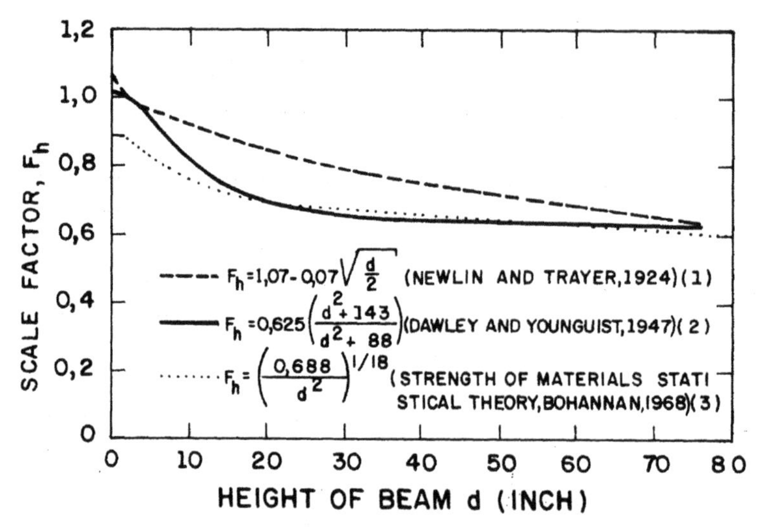

Fig.1. Empirical equations for the height factor Fh, Standard height 5.0 cm. (2"), MANTILLA CARRASCO (1989-a)

The researches developed by KEENAN (1972), MUTTAQIN and TAKEMURA (1981), and MAX and MOODY (1982) are in agreement with previous works. In these researches,various sorts of wood, number and width of the laminae were investigated.

Acording to the works discussed above, it is easily seen that not only height but also span and width influence the streqth of glulam beams. In order to evaluate the size effect on the behavior of glulam beams, a experimental program was conducted.

3 Instrumentation and models

The stantard beam is 11.70 cm x 39.05 cm x 860.00 cm, which was obtained splicing eleven layers, eachone 3.55 cm thick (figure 2). This beam was analysed extensively by MANTILLA CARRASCO (1989-a).

After testing the standard beam, it was divided in two small models, 1:2 and 1:3 scales. These small models were tested and then split in smaller ones, 1:4, 1:5, 1:8 and 1:10 scales. These models were prepared according to the arrangement of layers in the standard beams.

The wood species of the standard beam is the "Pinho do Paranã" |*Araucaria brasiliana*|. The moisture measured in the beams ranges from 12.25 % to 13.16%. The adhesive employed is the resorcinol resin. The Standard beam and the small models were fabricated according to recommendations presented by MANTILLA CARRASCO (1989-b).

To prevent buckling, each glulam beam was braced laterally in five points. The test was performed in two stages, first with the application of a single load at the middle of span and second with two concentrated loads (figura 2). In each stage, the test was performed up to the rupture. The load rate was 10 MPa/min.

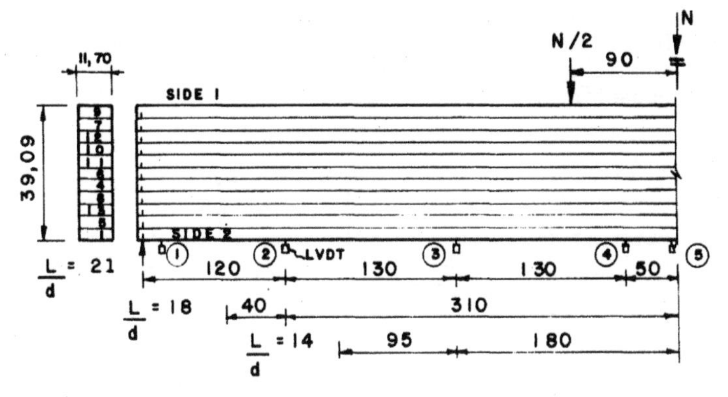

a) Model, scale 1:1 (11.70 cm x 39.05 cm x 880 cm)

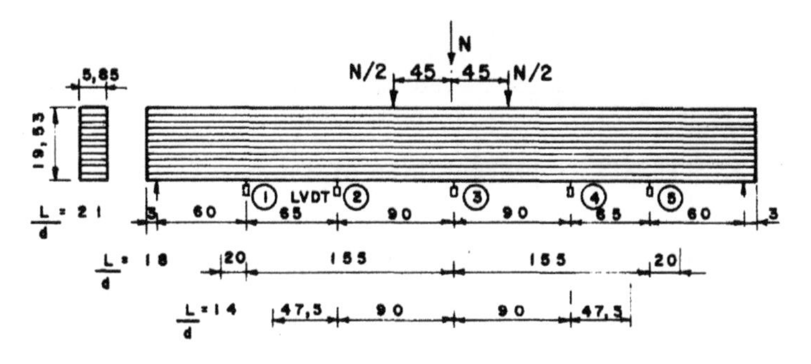

b) Model, scale 1:2 (5.85 cm x 19.53 cm x 436 cm)

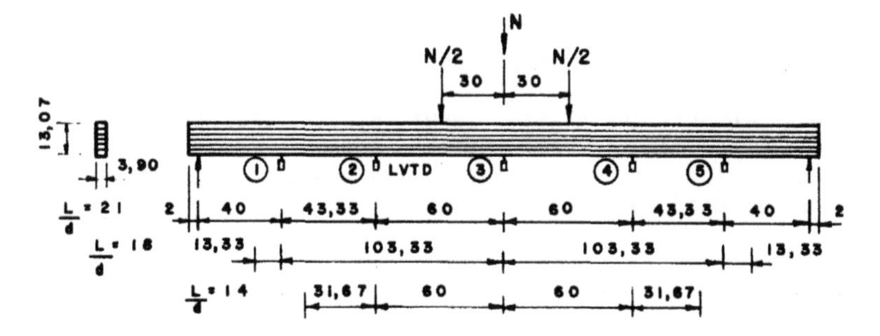

c) Model, scale 1:3 (3.90 cm x 13.07 cm x 290 cm)

Fig. 2. Schematic representation of the test

In order to determine the elastic constants, the rupture modulus and the elastic curves, the deflections were read off from displacement transducers (LVDTs). The LVDTs were set according to figures 2 and 3, i.e., 14 LVDTs in the standard beam; 5 LVDTs in 1:2 and 1:3 models; 1 LVDT in 1:4, 1:5, 1:8 and 1:10 models. The deflections were obtained up to the proportional limit.

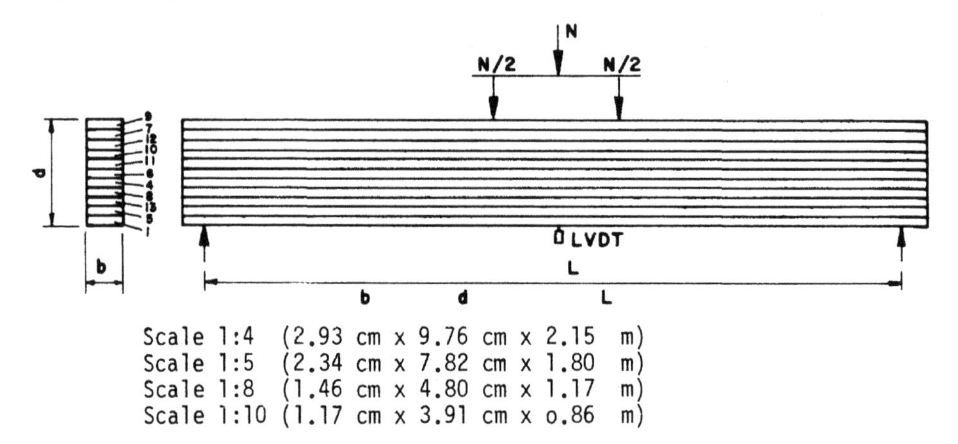

Fig.3. Schematic representation of the model's test

4 Data interpretation and discussions

The beams in 1:1, 1:2 and 1:3 scales were tested twelve times; six times with a concentrated load in the middle of the span, and six times with two concentrated loads. The l/d ratios adopted were 21, 18 and 14.

The beams in 1:4, 1:5, 1:8 and 1:10 were tested eight times; four times with a concentrated load in the middle of the span, and four times with two concentrated loads. The l/d ratios adopted were 21 and 14.

In each set of tests, half of them were performed changing the upper to the lower side position.

Figure 4 presents the elastic curves plotted from the deflections measured from the elastic curve of the beam with l/d equal to 21. In this calculus, the shear contribution was neglected, and a linear regression was used (equations 4 and 5).

$$v = \frac{L^3}{48EI}\left(3\left(\frac{x}{L}\right) + 4\left(\frac{x}{L}\right)^3 + \ldots\right)N \quad ; \quad v = \Theta N \tag{4}$$

$$\text{where:} \quad \Theta = \text{regression constant} = \frac{L^3}{48EI} \rightarrow E = \frac{L^3}{48L\Theta} \tag{5}$$

The shearing modulus of elasticity, G, was evaluated from the elastic curve of the beam with l/d equal to 14 and the elastic modulus E obtained from equation 4. In this calculus, a linear regression was used and the shearing coefficient was set equal to 2.387 (MANTILLA

CARRASCO, 1989-a), equations 6 and 7.

$$v = \left(\frac{L^3}{48EI}\left(3\left(\frac{L}{x}\right) + 4\left(\frac{L}{x}\right)^3 + \ldots\right)N + \frac{\alpha c}{2GA}\right) N \quad ; \quad v = kN \rightarrow N = \Theta v \quad (6)$$

For x= L/2, then:
for 1 load,

$$\Theta = \frac{1}{\frac{L^3}{48EI} + \frac{\alpha cL}{4GA}} \rightarrow G = \frac{\alpha cL}{\left(\frac{1}{\Theta} - \frac{L^3}{48EI}\right)4A}$$

for 2 loads,

$$\Theta = \frac{1}{\frac{L^3}{48EI}\left(\left(3\frac{a}{L}\right) - 4\left(\frac{a}{L}\right)^3\right) + \frac{\alpha cL}{4GA}} \rightarrow G = \frac{\alpha cL}{\left(\frac{1}{\Theta} - \frac{L^3}{48EI}\left(\left(3\frac{a}{L}\right) - 4\left(\frac{a}{L}\right)^3\right)\right)4A} \quad (7)$$

Equations 4 and 6 provide deflections as a function of a generic load N. Equations 8 and 9 are derived from these two equations for a single concentraded load in the middle span, and two symmetrical loads, respectively. The least regression coefficient obtained was 98%.

2 loads, l/d= 21, span 424 cm, small model scale 1:2

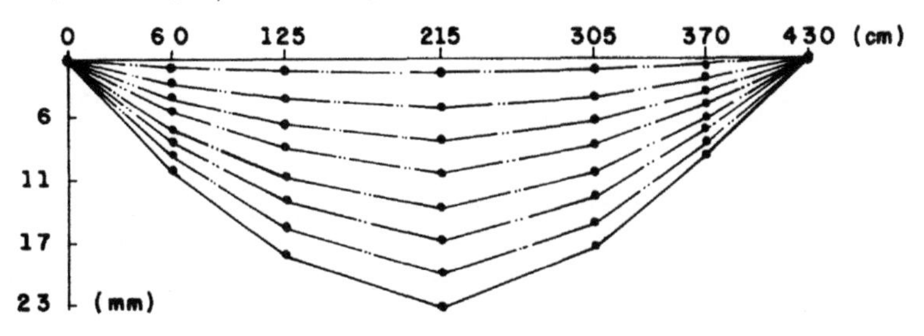

1 load, l/d= 21, span 282 cm, small model scale 1:3

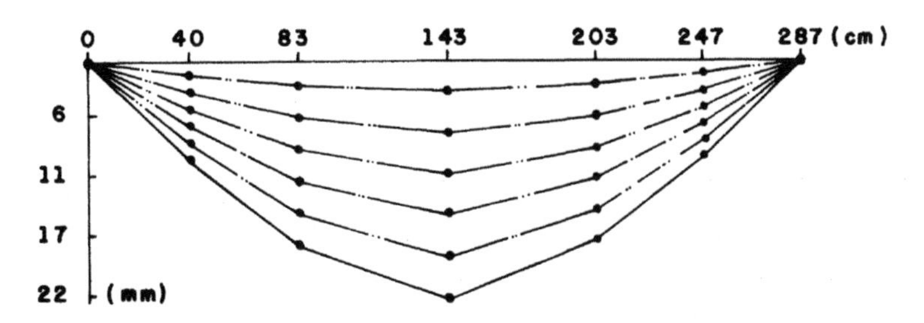

Fig.4. Elastic curves of the glulam beams

When a sigle load N is applied, the elastic curve is provided by the following equation:

$$\left.\begin{aligned} v &= \frac{NL^3}{48EI}\left(3\frac{x}{L} - 4\left(\frac{x}{L}\right)^3\right) + \frac{\alpha cN}{2GA}\,x & ; \quad & 0 \le x \le L/2 \\ v &= \frac{NL^3}{48EI}\left(3\frac{x}{L} - 4\left(\frac{x}{L}\right)^3 + 8\left(\frac{x}{L} - \frac{1}{2}\right)^3\right) + \frac{\alpha cN}{2GA}(L-x) & ; \quad & L/2 \le x \le L \\ v_{max} &= \frac{NL^3}{48EI} + \frac{\alpha cN}{4GA}\,L & ; \quad & x = L/2 \end{aligned}\right\} \quad (8)$$

When two symmetrical loads N/2 are applied, the elastic curve will be provided by:

$$\left.\begin{aligned} v &= \frac{Na^3}{12EI}\left(3\left(1+\frac{b}{a}\right)\frac{x}{a} - \left(\frac{x}{a}\right)^3\right) + \frac{\alpha cN}{2GA}\,x & ; \quad & 0 \le x \le a \\ v &= \frac{Na^3}{12EI}\left(3\left(1+\frac{b}{a}\right)\frac{x}{a} - \left(\frac{x}{a}\right)^3 + \left(\frac{x}{a}-1\right)^3\right) + \frac{\alpha cN}{2GA}\,a & ; \quad & a \le x \le a+b \\ v &= \frac{Na^3}{12EI}\left(3\left(1+\frac{b}{a}\right)\frac{x}{a} - \left(\frac{x}{a}\right)^3 + \left(\frac{x}{a}-1\right)^3 + \left(\frac{x}{a}+\frac{b}{a}-1\right)^3\right) + \frac{\alpha cN}{2GA}(L-x) & ; \quad & a+b \le x \le L \\ v_{max} &= \frac{NL}{48EI}\left(3\left(\frac{a}{L}\right) - 4\left(\frac{a}{L}\right)\right) + \frac{\alpha cN}{2GA}\,a & ; \quad & a \le x \le a+b \end{aligned}\right\} \quad (9)$$

Table 1 lists the elastic and rupture moduli evaluated in the experimental program. Analysing the results, it can be concluded that smaller the scale, greater the elastic and rupture moduli. Among the possible causes to this fact, we can cite:

Greater stiffness of adhesive.
Reducing the scale, the ratio adhesive/wood increases.
Greater homogeneity in the small beams.

5 Scale factor

The scale factor was computed by the ratio between the mechanical properties in the standard beam and the mechanical properties in the small model. These relationships were plotted in figures 5a, 5b and 5c, which represent the scale factor for the rupture, elastic and shear moduli, respectively.

The scale factor Ft is defined as the ratio between the mechanical properties in the small model and the mechanical propertie in the standard beam. This factor increases when the scale decreases, reaching 10% for E, 15% for G and 40% for MR.

Table 1. Mechanical properties of glulam beams, scales 1:1 to 1:10

Scale	Test	Side	L/d	1 Load		2 Loads		MR (MPa)
				E (MPa)	G (MPa)	E (MPa)	G (MPa)	
1:1	1	1	21	14950	--	15060	--	85.35
	2	1	18	14654	436	14654	583	--
	3	1	14	14222	1235	13521	1338	--
	4	2	21	14325	--	14932	--	--
	5	2	18	13927	629	14606	385	--
	6	2	14	13686	1321	13778	1219	--
1:2	1	1	21	15340	--	15712	--	89.51
	2	1	18	14409	635	14833	728	--
	3	1	14	12921	1235	13500	1267	--
	4	2	21	14481	--	14829	--	--
	5	2	18	14188	--	14847	--	--
	6	2	14	12795	1188	13911	1229	--
1:3	1	1	21	15430	--	15998	--	97.93
	2	1	18	15187	322	15935	115	--
	3	1	14	13586	1280	13995	1236	--
	4	2	21	15243	--	15505	--	--
	5	2	18	15620	496	14425	586	--
	6	2	14	13140	1190	13830	1135	--
1:4	1	1	21	15950	--	16130	--	97.35
	2	1	14	13546	1296	13135	1276	--
	3	2	21	14950	--	15936	--	--
	4	2	14	12196	1201	13785	1198	--
1:5	1	1	21	16223	--	16753	--	93.66
	2	1	14	14183	1263	14639	1218	--
	3	2	21	15439	--	15395	--	--
	4	2	14	13925	1225	13785	1242	--
1:8	1	1	21	16115	--	16453	--	115.16
	2	1	14	13295	1336	14103	1315	--
	3	2	21	15325	--	15135	--	--
	4	2	14	12986	1256	13985	1242	--
1:10	1	1	21	16438	--	16843	--	121.18
	2	1	14	14136	1395	14386	1388	--
	3	2	21	15935	--	16030	--	--
	4	2	14	13625	1199	13789	1256	--

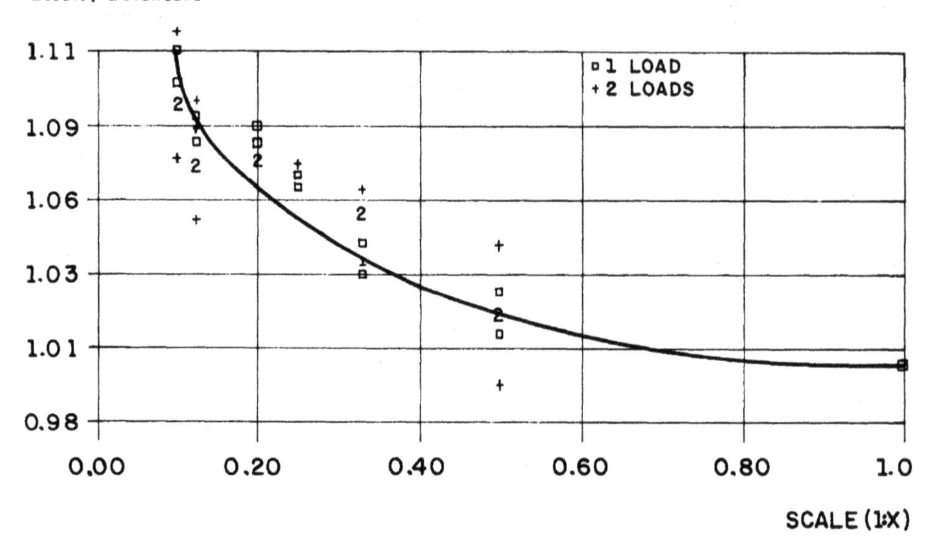

a) Elastic modulus (E) scale factor

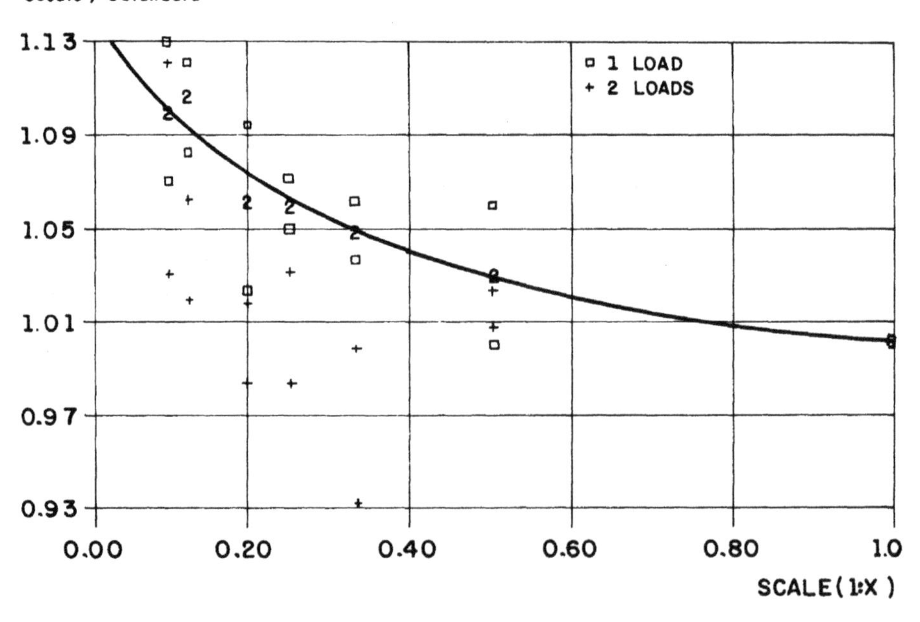

b) Shearing modulus of elasticity (G) scale factor

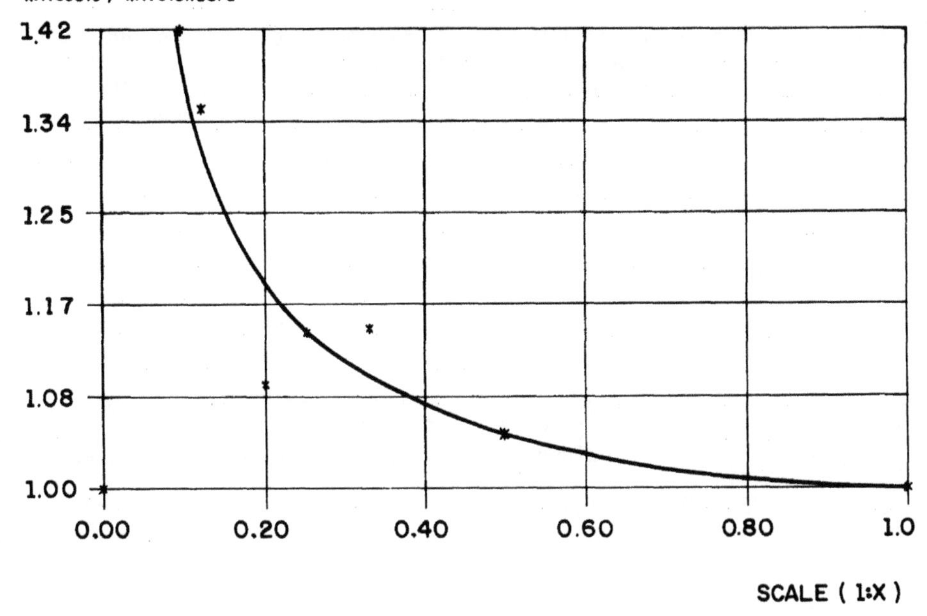

c) Rupture Modulus scale factor (MR)

Fig.5. Scale factor for the elastic constants

6 Conclusions

The mechanical properties of laminated timber beams depends on the specimen size in which they were evaluated. This effect is due to the adhesive stiffness and the greater homogeneity of small models. The size effect is measured by the size factor Ft, shown in figures 5a, b and c.

It is important to realize that these results are limited to the kind of timber and scales studied in this work.

7 References

ABBOT, A.R. and WHALE, L.R.J. (1987) An overview of the use of glued laminated timber (glulam) in the UK. **Construction & building materials**, 1 (2):104-110.

BOHANNAN, B. (1966) Effect of size on bending strength of wood members. **USDA-FS-FPL**. Mad., wis., (Res. Pap. FPL 56)., 26p.

______ (1968) Structural Engineering research in wood. **J. Struc. Div.** 94(02): 403-415.

CHUGG, W.A. (1964) **Glulam**: The theory and practice of the manufacture on glued laminated timber structures. London, Ernest Benn. 423p.

FREAS, A.D. and SELBO, M.L. (1954) Fabrication and design of glued laminated wood structural members. **USDA-FS-FPL**, Wash, technical bulletin 1069. 220p.

KEENAN, F.J. (1972) Shear strength of wood beams. **FPJ**. Mad. Wis. 24(09) 63-70.

MANTILLA CARRASCO, E.V. (1989-a) Resistência, elasticidade e distribuição de tensões das vigas retas de madeira laminada colada (MLC). **Tese Doutorado**, EESC-USP, 486p.

______ (1989-b) Fabricação de peças de madeira laminada colada. **III Encontro brasileiro em madeiras e em estruturas de madeira**, 35p.

MASCIA, N.T. (1985) Contribuição ao estudo da flexão estática em peças de madeira. **Dissertação de Mestrado**. EESC-USP., 308p.

MARX, C.M. and MOODY, R.C. (1982) Effect of lumber width and tension laminated beams. **FPJ.**, 32(01), 45-52.

MOODY, R.C. Improved utilization of lumber in glued laminated beams. **USDA-FS-FPL**. Mad., Wis., (Res. Pap. FPL 292), 48p.

MUTTAQUIN; Z. and TAKEMURA, T. (1981) Effects of beam size on flexural properties of glued laminated Jelutong beams: a case of compatatively uniform distribution of laminae. **J. Japan wood res. soc.** 27(04) 283-289.

29 DURABILITÉ DES COMPOSANTS EN BOIS DANS HABITATIONS À LOYER MODERÉ

(Durability of Wood Composites in Low Cost Housing)

A.C.Q. MASCARENHAS and M.J.A. SANTANA
Polytechnic School, Federal University of Bahia, Salvador, Brazil

Resumé
À Salvador, Bahia, il ya une réduite tradition de l'emploi du bois comme composant de cloison aux habitations et constructions en général et quand'il arrive c'est dans les oeuvres provisoires ou envisageant courte durée. Au délà de ces cas, dans sous-habitations aux bidonvilles ou "favelas", le bois et les panneaux à base de bois sont employés n'importe quelle façon, comme il arrive pour d'autres matériaux.

En fonction d'être une des rares expériences de l'emploi "rationel" du bois dans habitations chez nous, a été évalué l'ensemble des maisons du "Horto Florestal", au quartier "Retiro". Ont été observées les conséquences des attaques des xylophages au bois mis en oeuvre, dans différentes conditions.

Les conclusions se ont appuyées sur des aspects tels que la technologie du bois, des agents destructeurs, des conditions de mise en oeuvre et de conservation, des projets architectoniques et d'autres.
Keywords: Bois, habitation, durabilité, xylophages.

1 Introduction

La carence habitationelle au Brésil s'a agrandit année par année, sans que des mesures compatibles ont éte prises pour résoudre cette question. Un grand numéro de facteurs tels que la politique gouvernementalle, l'utilisation du sol, la spéculation immobilière, le finnancement, le niveau des salaires de la population moins riche ou pauvre, les matériaux disponibles, et les besoins ou éxigences des utilisateurs. L'interaction de ces aspects se pose dans le difficile processus de solution, des quels, les deux derniers attirent particulièrement l'attention dans cette analyse.

Sur les matériaux disponibles on pourait dire que le Brésil a des nombreux recours naturels possibles d'être appliqués à la production d'unités habitationelles, plusiers desquels avec technologie controlée. Le bois est d'entre eux, et même sans avoir un étude complet des essences disponibles avec un certain potentiel d'emploi, on a déjà quelque aptitude technologique à appuyer des actions dans ce sens. Malgré l'emploi de ce matériau dans les bâtiments, il est toujours limité dans les spécifications construtives, à cause des coûts, de la "disette" et de la durabilité.

Les éxigences des utilisateurs, selon la Norme ISO-DP 6241 (1), quatorze sont les conditions qui doivent être remplies par les bâtiments:

a) securité structurale
b) securité au feu
c) securité pour l'utilisation
d) confort hygro-thermique
e) pureté de l'air
f) étanchéité
g) confort visuel
h) confort acoustique
i) confort tactile
j) confort anthropodynamique
k) hygiène
l) adaptation à l'utilisation
m) durabilité
n) économie

D'entre celles-là, la durabilité, c'est à dire, la capacité qui a un produit, un composant ou encore un bâtiment de résister à la degradation, est fréquemment considerée comme l'éxigence plus faible remarquée aux éxperiences dans les maisons en bois bâties au Brésil.
La présente analyse qu'on a efectué sur des maisons, toutes ayant le même projet et employant les mêmes materiaux, bâties dans un esemble habitationel à 1964, a envisagé étudier la durabilité du bois mis en oeuvre comme composant de la habitation.

2 Caractéristiques des maisons

Les maisons, cent unités, ont été implantées sur terrain avec declivités, au quartier "Retiro", dans Salvador, capitale de Bahia. Les lots, de dimensions diférentes entre eux, abritent des constructions de 6,0m x 6,0m, à peu près, composées par deux chambres, salle, cuisine, salle de bain et véranda. (Voir Fig.1)
Les fondations, traverses de chémim de fer usagées, ont été les piquets qui supportaient les maisons ponctuellement. Les bâtiments gardent une séparation stratégique du sol, envisageant les aérer et les protéger contre l'humidité, au delà des éloingments entre eux, que les défendent contre la possibilité de propagation du feu, en cas d'un incendie.
Le plancher en bois est assis sur pièces en bois équarries, à l'exception de la cuisine et de la salle de bain, où on a employé le mortier de ciment teint de rouge, pour lui donner meilleur condition d'utilisation dans la zone hydraulique de maison.
Les parois ont été formées par un squelette de bois récouvert à l'extérieur par planches dans la direction horizontale de 13,5cm x 2,0cm et à l'intérieur par pièces de 13,5cm x 1,0cm. Toutes les planches se rejoignent par rainure et languette, de telle façon qui offrent etanchéité à l'eau. Dans la cuisine et la salle de bain les parois ont été bâties en maçonnerie de blocs de sable et ciment, jusqu à l'hauteur

Fig. 1. Vue partielle de l'ensemble

de 1,50m, complémentées par cloisons semblables aux autres décrits, en bois, jusqu'au plafond.

Le projet a defini un seul type de maison, couverte en deux pentes, en deux niveaux, sur le corps et la véranda. La structure portant de la toiture est composée par pièces de bois de différentes sections transversales soumises à flexion.

Les tuiles sont du type françaises ("francesas"), fabriquées d'une mélange de ciment et sable, à l'exception de la véranda, où elles sont du type ondulée, en amiante-ciment.

Toutes les maisons avaient plafond en planches minces (1,0cm) de bois, inclusivement les bords, qui se projetent 0,50m hors des parois sur la façade et les latérales.

L'essence de bois utilisée, le Pin du Paraná (Araucaria angustifolia), ne révèle pas des signes évidents d'avoir reçu traitement pour la préserver contre les xylophages soit par le créosote ou par solution hydrosoluble. Selon les plus anciens propriétaires, dépuis la construction à peine les parois, dans la face extérieure, ont reçu peinture de tinte huilleuse, ce qu'ils considèrent comme "l'unique traitement sur plusiers maisons jusqu'aujourd'hui".

3 Méthodologie d'évaluation

La méthodologie utilisée pour évaluation de la durabilité du bois employé dans la construction des maisons référées, s'a attachée à l'inspections sur terrain.

Le bois a été observé mis en oeuvre, dans la condition réelle d'emploi, c'est à dire, sujet à toutes variations qui ont arrivée aux maisons. Comme il ne s'agissait pas d'un test, on a pu vérifier "in loco" l'action de dégradation de diverses origines et entendre, des occupants ou propriétaires, presque tout qui s'est passé aux bâtiments.

On a consideré que la vie utile - 25 années - déjà ateinte par la majorité des maisons, confère à la methode d'inspection sur terrain confiabilité pour définir la durabilité du bois employé.

Seulement les aspects architectoniques et de dégradation par des agents biologiques ont été envisagés, parce que les incendies, et d'autres causes n'ont pas eu importance: une maison, selon les habitants, a brulé, en conséquence d'un problème dans un appareil de télévision, il n'y a pas longtemps.

4 Identification et analyse des problèmes

Les difficutés pour la conservations des immeubles ont eu plusiers origines, parmi lequelles, soit par sa fréquence, soit par sa intensité, d'aspects architectoniques et de destruction par xylophages ont mérité être détachés et évalués.

4.1 Aspects architectoniques

En dépit du parti architectonique, très usé au sud du pays, quelques détails méritent être commentés, parce qu'ils favorisent l'action des agents de dégradation.

Le bord prolongé de la toiture, qui entoure toute la maison, n'a pas bien accompli une de ses fonctions plus importantes: proteger les parois contre l'incidence de la pluie sur elles, ce que permet l'augmentation de l'humidité des pièces. En raison d'avoir 0,50m de largeur, il n'a pas pu défendre les parois, comme il fallait, permettant le dégât prématuré des pièce em bois. (Voir fig. 2)

La projection de la toiture principalle sur la véranda a offert aussi des facilités pour l'action des eaux pluviaux sur les composants des maisons.

On a remarqué aussi que la végetation ornamentalle, bien au goût de plusiers habitants, a contribué pour la destruction de l'immeuble, soit cassant des tuiles, soit humidifiant, par contact, las parois et encore étant chémin pour quelques infestations.

Un autre détail construtif important et prévu en projet a été l'éloignement des planches du sol, qui, au délà de l'aérer, distanciait le bois des termites souterrains.
Cependant, cet aspect a été négligé par les propriétaires, certainement par méconnaissance, les quels ont éléves maçonneries de blocs du sol jusqu'aux solives du plancher, cherchant profiter les espaces sous celui. De cette façon des conditions pour la propagation des xylophages ont été facilités. (Voir fig.2)

4.2 Dégradation par xylophages

Dans ce cas-là la durabilité de bois a éte profondément réduite par quelques agents biologiques par raisons qui seront exposées à la suivre.

Fig. 2 - Maçonnerie sous la maison, le bord endommagé et ampliation en blocs.

Champignons - ils ont actué fréquemment sur les zones inférieures des parois externes, du à la retention de l'eau là, où on vérifient diverses remplacements de planches par les maçoneries de blocs.

Sur d'autres régions on a pu remarquer l'action des champignons lignivores, comme sur les pièces de support et finition de la toiture, dans les bords, en conséquence des gouttelettes d'eau et aussi par casse des tuiles, ce qui met en évidence la précaire conservation des unités habitationnelles.

On n'a pas faite aucune identification des champignons trouvés.

Insectes - il y a eu des attaques generalisés par les deux types de termites de bois sec, representées par les Cryptotermes et souterrains, à travers des Nasutitermes. Ces genres ont une élévée incidence sur les bois à Salvador. Les infestations des termites du bois sec sont remarquables sur divers composants construtifs, tels que parois, planchers, menuiseries, planches des plafonds etc. Les galeries en direction parallèle à la axiale du bois, les excréments et l'absence des particules du sol ont bien evidencée cette infestation.

Les termites souterrains on été définies par ses "nasutes", par les

Fig. 3 - Des bords et tuiles detruits.

galeries extérieures au bois couvertes par grains du sol et excréments melangées, au délà des colonies secondaires, situées dans locales élévés des bâtiments.

Comparativement, d'une façon qualitative, les termites de bois sec ont été responsables par la majorité des dégats aux pièces des maisons, y provoquant des problèmes qui conduisent à l'éxigence de remplacement d'innombrables pièces et pour ça les habitants ont recouru à demolitions ou démontages d'autres immeubles du même ensemble. D'autre manière, le replacement est fait par materiaux différents, tels que des blocs, du mortier et du béton. Les termites de ce type sont plus difficiles d'être aperçues que de l'autre, surtout par des laïcs, parce qu'elles n' exibent pas des symptomes très évidents, ce que va contribuer pour sa dissemination et, conséquemment, plus forte destruction des maisons.

Les termites, par sa capacité de voler aux périodes de reproduction, augmentent leur rayon d'action et cette caractéristique contribue pour l'ampliation des endommagements et problèmes.

Il faut observer qu'il n'y a pas aucune orientation aux habitants pour conserver le bois, materiau prédominant dans les maisons, ça amène évidemment à une destruction plus accelerée.

5 Conclusion

Rares sont les applications , à Salvador, du bois comme materiau de védation dans les habitations et bâtiments em général. La pluspart des utilisations arrive aux solutions de caractère provisoire, comme dans les clôtures et chantiers de construction ou encore d'une façon desordonnée aux sous-habitations des bidonvilles. On peut affirmer que,

Fig. 4 - Destruition de pièces de la façade par termites de bois sec.

culturellement, Salvador n'a pas la tradition de l'utilisation du bois pour la production d'habitations, puisque on fait toujours l'association de ce type d'habitation à ceux des bidonvilles.

Face aux commentaires du paragraphe précédent, l'ensemble des cent maisons, des quels on a décrit aspects variés, présente une particularité: a été bâti pour résidence de fonctionnaires de la Mairie de Salvador, dans une expérience innovatrice et même, pourquoi ne pas dire, bien succedée.

Quoique projetée et executé sur bases technologiques, l'ensemble habitationel présente des défauts construtifs qui sont agravées par le fait d'être le bois le principal materiau. Nonobstant la vie utile que plusiers de ces maisons ont atteint (25 années jusqu'aujourd'hui), encore en bon état, réfléchie par le bon aspect et l'utilisation pleine, pourait être plus longue si des informations pour l'entretien fussent fournies aux propriétaires.

Par ce qu'on a pu vérifier visuellement et par entrevues avec les habitants, les pièces em bois n'ont pas reçu aucune protection par produits de préservation préventivemente, au cours de la production des maisons, ou même de façon curative dès les premiers signes d'infestation, soit par champignons, soit par insects. Ces défaillances ont contribué pour la réduction du période d'utilisation de l'immeuble.

La projection de la toiture du corps de la maison sur celui-là de la véranda, leurs bords étroits, le type des tuiles (lourdes et sans possibilité de réposition-on ne les trouve plus sur le marché) ont difficulté, d'un coté, la conservation de la couverture en bon état, et d'autre, ont facilité l'action des champignons qui endommageaient le bois.

Le vide sous le plancher, qui eloignait l'habitation du sol, en plusiers cas, a été éliminé, permettant l'indésirable accès des termites et le dévéloppment plus facile des champignons.

A dépit de la manque des recommendations techniques pour la conservation, il y a à considerer le désire de la population but, c'est à dire, rêver à sa maison de "construction", ce qui vaut dire, de maçonnerie de blocs, si possible y compris le béton armé. Ça a stimulé l'idée de plusiers habitants de ne pas entretenir leurs bâtiments en bois, envisageant les reformer avec d'autres matériaux. On a constaté, dans quelquer cas, le remplacemente partiel des pièces endommagées, sur d'autres, aucun remplacement et, à peu prés, 30% des maisons ont été déjà reconstruites en autres materiaux.

L'analyse des problèmes, evidencés à travers de l'inspection sur terrain, a amené à la conclusion que les composants en bois, employés dans la construction des cent unités, ont demonstré, dans conditions réelles d'utilisation, nonobstant les degradations sur quelques elements, que, s'ils eussent reçu des réparations, pouraient avoir vie utile plus longue.

D'autre coté, l'espèce forestière employeé, Pinho do Paraná (Araucaria angustifolia) a masse volumique moyenne (0,54 g/cm^3), est sensible aux agents xylophages et présente faible durabilité par rapport aux autres bois feuillus usés à Salvador. Tout ça suggère que des panneaux (contreplaqués et de particules) traités preventivement

Fig. 5 - Le premieré maison bâtie. Des signes d'attaques de termites sous fenetre vingt cinq annëes d'utilisation.

soient égallement viables et plus économiques pour l'emploi aux habitations.

Des exemples, comme ce qu'on présente, méritent être divulgués pour que au sein de la population puisse avoir un changement de pensée dévant l'immeuble en bois. Quelques familles, qu'aujourd'hui racontent l'histoire du dêbut de la construction de l'ensemble examiné, en rêvant à la maison de blocs, n'ont pas remarqué que, là, à maison en bois, ont vécu deux générations et que si entretenues eussent, ces maisons, sûrement, abritteraient la prochaine génération, comme la cas de M. Manoel Gomes, habitant de la première maison, que actuellement est résidence de ses fils et de de ses petit-fils. (Fig. 5)

6 References

ISO-D P 6241 London, 1970 apud Souza, Roberto in: **Avaliação de Desempenho Aplicada a Novos Componentes e Sistemas Construtivos para Habitação.** Tecnologias de Edificações, pg 531, Ed. Pini: IPT - São Paulo 1988.

30 RESEARCH ON ELEMENTS OF A ONE FAMILY LOW-COST BUILDING SYSTEM

Z. MIELCZAREK
Technical University of Szczecin, Szczecin, Poland

Abstract
This paper gives the short recapitulation of the researches on prefabricated wall panels designed for one - family low cost building system in Poland. The concept of this system assumes maximum ulitization of non - classified wood and wood waste. Wall elements designed as panels of wooden framework made of small dimension wood. Several hundreds of houses have been constructed in this system in Poland. The tests proved satisfactory properties of plates as far as thermic insulation is concerned. Load bearing capacity and stiffness of plates depend on their type. Everage ultimate loads under compression where 591 kN for full plates, 365 kN for plates with window openings and 273 kN for door openings. We found out stiffness of plates with door openings was only 2,2% and with window openings 14,1% in relation to full plates.

1 Introduction

Difficult economic situation in Poland and shortage of popular building materials such as wood, bricks and light concrete elements make it necessary to search for new, cheap technologies which make use of local raw materials and industrial wastes. One of the attempts to solve the problem is the conception called the, Domino system. The system assumes maximum utilization of small dimensions wood and wood wastes.

Several hundreds of detached houses have been constructed in this system in Poland. Visual inspections of the houses having been used for several years not disclosed any significant deficiencies, neither the dwellers have any complaints as to the living conditions and besides they are of opinien that these are houses with good thermic insulation.

As this type of building is implemented in Poland by a small company, which is not provided with own material testing laboratory and neither has higly qualified technical engineers, there arose necessity of extensive research work which might enable to eliminate possible mistakes, the system modification and prolonged the object durability.

2 Characteristic of the System

Wall elements have been designed as panels with wooden framewark made of small dimension wood of diameter 8 to 12 mm. Practically those are young trees obtained during making glades in young forets. This small dimension wood are obtain very inexpensiwely, so that the framework price is determined mainly by transport and labour costs. This framework is, after impregnating, filled with sawdust or wood chips concrete, while mixing the compound, calcium chloride is added to concrete mixer to impregnate the abatement and sawdust. The framework of the wall panel with window opening is shown in Fig.1a and with door opening in Fig.1b.

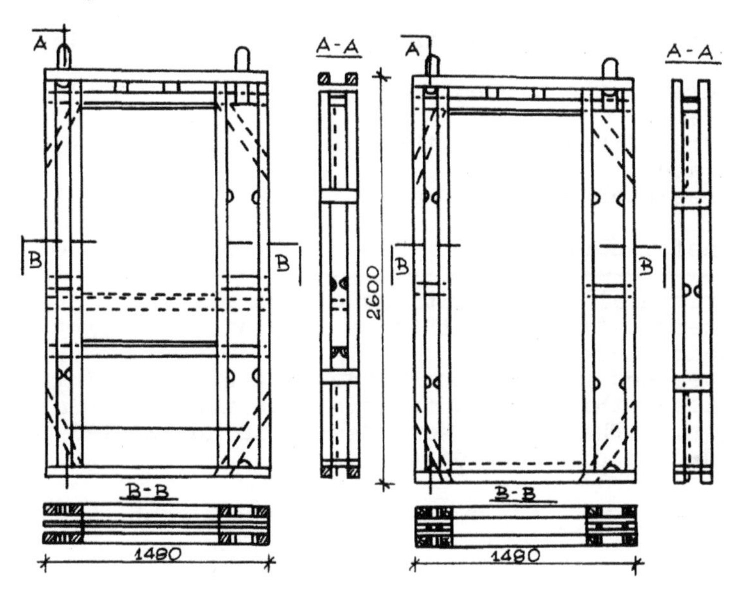

Fig.1. Plates with window and door openings

Roof construction consist of wood - board trussed rafters joined by means of nails.

3 Investigations on some properties of small dimensions wood

At the Technical University of Szczecin a great number of tests of strength - elasticity properties have been car-

per - constantan composite.
Experiments lasted for 10 days. Ten day period was indespensable to establish the flow of heat flux through the partition. During this time the temperature of the inner and outer surface was registred. Wall elements where investigated at the same humidity level e.g. after partly drying for about for weeks in temperature 60 C. After completing the experiments some specimens of saw-dust and chip concrete were exstracted from the plates to determine their density and humidity.

Results of the tests are presented in table 2.

Table 2. Results of investigations of heat transfer coefficient

Specification	Wall element No	
	I	II
Temperature of the air on the warm side, C	21,3	22,1
Temperature of the air on the cold side, C	-16,2	-16,4
Temperature of the air on the warm side surface	17,9	18,3
Temperature of the air on the cold side surface	-13,7	-13,7
Thermal resistance od the wall, m^2 k/W	1,62	1,54
Heat transfer coeficient	0,56	0,58

4 Experimental investigations of the plates

4.1 Determination of stiffness of plates shearing in their plane.

Full plates and plates with window and door openings were investigated, three units of each type i.g. 9 plates altogether. The plates were tested in horizontal position loaded in diagonal direction

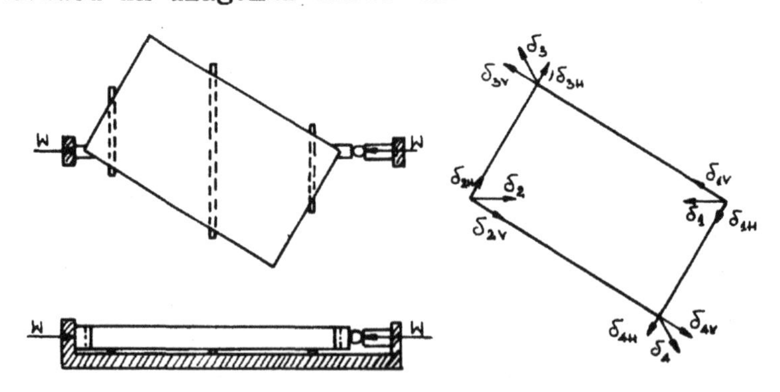

Fig. 2. Scheme of experiments

ried out. We have investigated the following basic properties
density with some moisture content
compression strength parallel to the grain
bending strength
modulus of elasticity for tension acting parallel to the grain
modulus of electicity for tension acting parallel to the grain modulus of elasticity for bending

Results of these experiments were presented at the International Sympozjum in Bagdad in 1986 . The work was continued. Such properties as small dimensions wood as shear strength, strength and elasticity when compression is acting perpendicularly to the grain were tested. Results of these tests are shown in table 1.

Table 1. Strength of nonclassified wood on shear and compression acting perpendialarly to the grain

Kind of properties	Everage value	Boundary value		Interval of confidence	standarts deviation
		min	max		
Shear strength parallel to the grain R_{dv} MPa	8,16	5,1	11,0	/7,33;8,59	1,212
Compression strength perpendicularly to the grain $R_{dc\ 90}$	4,98	2,6	7,9	/4,49;5,47	1,472

4 Tests of thermic insulation of wall panels
Two wall panels where prepared for the tests. Each panel consisted of wooden framework filled with sawdust or chip concrete. Dimensions of the plates were: 1500 x 250 x heigh of the story. It was necessary to cut the height of plates to 1500 mm due to the size of conditioning room. The elements were marked as I and II.

Investigations on thermic insulation of wall elements where parformed in conditioning room with constant air temperature. The measurments were made on the cold and warm side of the panel.

Tests of heat flux density where carried out in conditions of stationary heat flow using 50 x 50 cm thermic mea surment facilities. Temperature of the air and panel sur faces was measured by thermoelectric couples made of cop-

The loads for the purpose of experiments were matched so, as to make the material work within the elasticity limits. It make possible to use the same plates to further tests under vertical load. Diagrams in Fig.3 show relation between the compression force H acting along the diagonal line and deformations at the some points of the plate.

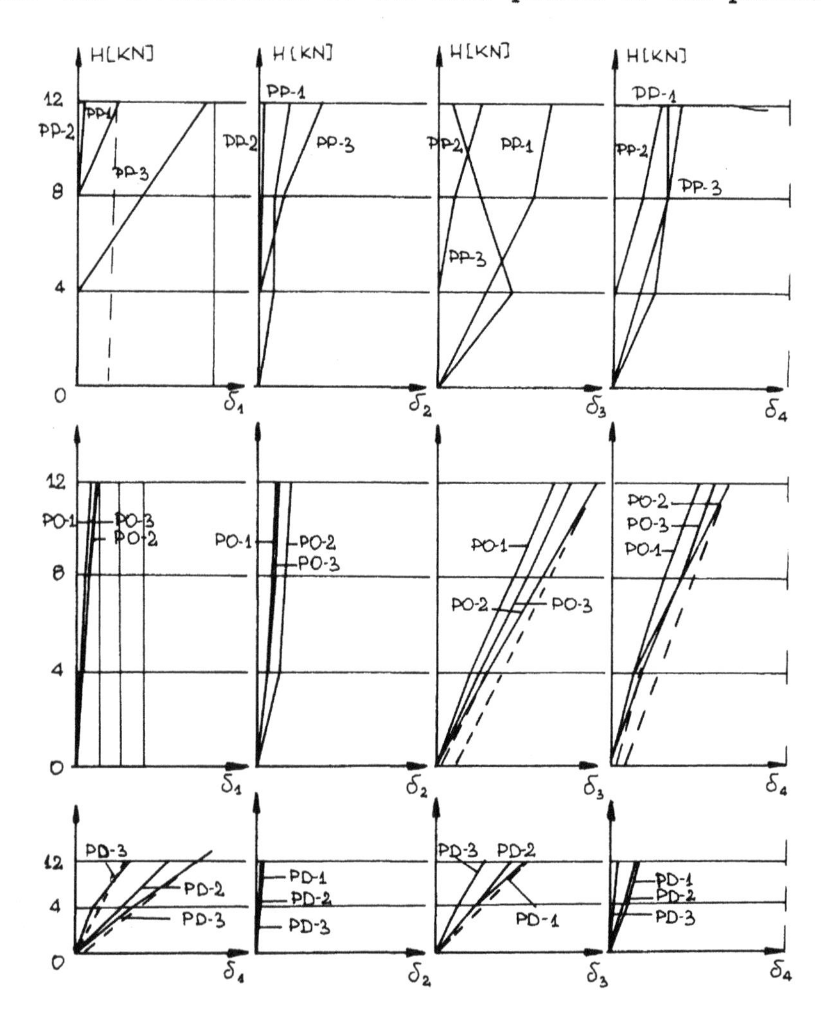

Fig.3. Relations between loads H and deformations δ_i

Average values of deformations of full plates and plates with door and window openings are presented in table3

Table 2. Mean values of deformations

Type of plates	Measured deformations mm			
	1	2	3	4
Full plates [1/]	0,42	0,22	0,70	0,42
Plates with window openings [1/]	0,76	1,02	6,08	4,6
Plates with door openings [2/]	12,57	0,056	10,35	2,83

1/deflection caused by load H = 12 kN

2/deflection caused by load H = 4 kN

Values of the modified shearing modulus are setting in the table 3.

Table 3. Mean values of the modified shearing modulus

Type of plates	Modified shearing modulus G	
	G MPa	G %
Full plates	23,440	100,0
Plates with window openings	3,312	14,1
Plates with door openings	0,533	2,2

4.2 Determination of vertical ultimate load

In Fig.4 there are schematically presented types of plates investigated on the vertical load. On this figure there are marked points /and directions/ where deformations of plates caused by compression vertical load were measured

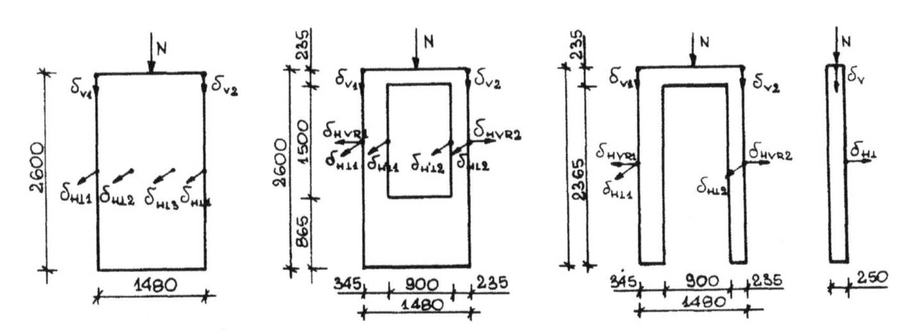

Fig 4. Points where deformations of the plates where measured.

Values of ultimate loads for each plates are given in the table 4.

Table 3. Ultimate loads of tested plates

Symbol of the plate	Full plates kN	Symbol of the plate	Plates with window openings	Symbol of the plate	Plates with door openings
PP-1	507	PO-1	345	PD-1	285
PP-2	667	PO-2	420	PD-2	272
PP-3	600	PO-3	330	PD-3	261

5 Conclusions

On the basis of carried out tests the following conclusions can be drawn

strength of small dimension wood is about the same as that of classified wood except that compression strength under loads acting perpendicularly to the grain is remarkably lower

stiffness of plates with door openings is only 2,2% of full plates, and plates with window openings - 14,1%.Thermic insulation of plates is relatively good and is up to present day standards

ultimate compression strength of plates is significant and the average for full plates is 590 kN, with window openings 365 kN and door openings 270kN.

PART SIX
BUILDING WITH CULMS AND STICKS

31 THE USE OF FORESTRY THINNINGS AND BAMBOO FOR BUILDING STRUCTURES

P. HUYBERS
Civil Engineering Faculty, Delft University of Technology, The Netherlands

Abstract
At Delft University of Technology a research programme has been initiated, aiming at the employment of thin roundwood poles and bamboo culms in primary structural building applications. For this purpose new connection techniques have been developed, making use of steel wire that is laced around the members to be connected, with the help of a special tool. This tool is operated by hand and produces tight and strong bindings, so that fully reliable connections are obtained although both roundwood and bamboo show a tendency to form shrinkage cracks.

The wire lacing method has been or is being used for a number of structures in The Netherlands and in a few other countries. As the tool and the designed construction and connection techniques offer opportunities for the employment of this group of cheap - yet powerful but generally under-estimated - vegetable materials, these methods could contribute to the mitigation of the materials shortness problem that is becoming very urgent.
Keywords: Roundwood, Thinnings, Bamboo, Wire lacing tool, Connections, Space structures, Trusses.

1 Introduction

Building materials tend to get scarce in many places of the world. The present sources become exhausted and the ecologic systems are distorted. This arouses the necessity to make as good use as possible of the readily available materials. Vegetable materials are self-replenishing and play therefore a very important role in this respect. Thin roundwood, as a maintenance product of the woods, and bamboo are available almost everywhere.

They are however not always considered as to be a sophisticated material for building structures. They both have the image that it is not well possible to build structures out of them that are acceptable from the engineering point of view, i.e. strong, stiff and reliable.

This is not altogether true. On the contrary, they have

properties that make them very suitable in many respects. They have a great normal strength, they are light, are relatively cheap and require little energy during production. They both have one great disadvantage to overcome: they have a tendency to split axially, so that it is difficult to form reliable connections between the structural members.

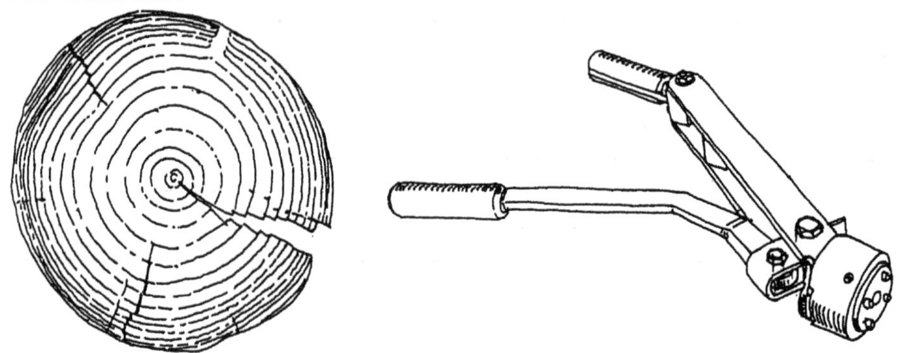

Fig. 1. Typical cross-section of roundwood.
Fig. 2. Lacing tool for 2 to 4 mm thick steel wires.

2 Wire lacing tool

For this purpose a hand operated tool has been developed, that facilitates the production of tight and strong bindings of galvanized steel wire - basically around any object - but particularly as a jointing method reminding of the conventional rope or sisal lashings in building construction.

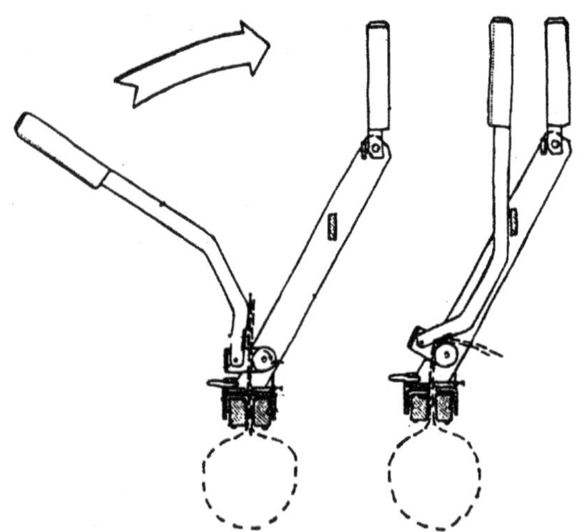

Fig. 3. Stretching of wire by horizontal handle motion.

The tool is shown in cross-section in Fig. 3. It has a hollow cylindrical base in which a smaller cylinder - with a central opening - rotates. On this base a handle consisting of two parallel bars is welded under an angle. Between these bars a rotating wheel and a hinged bent lever, provided with saw-teeth, are positioned.

The lever is connected to the wheel by a knee hinge rotating the wheel when the lever is pulled towards the fixed handle. At the same time the saw-teeth grip the wire, pressing it firmly against the wheel. With each pull the wire is stretched a bit more over the wheel, preventing it from slipping back. When a proper stretching force in the wire is reached, the tool as a whole is rotated in order to twist the wire. Between the wheel and the base there is a cutting device, consisting of two hardened plates. These plates indent the wire during the twisting and finally break it, when the twining goes on.

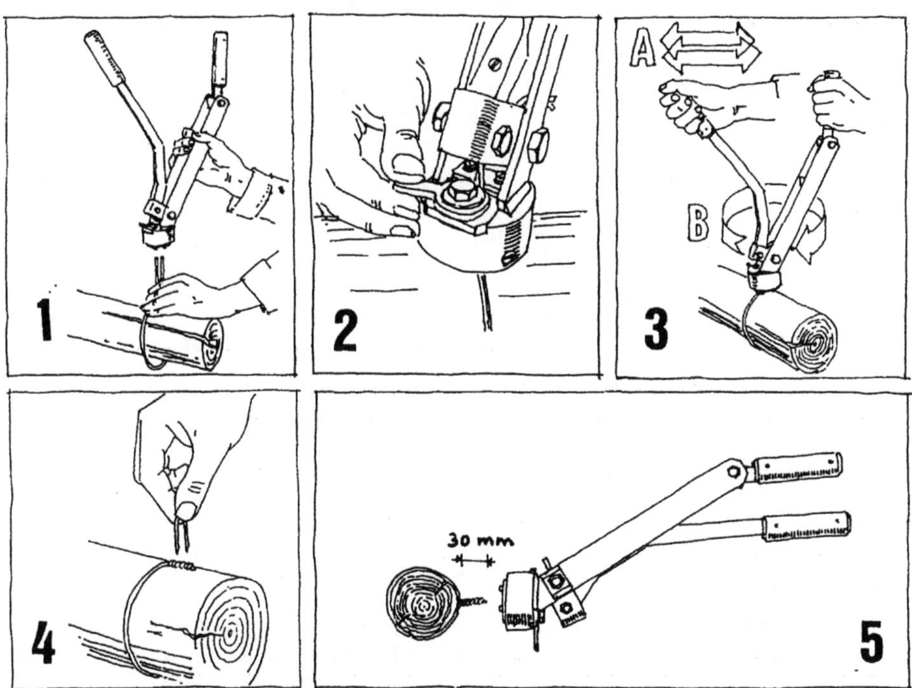

Fig. 4. Various phases in the production of wire lacings:
1. The wire is bent in shape and the tool placed. This has one fixed and one movable handle, the latter with saw teeth.
2. The tool has an automatically working cutting device.
3. The wire is stretched with the movable handle (A), the cutter is closed and then the wire ends are twisted by rotation of the whole tool around a vertical axis (B).
4. The wire ends are hammered flat and fixed with a staple.
5. The twisted end is about 30 mm long.

3 Roundwood

Almost half the production of timber has diameters of less than 15 cm and is not suitable for further sawing into rectangular cross-sections. This thin roundwood is therefore generally considered as to be of little economic value and is used for secundary purposes only, notwithstanding the fact that their structural potentials are undeniable.

3.1 Steel plate connections

For the construction of larger span load bearing building structures a solution has been developed for the interconnection of the members. To this purpose poles are provided with steel plates that are inserted in slots at both ends. Galvanized wires are laced around the timber and pass through holes in pole and plate, containing a short piece of steel tube. Wires and tubes act together to transfer force from the pole to the steel plate.

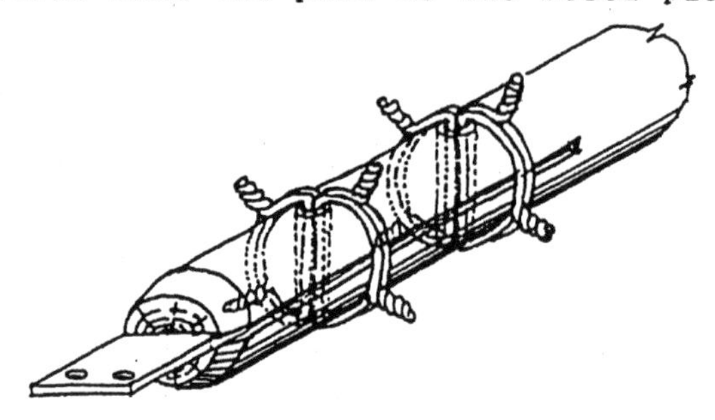

Fig. 5. Schematic detail of the plate connection.

This solution consists of the following aspects:

- An approx. 6 mm wide clearance slot is made in the end of the pole.
- 2 holes are drilled, in row and perpendicular to the clearance slot.
- A 6 mm thick steel plate is placed in the slot.
- 2 tubular dowels of 17 mm dia, 3 mm wall thickness and 90 mm length are then placed in the predrilled holes at each end.
- These dowels are held in place by steel wire lacings. Generally, 4 pieces of wire are led through each dowel. Each wire is bent around the wood and then secured, two wires at both sides. Generally a wire thickness of 4 mm is used.

Numbers of loading tests - mainly under tension - were executed on this connection method and these proved, that in this way an economic introduction of the axial forces from one member into another is obtained. This leads to a comparatively great strength and reliability, as the wires exert a pretensioning force which reaches its maximum values in the collapse phase of the connection.

3.2 Roundwood space structures

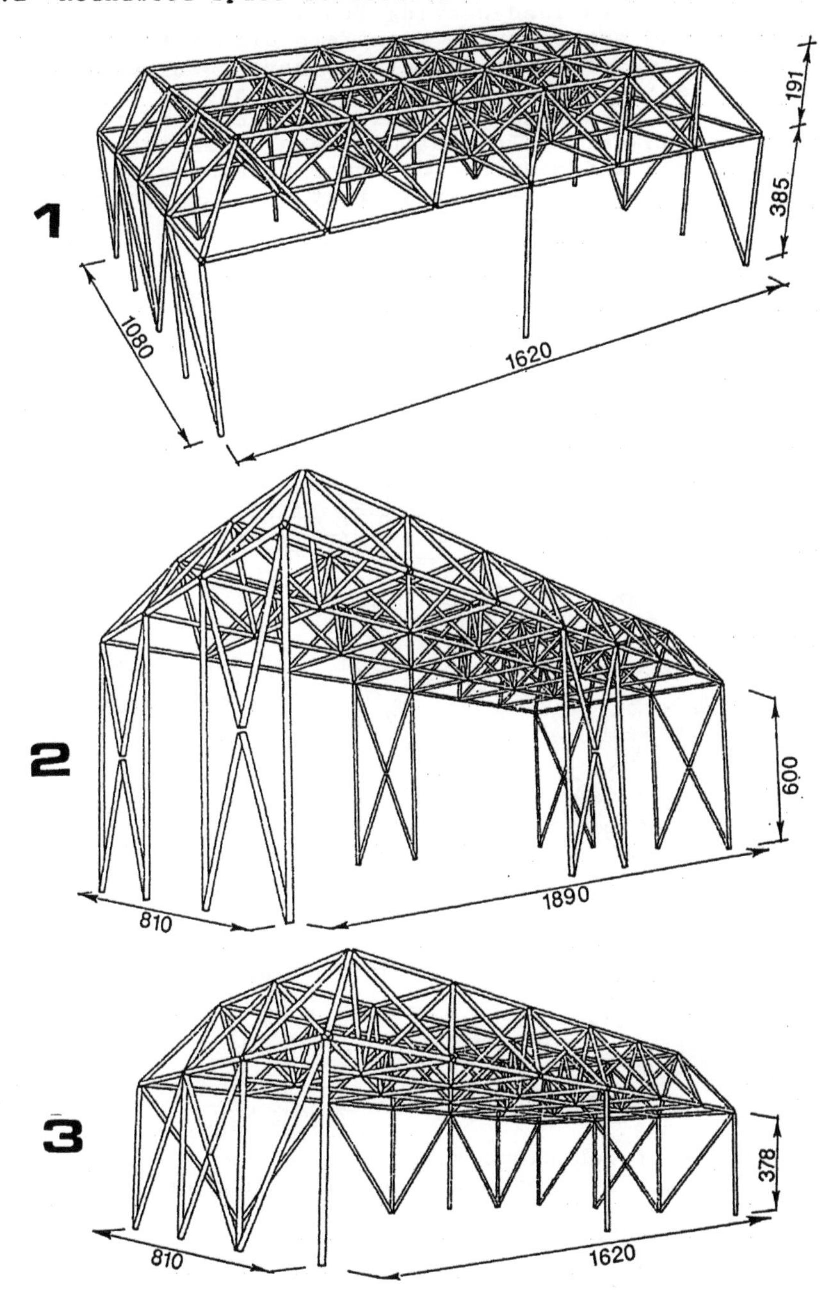

Fig. 6. Realized roundwood space structures in The Netherlands (1 and 3) and in England (2).

This connection method facilitates the construction of relatively large span load-bearing frames for buildings, made of roundwood members with diameters of not more than about 10 cm. A few of these structures have been realized during the recent years in the Netherlands and in England. Their structural behaviour is promising; they tend to behave strong and stiff. In most of these cases the wood was debarked only and required no other machining apart from the provisions at their ends, that were necessary for the connections.

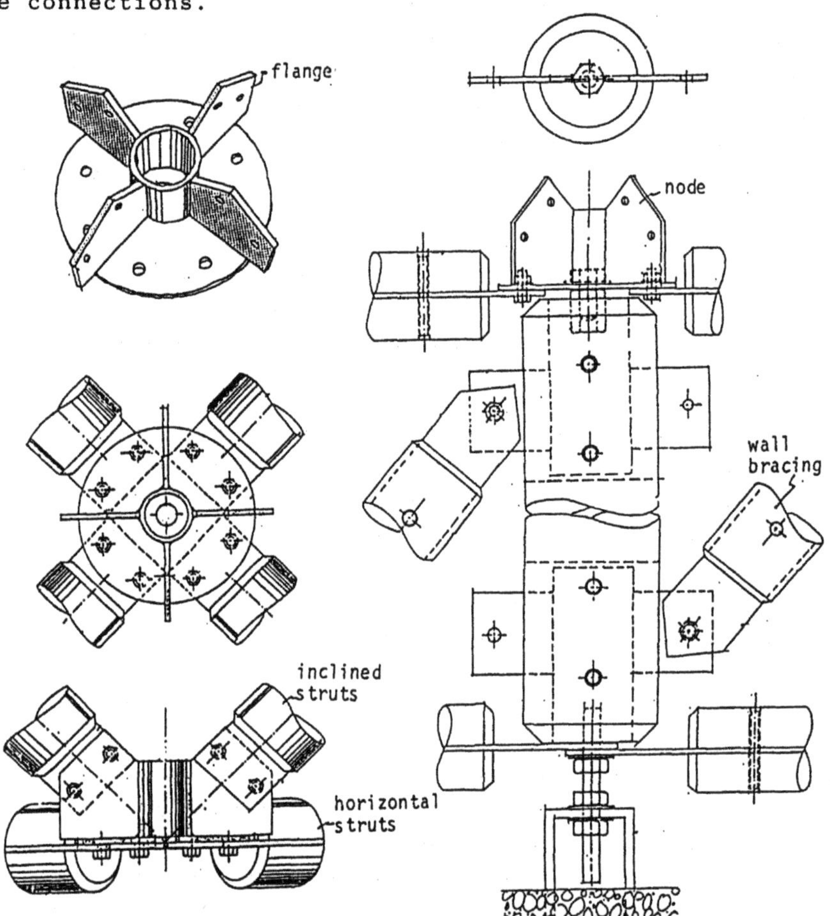

Fig. 7. Details of roundwood space structures:
1. Steel node with 1, 2 or 4 flanges depending on place in system
2. Typical standard detail in space deck.
3. Column detail with wall bracing. The steel plates in the columns are provided with welded-on long nuts or threaded ends for their connection to space deck or foundation and with occasionally added steel plates for bracings.

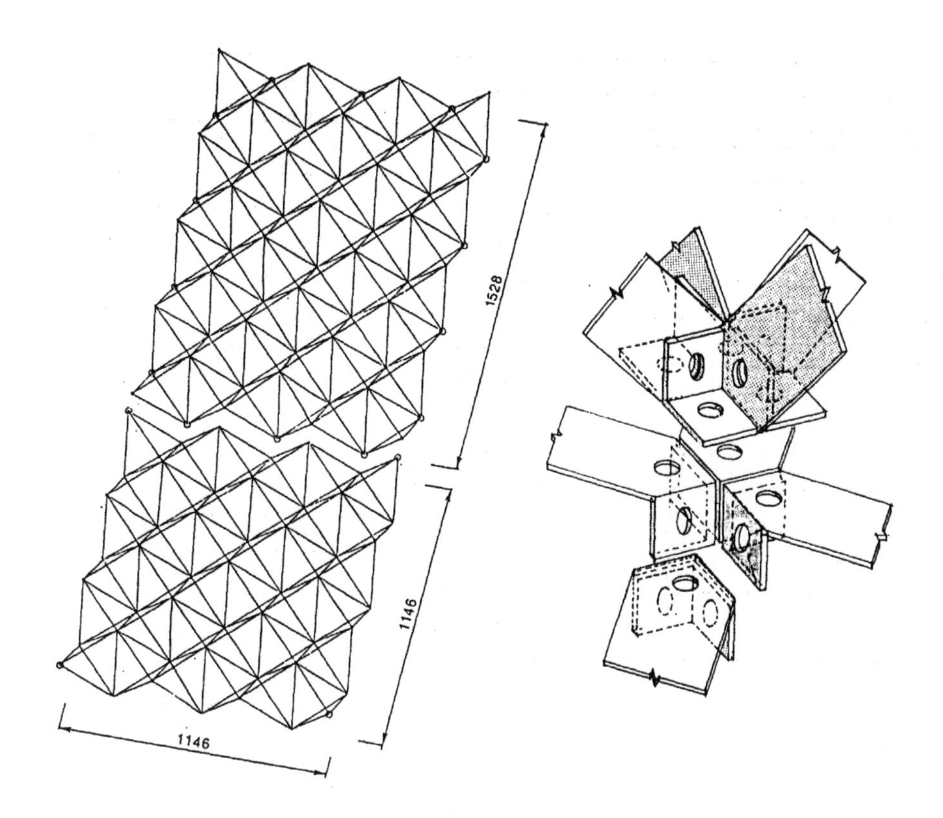

Fig. 8. Lay-out and standard detail of diagonal space structure to be erected in April 1990 in Rotterdam.

3.3 Simple lap joints

With simple lap joints trusses can be constructed out of roundwood, that meet the normal requirements for structures. A design has been worked out for a school building structure, making use of these techniques. Although the designs reminisce conventional truss shapes, it should be kept in mind, that the meetings of the poles are essentially different from those used in sawn timber with rectangular cross-sections.

E.R. van den Ham did a special study on this subject and compared two alternative plans, one by the author and one by himself.

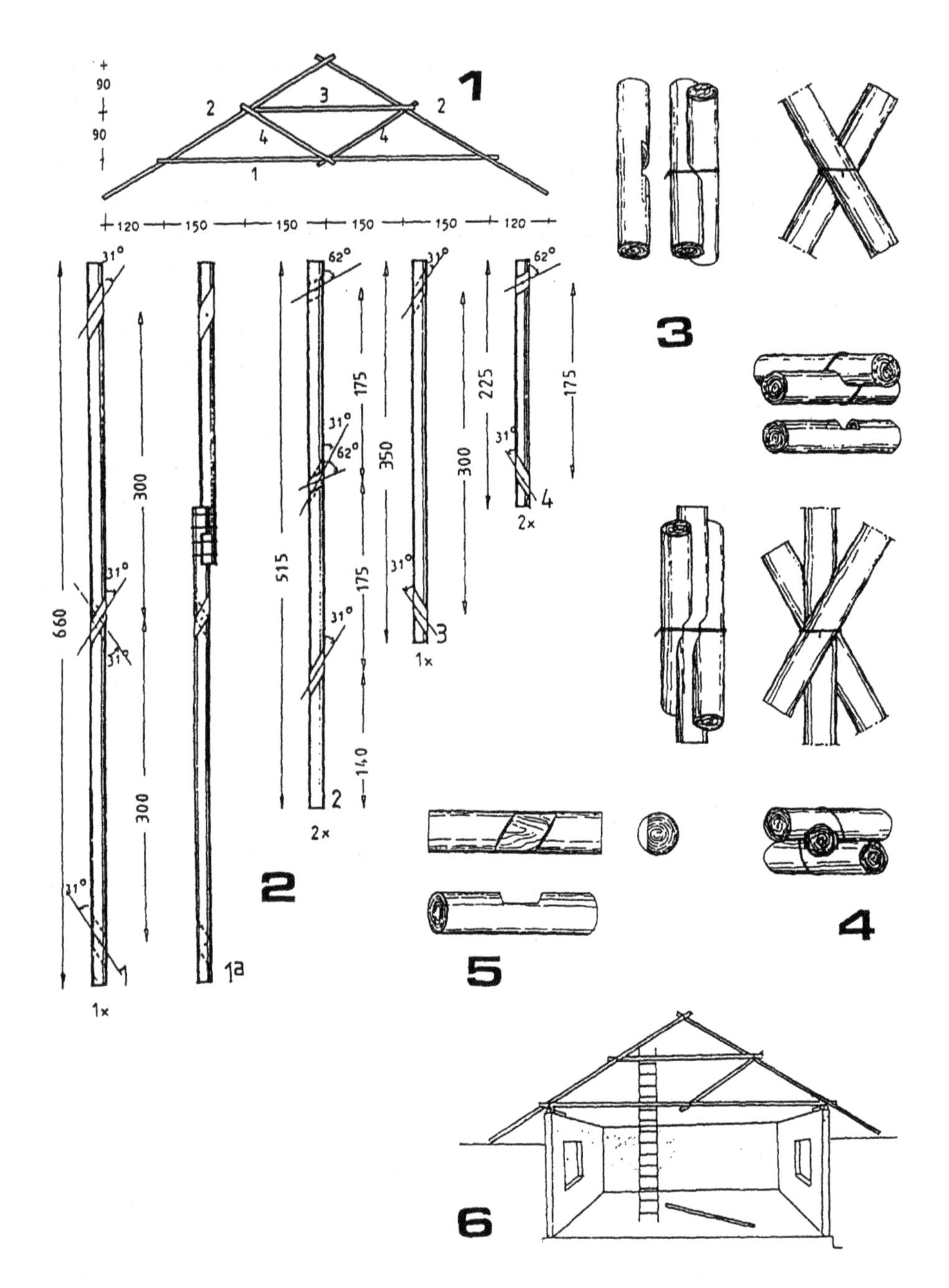

Fig. 9. Components and details of roundwood trusses for school building project.

4 Bamboo as a structural material

Similar solutions have been worked out for the construction of bamboo frames. The tool was used for a bamboo dome in India and for houses in Colombia and Costa Rica. Its suitability is being studied in many other countries all over the world.
A few alternatives for connections in bamboo trusses are being studied:
- a solution comparable to the simple lap joint in roundwood. In the present case a peace of wood is included in the joint in order to prevent the hollow culm from being crushed by the stretching force in the wire, which can under certain circumstances reach a considerable value.
- A plywood plate is inserted in slots in the ends of the culms, fixed with hardwood dowels and circumferential wires.
- Two bent thin steel plates are welded together cross-wise. This is used as a connector between two culms, that are fixed by wires and a hardwood or steel dowel through holes in the poles and the connector. It is advised to use rubber sheets - presumably made of second-hand inner car tube - as an intermediary between poles and connector in order to avoid loss of prestress by dry-shrinking. This is advisable in the first solution also, as well as in the simple lap roundwood joints.

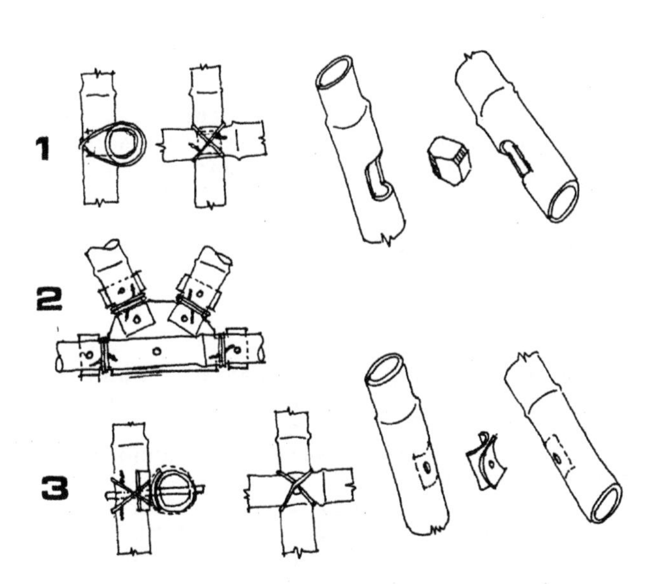

Fig. 10. Three solutions for connections in bamboo.

In Jamshedpur, India, a bamboo dome structure having a diameter of 18.00 m has been built by V.R. Sonti of the firm ASCU with the help of the tool, which is claimed to be the largest bamboo dome ever.

5 Scope

Generally spoken, the wire bindings can replace any kind of conventional laced joints. Lacings are usually of sisal or hemp rope. Strings of sliced bamboo or rattan are also used. A more modern material is thin steel wire. The common application method is with the help of a pair of pliers.

There is a number of basic and very important differences if compared to the wire lacing method which make much less dependable:

- the thickness of the wire is very limited, max. 2 mm
- no compression force between the members is exerted, so that the joint is not tightly packed
- the wire ends are gripped near to the material to be connected. This means that the wires are indented at a point were they should actually have their total strength available, but therefore there is a tendency to break just before the place where the twisting begins. In the tool the ends of the wire are gripped at 30 mm behind the beginning of the twisted area: the actual seal of the wires is undisturbed.

6 Acknowledgements

The wire lacing tool has been developed by J. Lanser in collaboration with S.Th. van der Reijken. The measurements were done by G. van der Ende. J. Blom evaluated a number of the test results as a part of his final year study at the Civil Engineering Faculty. H. Bijsterbosch and D. Bijl made the 3D-sketches with the MEDUSA-CABLOS system. The latter also did the computations for some of the projects.

7 References

Huybers, P.(1984) Farm building structures of round timber, in **Proc. of Conf. on timber constructions for farm buildings**, Ronneby, Sweden, Ch. 2:5

Darby, H.J. (1987) Building with home-grown round timber, **Farm Building and Engineering**, 3, p. 18-21

Sonti, V.R.(1988e) Delft wire lacing tool and a unique application - in making a geodesic dome 60 Ft. in dia., International Bamboo Workshop, in **Proc. of Conf. in Peechi, Kerala.**

Nimpuno, S. (1988f) **Manual on the Delft wire lacing tool**, Delft University of Technology, Fac. of Civ. Eng.

32 THE USE OF TIMBER AND BAMBOO AS WATER CONDUITS AND STORAGE

T.N. LIPANGILE
Ministry of Water, Iringa, Tanzania

Abstract
Forest products such as timber and bamboo are not much known as useful materials in hydraulic engineering construction.

Our country Tanzania has 15 years experience on the use of timber and bamboo as suitable materials for application in pipe lines, culverts, water storage and food storage constructions.

This paper explains briefly about this technology which has proven as very suitable appropriate technology for developing countries. The paper outlines research findings, pilot projects achievements, cost benefits and suitability of the timber and bamboo technology in water conveyance and storage. Today this technology is broadly applied in Tanzania only as a National Scale Project application and is likely to spread in third world faster.

The technology of using timber and bamboo as water conduit and storage has been well reisearched tested and successfully applied under Tanzania's condition in the following areas of application: water supply, irrigation, road culverts, sewerage disposal and drainage, food storage.

A population of about a quarter million people are today benefitted from this technology in Tanzania. Several International Symposia including that of Rilem, held in Iraq 1986 have discussed and approved our Timber and Bamboo Technology uses.

1 Introduction

In the past few decades the economies of Third World Countries have declined sharply and these are no signs of recovery for the near future. Because of their economic resource problems third world countries face acute shortages of a number of construction materials. Among materials that are in short supply are those used for water conduits, e.g. pipes and tanks. Conventional materials such as plastic, steel and concrete must often be imported. These materials are very expensive and, therefore, often not feasible alternatives for many of the local constructions.

In many third world countries (particularly in tropic) there are abundant forest reserves and vast land areas that could be forested. For the past decade researchers in Tanzania have studied the possibility of utilizing forest products, e.g. bamboo and timber, to construct pipes for water conveyance and tanks for storage. The water made available will be used for irrigation, sewerage culverts and water supplies.

The results achieved so far are positive and show much promise. The techniques used have major potential as alternative appropriate technologies for the Tanzania rural and urban areas and other third world countries.

Tanzania now has more 200Km of timber and bamboo pipelines in 40 villages and about 250,000 people are being supplied with water from these. In addition some major engineering constructions e.g. irrigation and road culverts have successfully been constructed in some parts of Tanzania. All schemes are operating very satisfactory.

2 Availability of Raw Materials Bamboo and Pine Wood

Bamboo is a unique group of gigantic grasses the culm of which originates in underground rhizomes. It grows naturally in many parts around the world country but some species are artificially planted.

Bamboo forests are found across tropic and subtropic zones between latitudes of about 40° South, i.e. areas with mean annual temperatures of from 20^oC to 30^oC. The density of Bamboo forests is 5,000 to 10,000 stems per hectare.

Bamboo suitable for water pipes grows at altitudes from 20 to 3000 meters. The plant is fully mature at an age of three to four years. The inner core diameter varies from species to species, the minimum being 3.5cm and the maximum 12.5cm.

The largest bamboo forest reserves around the world are found in Asia (India, Burma, Thailand, Malaysia, Bangladesh, Hongkong, Indonesia, China, Nepal, Japan, Sri-Lanka and Central America). The largest species of bamboo (200 cm inner diameter) Dendrocalamus giganteus is also found in these countries.

In Tanzania two types of bamboo are available, Arundinaria Alpina (Green) and Bambusa Vulgaris (Yellow). Suitable forests are found in Southern Tanzania. Both of these species are suitable for use in constructing water pipes. Timber is also used to contruct water pipes and tanks, pine being the pricipal species used. Pine grows rapidly and is readily permeated by preservatives, which provides for a long life span.

Pine forest reserves are located between latitudes 30° North and 30° South in the tropical and subtropical zones that are favored by high rainfall and cold climates. Pine reaches full maturity at about 12 years of age.

In Tanzania pines are plentiful in the Southern and Northern parts of the country. All pine forests in Tanzania are government controlled and artificially planted, thus, there is limited danger of deforestation resulting from over-utilization of this resource.

3 Historical Background

The use of bamboo and timber piping materials is an ancient practice. However, modern scientists have not been able to benefit from earlier practices because of the lack of documented scientific and technological information. The modern world possesses all the facilities, as well as the scientific and technological knowledge, necessary to develop products from bamboo and timber. These materials can be developed to a level where they can compete with conventional materials such as plastic, steel and concrete, in terms of durability, quality and cost. Many local and external scientific/technical institutions are supporting Tanzania Wood/Bamboo Technology today.

4 Bamboo Piping Technology

4.1 Desingn and construction of a bamboo water supply system

After harvesting, the bamboo canes are cut into four-meter sections and are drilled through the core pierce the natural internodes.

To improve the strength of the bamboo pipe and to enable it to tolerate high water pressure, the outer surface must be reinforced with bands knotted at about 5 cm intervals. The capacity of a reinforced bamboo pipe is 0,3 MPa water pressure. The end of the pipe section is sharpened into a pencil shape to allow joining by means of a polythylene socket 20 cm in length. System construction and maintenance is carried out by local people who have been trained for the job. A crew of about six men can lay a pipeline of 1 Km in one day. Maintenance can be carried out by two men, usually village residents.

The normal procedure for designing and constructing conventional water systems is generally followed when constructing bamboo pipeline systems. However, the diameter of bamboo pipes has to be slightly larger compared to conventional pipes used for similar purposes, e.g. 7.5cm bamboo pipe will be necessary where a 5cm plastic pipe is normally used. The most commonly used diameters for bamboo pipes are 3.5cm, 5cm, 7.5cm, 9cm, 10cm, and 12.5cm. The use technique adopted to preserve the pipes determines the overall cost and life span of the system. The hydraulic design C-factor 75 and 95 Hazen Williams formulae are applied for non-inner lined bamboo pipes respectively. Bitumastic paint is normaly use for inside lining of bamboo pipes.

5 Durability

Bamboo is an organic material and therefore prone to rapid decay after harvesting. Its natural life-span after cutting varies from one to three years. In order to make the material viable and economically competitive bamboo piping must be treated with preservatives in order to ensure a long service life. The most known enemies to bamboo plant life are termites, bacteria and fungi. Both insects and soil bacteria are harmless to human beings. They are all treatable by various established procedures of preservation techniques.

5.1 Preservation Techniques for Bamboo Pipes

Research during the last decade has shown that a service life of 10-20 years can be achieved by using special preservation techniques. These techniques can vary depending on the source of the water that will run through the pipeline, environments and climates of the areas.

The most important business in bamboo preservation technology is first to ensure that bamboo pipes freshly cut from forest and transported to pipe processing centres are protected from contacting any infectious disease e.g. fungi and bacteria attack by dipping in pure clean water pools or tanks. If necessary these pools can be treated with non hazard preservatives which have fungistatic and bacteriostatic properties e.g. Chlorine, Borac, Moncozeb (Dithane M 45), Maneb, Kocide 101.

5.2 Clean water Fungicide and Tar Uses

Clean water free from fungi and bacteria (naturally purified) or artificially purified is the best protection against bamboo pipe decay in the interior of the pipe conveying water when constantly saturated with water preventing air attack. Prolonged periods of intermittent sterilization by flushing the whole system guarantees extra protection. The outer surface of pipe can also be coated with tar/bitumen to prevent decay caused by contact with ground.

Where conveyance of pure clean water in not feasible bamboo pipes are impregnated with fungicides which are harmless to human beings e.g. boron based compounds (Borax) and copper based compounds (Kocide 101 and CQA). The pipes are sealed inside and outside by means of bituminous paints (two coats) to prevent chemical leaching and to ensure fixation. All these process prevent rotting.

In less termite - infected areas tar treatment alone is sufficient to prevent termites attack but in areas where termites are abundant treatment with insecticides Permethrin and Deltamethrin admixture with tar is most suitable when coated on the pipe outer surface. Expected life span using this system is 10 - 15 years where termites and fungi cases are severe. In areas with less aggression of termites and fungi a life span of up to 20 years is very possible to achieve.

Under normal circumstances bamboo pipes are stored in clean water pool or (tank). Treated bamboo pipes are buried in the ground in bulk for future uses.

For proper management of manufacture and treatment of pipes it is recommended that the work should be centralised.

6 Health and Environmental Aspects

Water carried by bamboo pipe systems is purified to meet the requirements of normal public water supplies. The interior of the pipeline is coated with lining approved for contact with drinking water whenever necessary. Direct contact between water and chemically preserved bamboo pipe surfaces is not allowed. The chemicals used e.g. Permethrine, Deltamethrine, Borax, Mancozeb, Maneb, Chlorine etc. etc have the World Health Organisation approval. All both chemicals are on the pipe by tar thus preventing ground chemical movements. This action prevents environmental pollution.

Water analysis conducted in Tanzania by the University of Delft, the Netherlands and U.K. showed that water quality at the distribution points is within the approved standards of the World Health Organization.

7 Economics of bamboo pipes

Besides the bamboo itself the main product used in manufacturing bamboo pipes is galvanized steel wire. Other relatively cheap materials such as joints and preservatives are also required.

The manufacture of the pipes is labor intensive, but the low-skilled labor required is locally available. The fig. 1 shows bamboo pipe in operation.

The cost of bamboo pipe is between four and ten times cheaper than conventional steel and plastic material, even including cost of transport. The cost of importing these materials to Tanzania was investigated recently and it was found that a twenty kilometer long plastic pipeline would require thirteen containers carrying polythylene rolls. The same length of bamboo pipeline would require only one container for the supplementary products such as galvanized steel wire, joints and preservatives. The installation cost per capita for a village bamboo water supply system, utilizing a timber tank, is between USD 2-4.

8 Timber Technology: design Manufacture and Construction

Timber tanks and pipes are manufactured at sawmills. Tanks are constructed for ground and above ground use. The whole structure is manufactured from timber pieces of 10cm x 5cm or 5cm x 5cm. Timber pipes are similarly constructed. Timber for both tanks and pipes is machined to obtain a tongue and groove along the edges of the joints.

Fig. 1 - Bamboo pipe

Fig. 2 - Timber conduit structures

Fig. 3 - Timber pipeline irrigation pipeline

The timber pieces are held together firmly by means of galvanized steel bands. The inner pressure capacity of the materials (pipe/tank) is determined by the spacing of the steel bands. The economic capacity of a timber pipe is 6,0 MPa water pressure. The soil backfill for timber constructed road culvert is half diameter of the culvert e.g. for 1 m diameter culvert the depth (minimun) of backfill will be 0.5 m. For design purpose refer to (appendix 1) structural design data. Fig. 2 & 3 are showing timber conduit (pipes and culvert) structures and a 60cm diameter irrigation pipeline in operation respectively.

9 Durability Health and Environmental Protection

Durability of materials is guaranteed for 50 years by impregnating the wood with a waterborne or oilborne preservative, chemical fixation is ensured so that no environmental pollution occurs. For drinking water

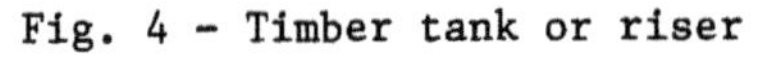

Fig. 4 - Timber tank or riser

Fig. 5 - Timber Tank on ground

system the treated surfaces of the materials used are sealed when necessary to prevent water contamination. The suitable preservation chemicals are Copper Crome Arsenic (C.C.A.) (see appendix I) and Creosote.

Water analysis conducted locally in Tanzania and abroad have shown positive good results.

10 Hydraulics and Economics of Timber

The flow of the water through timber pipes is similar to that for plastic pipes since the inner surface is lined with plastic film, coated or has been planed. In Tanzania pipeline up to 60cm in diameter, used as water conduits for irrigation, road culverts and water supply systems, are all operating successfully. Similarly, water tanks with a capacity of up to 45 cubic meters have been constructed in a number of rural villages. All materials have been fabricated and constructed by Tanzanians locally. The cost of timber materials is about 3 to 12 times cheaper to conventional. Figures 4 & 5 are showing timber tanks on riser and on ground respectively.

11 Evaluation of Timber Bamboo Technology

In 1983 the Swedish and Tanzania Government appointed a Swedish Consulting from M/S BROCONSULT to evaluate Tanzania Wood/Bamboo Technology. The conclusion reached were that both timber and bamboo technologies were feasible for Tanzania conditions. Bamboo pipe systems are cost competitive if the systems can be operated continuously for a period of ten years or more. Timber systems are competitive if they can be operated continously for up to 20 years.

12 Market Study for Timber and Bamboo Technology in Tanzania

In 1988 again the Swedish and Tanzania Government appointed a Swedish Consulting firm M/S SwedForestry Consult the Study Mission was headed by a Swedish Forestry Consultant Dr. Roland Palm. According to a published report the following sectors of applications were recommended:

a) Water Supply and Sewerage Disposal
b) Road Culverts
c) Irrigation and Food Storage
d) Food Storage
e) Sewerage Disposal and Road Culverts
f) Hydropower Penstock
g) All Principal Donors Agencies (e.g. SIDA, DANIDA, FAO, UNDP, REDCROSS etc.) water, irrigation, food storage, culverts etc. can use timber technology.

13 References

Lipangile, T.N. (1984) Wood/Bamboo, in **Rural Hydraulic Development,** A.P.C. Countries, Marseille, Comission I.

Lipangile, T.N. (1985) Use of Bamboo Water Pipes, in **The International Bamboo Conference,** Honzou, China, p. 315.

Lipangile, T.N. (1986) Wood and Bamboo in Water Conveyance in **The International Water and Sanitation Conference,** Calutta, India, p.87.

Lipangile, T.N. (1986) Use of Bamboo Pipe in Water Supply System in **Use of Vegetable Plants and their Fibres as Building Materials,** Baghdad, Iraq, p. D 37-40.

Lipangile, T.N. (1986) Development of Bamboo Water Piping Technology in Tanzania, in **Bamboo Production and Utilization IUFRO World Congress,** Ljubljana, Yuguslavia.

13 Apendix I

Mean Strength Properties of Pine Timber (0,05m - standard)

Properties	Pinus Patula (Pine 12 Yrs)		Pinus Radiata (Pine 33 Yrs)	
	Green	Air Dry	Green	Air Dry
Moisture Content (%)	50	12	84	12
Weight per m^3 (Kg)	530	380	800	610
Maximum bending strength (e.f.s. at max. load) (MPa)	24	48	42	84
Stiffness (Modulus of Elasticity) (MPa)	5300	5850	9700	13008
Energy consumed in bending:				
a) To max. load (KNm/m^3)	-	-	4,3	7,9
b) To total fracture (KNm/m^3)	-	-	12,8	10,5
Compression (max. compressive atrenght parallel to gain) (MPa)	14	28	24	51
Hardness shear:				
a) resistance to indentation on side grain (N)	896	1112	-	-
b) Max. shear strenght parallel to gain (MPa)	46	70	-	-
Resistance to Spliting :				
a) in radial plane (KN/m width)	23	35	-	-
b) in tangential plane (KN/m)	27	56	-	-

33 USE OF VEGETABLE PLANTS IN HOUSING CONSTRUCTION IN NORTHERN IRAQ

M.R.A. KADIR
University of Anbar, Ramadi, Iraq

Abstract
Vegetable plants and their fibres have been used in the construction of the traditional house buildings in the northern parts of Iraq in many ways. Wood and canes are used for roof construction, soil stabilized with hay is used for brick making, wall plastering and roof coverings. Buildings constructed using these materials have lasted for many years and have provided comfortable environment for people despite the severe local climatic conditions.
In this study the use of vegetable plants and their fibres have been described and discussed and suggestions for improvement have been given.
Keywords: Traditional Buildings, Sun-burned Bricks, Hay, Reed, Vegetable Plants.

1 Introduction

Vegetable plants and their fibres have been used as building materials in Iraq since ancient times and are still used. Low cost house buildings are built using reed at the southern parts in the marsh areas where they occur in immense amounts, Raouf (1986). Reed panels have been used for interior walls of housing with both sides gypsum plastered. For better housing construction employing reeds many improvement have been suggested concerning their usage and their protection for longer life, Samarai (1986).

In the northern parts of Iraq vegetable plants and their fibres have been used in building construction in many ways. Wood, reed or cane and vegetable leaves are used for roof construction, reinforced soil with hay is employed for making sun-dried bricks and for plastering walls and roof coverings. Since no reported study is available it is the aim of this paper to describe and discuss the construction of the houses built employing vegetable plants and their fibres.

2 Environmental requirements

Most traditional building forms have evolved gradually by a process of trial and error so that they eventually provided homes that are considered to be the most suitable for local conditions.

The climate and weather condition of the northern part of Iraq -Kurdish region- is breifly as follows. Most of the northern areas experience rainfull for more than six monthes ranging from mediam to very heavy rainfull at the mountaineous regions. Villages at the foot of the mountains will be covered with snow in winter for many days. Temperature variation may range from -10C or less in winter to +45C in summer in some areas.

The principle requirement of the roof and the walls are that they should provide comfortable, safe and healthy conditions within the building. They should,therefore, keep out the wind and the rain and prevent the penetration of excessive heat or cold. The severe climatic condition have required heavy wall and roof construction using local materials available in the area. Most of the traditional house buildings in the urban areas and almost all the houses in the rural areas satisfy most of the local environmental requirements.

The houses are usually built facing south and south-east that is facing the sun for most of the day. Windows are usually small in area and on southern walls mostly. The roof of the house provides also for other amenities including provision for sleeping in summer or recreation. In some cases the roof serves as a floor or an open area for the house built above especially where the houses are built on steep hills or at mountain foot.

3 Wall construction

The walls of the traditional house buildings are built using sun-dried bricks or stones depending on the availability of the material. Near mountains and where no distant transportation is required stones are used for the walls. The footing of the wall is usually built of stone laid on clayey soil mixed with lime or gypsum if available in the near area or laid on clayey soil reinforced with hay. Hay is the brocken stems of wheat and barley. It is known that lime and hay both can be used for soil stabilization and strengthening, Bhatnager et al. (1988), and Al-Layla and Al-Saadi (1984). Studies have shown that the treatment of clay with hay causes a decrease in the potential of clay to volume changes upon water content variations.

The wall above the footing is made of sun-dried bricks which is made at the building site in the following way, Sujadi (1974). A clay soil which is neither too coarse nor too fine is thoroughly mixed with water then hay is spread over the mixture by hand while mixing continues. Mixing is done usually by treading with human foot. The mixture is left to ferment for some time usually till the next day and then remixed again. The mixture is then cast into a wooden mold of approximatly 250x250x70 mm. and the top surface is leveled with a straight wood. The mold is raised directly after casting and another brick is cast. After three days exposure to sun heat the bricks are turned over and stacked each two together and left to dry out for a few days more, then used for wall construction using a morter of similar mixture of clay and hay.

The wall thickness varies from 2 bricks upto 5 bricks in some cases where two floors are to be built. Some times large pieces of wood of length about one meter or more are provided in the wall construction to serve as tie member specially at the corners. After completion of the house both sides of the walls are plastered with the same mixture of soil and hay possibly a larger amount of hay is added.

In the traditional houses built in the towns for those who can afford it a layer of thin burned clay brick 250x250x40 mm. is laid on edge at the outer face of the external walls, and the joints are filled with soil, sand and lime mortar, thus providing a good durable protective layer with a pleasant appearance.

4 Roof construction

The roof in the traditional buildings is made of several layers. The roofing layers are put directly on the finished walls for spans up to 3-4 meters, for larger spans columns and beams are provided at mid spans. Oak or walnut trees are used for columns and beams. The roofing layers consist of joists which are plane trees circular in section of diameter about 70-100mm. spaced at 200-500mm. center to center depending upon the span. No kind of any protective treatment is applied, however the woods are left for natural seasoning to take place before using them.

The next layer is a mat of reed or cane fastened together on site on top of the joists and perpendicular to the direction of the joists, some times another layer of cane perpendicular to the first is used. At the level of the joists extra pieces of hard wood are put over the exterior walls extending beyond the wall for about 1000mm. protecting the wall from rainfall. The cane mat is then covered with bushes and leaves followed by layers of clean soil, the total thickness ranges from 300 to 600mm., and the slope of the roof for drainage purposes is provided for in this layer.

The final layer is a hard well compacted rendering of clay and hay mixture about 60mm. in thickness prepared the same way as described for making sun-dried bricks.

5 Maintenance and improvements

With the introduction of modern building materials people now are rarly using sun-dried bricks for wall construction except at very distant areas where cost of materials transportation is too high. However the roof system described above is still in use although the price of wood joists is no longer low.

The roofing system requires contineous maintenance, at the end of almost every heavy rainfull or snowfall its top layer is recompacted using a hand driven cylindrical roller made of stone weighing about 30 to 40 kgs. This compaction reduces voids and cracks in the top layer and reduces the possibility of rain penetration in to the layers below.

Heavy rain and wind cause errosion of the top layer and the plastering of the walls, moisture and drying movements cause cracks which are unrepairable, therefore every three or four years it is necessary to remove the top layer of the roof and the external plastering both made of clay and hay, and apply a new layer.

Most of the traditional houses built the way described above have lasted for more than 50 years. The durability of these houses or the materials used in them can be improved at the expense of some extra initial cost but the future maintenance cost would be greatly reduced. The external walls can be plastered from outside using cement, lime, and sand mortar on a metal lathing fixed to the wall.

Wood and reed are easly attacked by insects, rot fungi and fire, these disadvantages can be overcome to a great extent by treatment with suitable preservatives and coating with protective materials. Applying a water proof coat or membrance such as polythelene sheet below the top layer prevents moisture from penetration inwards. Also a thin coat of steel mesh reinforced concrete or mortar on top of the final layer of the roof provides best protection and minimizes maintenance costs.

6 Concluding remarks

Houses are built for the comfort of the people who live inside of them. Thermal comfort depends largely upon the thermal properties of the walling and roofing materials. The traditional house buildings in northern Iraq which have employed the wall and roofing systems described are much more comfortable than most houses built using the common modern wall and roofing materials. Durability of these traditional buildings can be improved at the expense of some extra initial cost.

7 References

Al-Layla, M.T. and Al-Saadi A.H. (1984) Soil stabilization with hay. J. Building Research. Baghdad. 3, 2, 17-32.

Bhatnagar,J.M. Goswami, N.L. and Singh, S.M. (1988) Stabilized soil for brick making. J. CIB. Building Research and Practice. 16, 3, 177-181.

Raouf, Z.A. (1986) Examples of building construction using reeds. in Use of Vegetable Plants and Their Fibres as Building Materials. RILEM/CIB/NCCL. Symposium, Baghdad.

Samarai, M.A. Al-Taey, M.J. and Sharma, R.C. (1986) Use and technique of reed for low cost housing in the marshes of Iraq. in Use of Vegetable Plants and Their Fibres as Building Materials. RILEM/CIB/NCCL. Symposium, Baghdad.

Sujadi, A.A. (1974) Kurdewari [Typically Kurdish- in Kurdish Language] Al-Maarif Publication, Baghdad.

PART SEVEN

RECYCLING OF AGRICULTURAL WASTE AND RELATED TOPICS

34 STUDY ON THE USE OF ROUGH AND UNGROUND ASH FROM AN OPEN HEAPED-UP BURNED RICE HUSK AS A PARTIAL CEMENT SUBSTITUTE

G. SHIMIZU and P. JORILLO Jr
College of Science and Technology, Nihon University,
Tokyo, Japan

Abstract
This is an experimental study on the use of the rough and unground ash from an open heaped-up burned rice husk as a partial cement substitute in concrete. The open heaped-up burning method is the most convenient way to dispose rice hulls, thus the ash produced is abundant and readily available especially during dry season. In this study, various percentage replacement of ash to cement were made, and were respectively introduced to four grades of concrete. Three different degrees of fineness of ash were introduced to HG1 and LG series. Fresh and mechanical properties at three curing ages were measured to evaluate the effect of this kind of ash to concrete. Test data revealed the feasibility of using this kind of ash as a construction material. At a proportion of 80:20 by volume (or 87:13 by weight) the compressive strength of a low grade RHA concrete was 93% - 100% to that of plain portland cement. The equations for the tensile property and modulus of elasticity of RHA concrete did not significantly vary compared with other experimental equations on plain portland cement concrete.
Keywords: Concrete, Compression, Elasticity, Tension, Rice husk ash, Water-cement ratio, Pozzolan.

1 Introduction

There is a growing awareness among researchers and engineers all over the world, especially in the developing countries, to use an indigenous material in the development of a low-cost but efficient and effective building material. Abundance and availability of raw materials are the most important factors which make a product inexpensive. In the case of concrete, an indigenous and readily available material which can save cement without sacrificing the strength and the durability will surely reduce the costs. A potential for this material is the ash from the rice husk which abounds in the countryside as an agricultural waste. From this point of view, it is the objective of this study to examine and discuss this kind of material as a partial-cement substitute in concrete for low-cost construction work.

1.1 Background

Previous studies by Mehta (1975), Loo and Weragama (1978), Chopra (1978) and Dass (1978) reveal that the ash from the burned rice husk contains 70 to 98% silica, which is a very good index of its potential as a cement extender and pozzolanic material. Pozzolans are materials containing high percentage of reactive silica which reacts with other binding medium like lime or cement in the presence of water, forming compounds possesing cementitious properties (ASTM,1988). At present, most of the data available about the effects of rice husk ash (RHA) on the fresh and mechanical properties of materials are very limited, in a sense that, most of these data on RHA with mortar (Dass, 1978), grout (Chopra, 1978), concrete (Mehta, 1975) used the ash which has undergone a process like controlled ashing or ballmilling.

The quality or the reactivity of the RHA depends on many factors and one of the most critical is the method of ashing, because this affects the resulting fineness of ash, chemical composition and quality of the silica content (Mehta,1975). In this particular study, the effect of fineness, quality of ash from the open heaped-up burned rice husk to portland cement concrete was further examined.

2 Experimental Investigation

2.1 Materials

Table 1 and 2 show the properties of the Type 1 portland cement and the aggregates, respectively. Figure 1 shows the grading of aggregates used in the tests. In order to ensure the uniformity in grading, all particles of fine aggregates retained in a 5 mm sieve and particles of coarse aggregates passing the 5 mm sieve were discarded.

The hulls of the Philippine variety of rice (e.g. C-4) were heaped-up and burned in an open field by pouring 1 to 2 liters of kerosene at the edge of the heap to start combustion. It took 4 to 6 days before the ash could be collected. This method of ashing is a very slow process and it is quite dependent on the windflow. The collected ash was cooled to room temperature before stocking it in plastic bags. The physical and chemical properties of the ash are given in Table 3. It can be observed from the given table that inspite of the relatively low pozzolanic index of the ash, it was still used in the experiment with the aim of providing information for such kind of ash and its effect to concrete for its utilization in the country-side construction works.

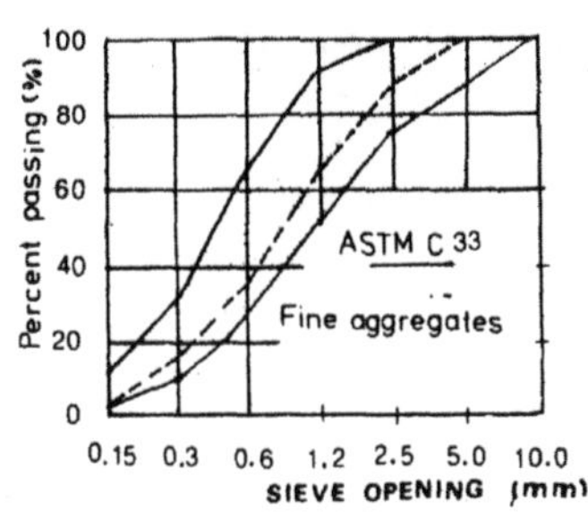

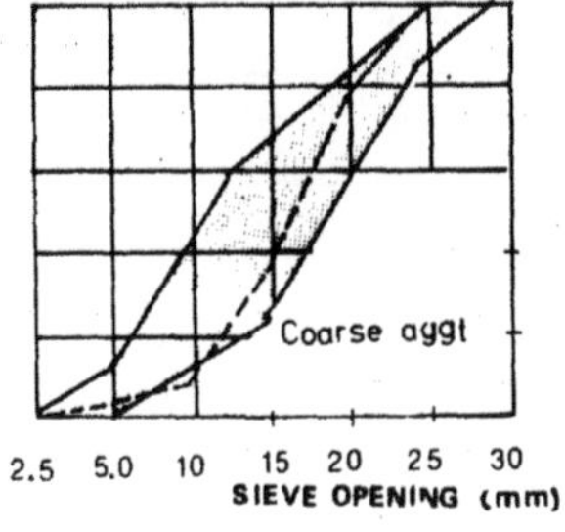

Fig. 1 Particle size distribution of aggregates

2.2 Outline of Experiments

Four grades of plain concrete with a nominal 28-day strength of 5300, 5000, 4000 and 2800 psi were cast, and were designated by mix code HG1, HG2, MG and LG series, respectively. To determine the optimum proportion of ash and portland cement, different percentages of ash were introduced as cement extender, namely 13.6%, 21.1%, 29.5% and 38.7% by weight of cement (or 80:20, 70:30, 60:40, 50:50 by volume). Three different degrees of fineness of ash by dry sieving were introduced to HG1 and LG series to determine its effect to the setting time and the strength development of concrete. RHA-N refers to ash at its natural fineness, and RHA-B and RHA-C refers to ash passing the 0.30 mm and 0.15 mm sieve, respectively. Each mix was designated by a code as:

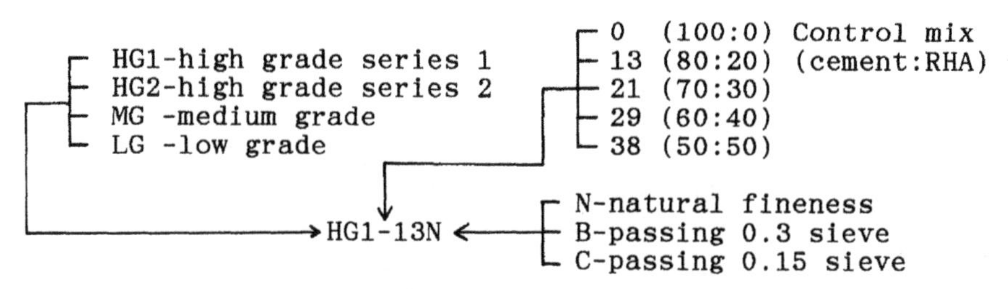

Each mix had a controlled consistency of 10 cm slump, since no water reducing admixture was used, it was necessary to increase the quantity of water in order to satisfy the slump requirement. A series of trial mixes were carried out to determine the required slump and finishing characteristics. Final mixing by a tilting type drum mixer was done in the laboratory where the average temperature and humidity was 29 C and 85%, respectively. The quantities of mix proportion are given in Table 4, all in all there are 30 mix batches. For the fresh properties, the concrete mixture temperature, air content, plastic density and setting time were measured in accordance with ASTM. Specimens were cured in air for 24 $\pm$ 4 hours and placed in a lime saturated water tank for the rest of the curing days.

Table 1. Physical properties of Type 1 cement.

ASTM	Physical Test	Value
ASTM C-184	Fineness No.100 (%)	99.12
	Fineness No.200 (%)	97.26
ASTM C-188	Specific Gravity	3.075
ASTM C-187	Normal Consistency (W/C %)	30.10
ASTM C-191	Initial setting (min)	142.0
	Final setting (min)	210.0
ASTM C-109	Compressive Strength	
	7 days (kg/cm)	152.2
	28 days (kg/cm)	258.3

Table 2. Physical properties of aggregates.

Classification/test	Fine aggregate	Coarse aggregate
Source	Porac river	Bulacan river
Max particle size (mm)	5.0	20.0
Specific Gravity	2.51	2.78
Absorption (%)	1.28	2.40
Fineness Modulus	3.0	6.80

Table 3. Physical properties and Chemical Composition of rice husk ash.

Property Test	RHA-N	RHA-B	RHA-C
Fineness No.100 (%)	47.67	64.01	99.89
Fineness No.200 (%)	25.29	31.59	68.48
Specific Gravity	1.944	1.948	1.956
ASTM C-311			
Relative water requirement (%)	162.8	156.2	143.7
Pozzolanic Index with Type 1 cement (%)	46-53	44-50	41-56

Chemical composition of RHA-N ash

Si_2O - 92.39%	Al_2O_2 - 0.185%	Fe_2O_3 - 0.040%
TlO_2 - trace	CaO - 0.310%	MgO - 0.420%
Na_2O_2 - 0.030%	k_2O_2 - 0.980%	P_2O_5 - 0.460%

The mechanical properties were measured in accordance with ASTM and as follows:

(a) Compressive test of 10 Ø cm cylinder at 7, 28, and 56-day curing ages.
(b) Stress and strain measurement for the 28-day age specimen for the modulus of elasticity.
(c) Split tension test of 10 Ø cm cylinder at 28 days.
(d) Third-point flexure test on 10 x 10 x 40 cm beam specimen at 28 days age.

3 Discussion of Result

3.1 Effect of RHA to the Fresh Properties of concrete

It can be noted in Table 4 that the amount of water needed to maintain a constant consistency of 10 cm slump increases as the percentage of ash increases. This is due to the crystalline structure and absorbent form of the RHA. The fresh concrete appeared very dry during the first couple of minutes of mixing and the general tendency was to add more water to increase its consistency, but experience proved that the mix required only a little addition of water and a longer mixing time. It is rather difficult to keep the consistency and the W/C ratio constant simultaneously without the aid of admixture, because the incorporation of RHA will surely result to a decreased workability.

As the percentage of ash increases the yield of resulting mix also increases, and mixtures with relatively high yield possessed harsh characteristics. The air content on the other hand, tends to be reduced with the increasing percentage of ash. This can be due to the overall decrease in the amount of the binding medium, change in overall fineness of cement and change in the water requirement. These behaviors were also observed on the fresh concrete containing RHA-B and RHA-C ash. Slight improvement in the workability was noted, resulting to the reduction of as much as 1.2% to 3.2% in the water requirement.

3.2 Setting Time of the Concrete

The processes of setting time of concrete mixtures with different ashes are given in Figure 2. Data were curve-fitted using the method of least square. It can be observed that regardless of fineness both the initial and final setting time of concrete retards as the percentage of ash increases. This result has a beneficial effect to concrete in a sense that a longer initial setting time allows longer handling of fresh concrete especially at high temperature. In the case of RHA-B and RHA-C concrete with controlled ash fineness, these show a setting time which is more or less predictable and consistent compared with the RHA-N concrete.

Table 4 Mix Proportion and Fresh Properties of test batches

Mix code	Cement (kg)	RHA (kg)	Water (kg)	W/C (%)	Slump (cm)	Air (%)	Temp °C	Unit Wt (kg/m³)
HG1-0	484	--	193	40.0	11.4	1.40	30.0	2433
HG1-13N	387	61.09	228	50.9	11.6	1.10	29.0	2347
HG1-13B	-do-	-do-	224	50.1	11.0	0.35	29.0	2343
HG1-13C	-do-	-do-	222	49.5	9.8	0.20	28.5	2368
HG1-21N	339	91.65	243	56.5	11.2	0.90	32.0	2288
HG1-21B	-do-	-do-	234	54.3	9.4	0.20	29.0	2307
HG1-21C	-do-	-do-	240	55.7	11.2	0.15	29.0	2325
HG1-29N	290	122.19	271	65.7	11.8	0.25	32.0	2219
HG1-29B	-do-	-do-	267	64.7	11.9	0.25	29.0	2261
HG1-29C	-do-	-do-	268	65.0	11.2	0.10	29.0	2268
HG1-38N	242	152.75	325	82.4	11.2	0.10	30.0	2209
Coarse Aggt.- 1073 kg; Fine Aggt. - 607 kg								
HG2-0	356	--	178	50.0	10.0	1.20	30.0	2397
HG2-13N	285	44.98	195	59.2	9.2	1.20	30.0	2342
HG2-21N	249	64.45	231	73.7	10.6	1.30	30.5	2290
HG2-29N	214	89.96	244	80.5	12.0	0.70	31.5	2276
HG2-38N	178	112.45	252	86.9	12.3	0.75	30.0	2210
Coarse Aggt.- 1075 kg; Fine Aggt. - 788 kg								
MG-0	313	--	181	58.0	10.3	1.40	32.0	2376
MG-13N	250	39.49	197	68.2	10.2	1.20	30.0	2340
MG-21N	219	59.23	230	82.9	9.7	1.30	31.0	2289
MG-29N	187	78.97	247	92.5	10.3	1.10	31.0	2262
Coarse Aggt.- 1033 kg; Fine Aggt. - 848 kg								
LG-0	260	--	172	66.0	11.0	2.20	30.5	2369
LG-13N	206	32.89	195	81.6	9.2	1.10	30.0	2319
LG-13B	-do-	-do-	196	82.1	9.3	0.50	26.0	2347
LG-13C	-do-	-do-	194	80.9	9.0	1.00	27.0	2339
LG-21N	182	49.34	223	96.5	10.7	0.90	31.0	2302
LG-21B	-do-	-do-	223	96.3	10.3	0.70	27.0	2297
LG-21C	-do-	-do-	221	95.3	11.0	0.60	27.0	2297
LG-29N	156	65.78	227	102.1	10.5	1.00	30.0	2276
LG-29B	-do-	-do-	229	103.0	9.4	1.00	27.0	2268
LG-29C	-do-	-do-	230	103.5	10.0	0.60	28.0	2279
Coarse Aggt.- 1013 kg; Fine Aggt. - 911 kg								

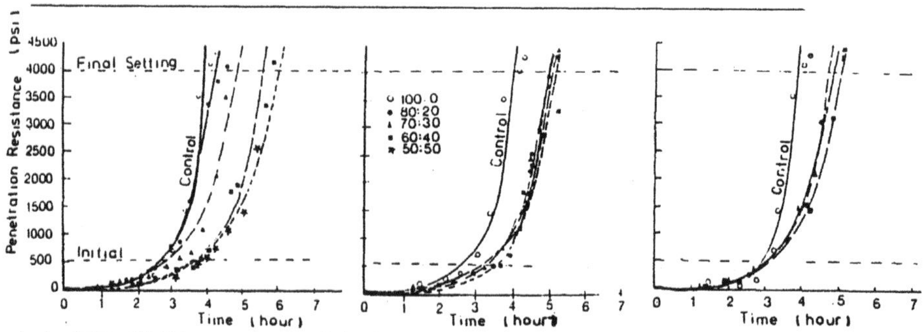

(a) RHA-N Concrete (b) RHA-B Concrete (c) RHA-C Concrete

Fig.2 Process of setting of Concrete mixture

3.3 Effect of W/C ratio to the strength of RHA concrete
Figure 3 shows the results of the regression analysis of the relationship of W/C versus the 28-day age compressive strength. Increase in the water requirement for a constant consistency of RHA concrete mix has undoubtedly resulted in the decrease of its strength. It can be noted that for a 28-day strength of 3000 psi, the mix requires a W/C value of 75, 72, 67, and 58% for a 100:0, 80:20, 70:30 and, 60:40 cement to RHA by volume proportion. Thus, the W/C ratio has to be reduced in order to produce the required 3000 psi strength. Fig. 3 can be used as a basis in choosing the W/C ratio for a certain strength requirement, the strength value in the figure act as an upper bound in the practical sense; if at a fixed consistency the water requirement can be lowered due to the use of a water-reducing admixture, this can only help to improve the resulting quality of concrete.

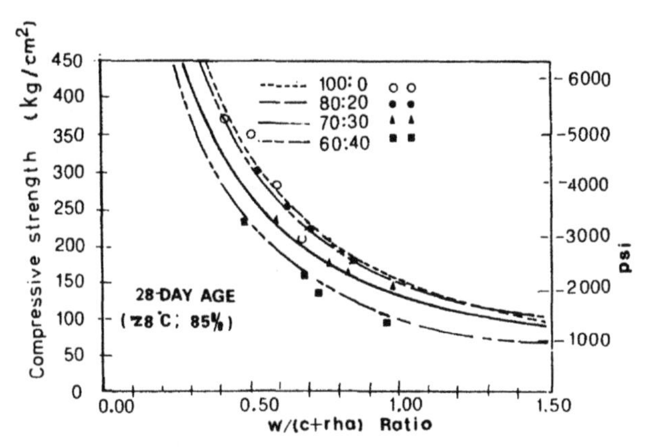

Fig.3 Strength and W/C ratio relationship

3.4 Strength development of plain and RHA concrete
Fig. 4 shows the development of strength of RHA-N concrete up to 60 day age. Average coeffecient of variation for all the data of compression test was 5.20%. Data were also analyzed and curve-fitted by the method of least square. The higher the strength like the HG1 series was, the larger the drop in strength become in proportion with the amount of ash. But in the case of LG series, it was comparatively small even though the percentage of ash was as high as 29.5%.

Fig. 5 shows the strength development of the three kinds of RHA introduced to HG1 and LG grade series. It is notable that RHA-B and RHA-C concrete resulted to 8% and 12% higher strength compared with the RHA-N, that is, the strength of RHA concrete also depends on the fineness of ash. The finer the ash, the better the strength quality of concrete becomes.

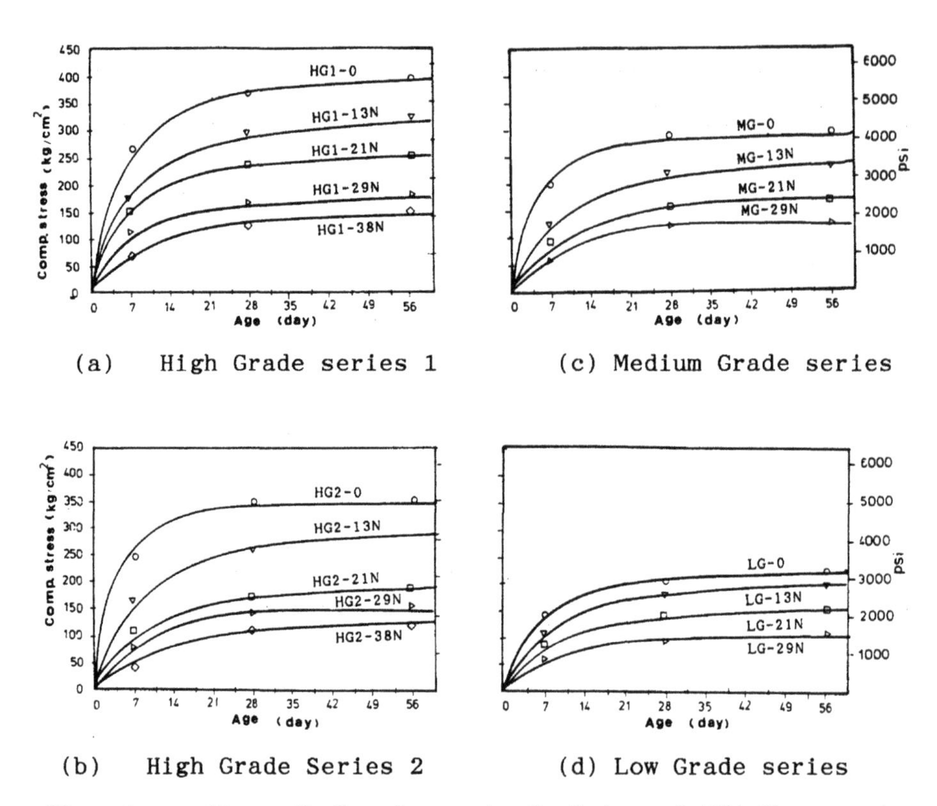

(a) High Grade series 1 (c) Medium Grade series

(b) High Grade Series 2 (d) Low Grade series

Fig. 4 Strength Development of plain and RHA-N concrete

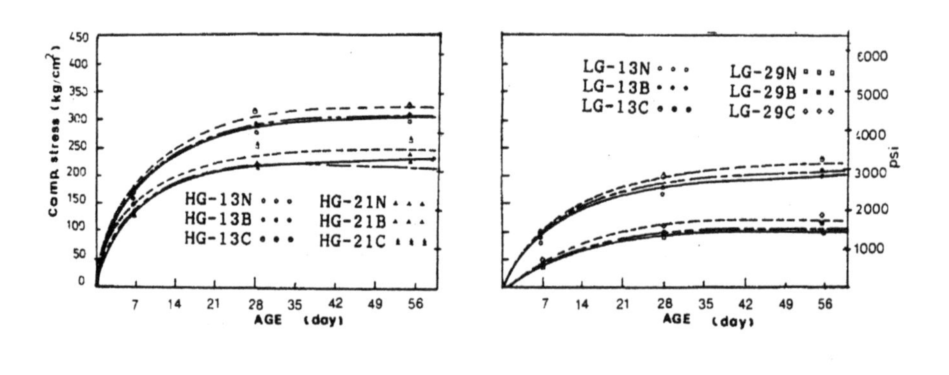

(a) High Grade series 1 (b) Low Grade series

Fig. 5 Strength Development of Three RHA concrete

3.5 Mechanical properties at 28-day age

Table 5 shows the 28-day age mechanical properties of both the plain and RHA concrete. The indirect tensile or split tension test results were likewise plotted and compared with other empirical equations, as shown in Figure 6. The tensile strength (Ft) of RHA concrete ranged from 10 to 15% of its ultimate compressive strength (Fc'), and these data were curve-fitted by using Equation (1) as the function. The equation of the curve fit does not significantly vary in terms of the values of C_1 and C_2 when compared with the Chen (1982) and the CSA-ACI (1983) suggested equations.

$$Ft = C_1 Fc^{C_2} \qquad (1)$$

Table 5 Mechanical Properties at 28-day age

Mix code	Comp Strength (kg/cm^2)	Split Tension (kg/cm^2)	Mod. of Rupture (kg/cm^2)	Modulus of Elasticity 33.3% Fc (kg/cm^2)	40% Fc (kg/cm^2)
HG1-0	371	37.7	37.9	285,380	278,474
HG1-13N	293	31.0	31.9	234,557	228,475
HG1-21N	230	28.8	26.0	222,605	216,856
HG1-29N	161	22.0	22.2	184,859	179,648
HG1-38N	129	14.6	19.7	---	---
HG2-0	352	36.5	35.1	259,652	254,423
HG2-13N	259	29.7	27.4	249,074	245,067
HG2-21N	169	22.6	25.9	200,249	193,440
HG2-29N	144	20.1	21.9	201,844	190,871
HG2-38N	106	15.5	19.4	---	---
MG-0	283	30.1	28.5	256,433	250,435
MG-13N	214	23.8	22.1	241,369	235,143
MG-21N	156	19.2	23.3	222,872	216,523
MG-29N	116	16.8	15.5	187,488	183,088
LG-0	207	26.4	29.5	220,216	212,328
LG-13N	180	24.6	22.9	216,519	208,091
LG-21N	143	18.7	22.2	198,080	189,879
LG-29N	91.6	13.3	11.9	172,945	160,569

The modulus of rupture (Fr) of RHA concrete on the other hand seems lower when compared with the value suggested by the ACI Equation. These tensile behavior findings are still inconclusive, in a sense that the number of specimens and the variables taken are very limited. Thus, it is more adequate to summarize it by a curve-fit equation.

The static modulus of elasticity of RHA concrete was estimated as secant modulus at 33.3% and 40% of the ultimate compressive strength. The results were compared with the established ACI Equation (see Fig.7). It can be observed that the modulus of elasticity is relatively high, and does not significantly differ. Thus, the simpler ACI Equation can be used to conservatively estimate its static modulus of elasticity.

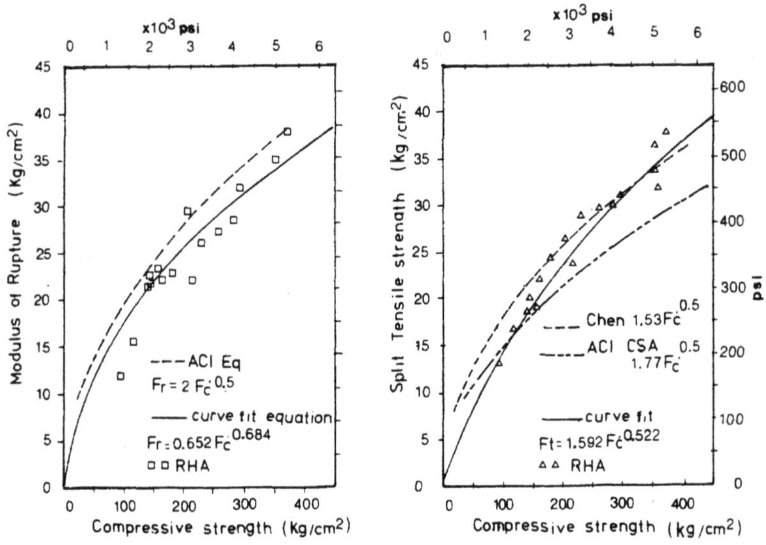

Fig.6 Tensile Properties of RHA Concrete

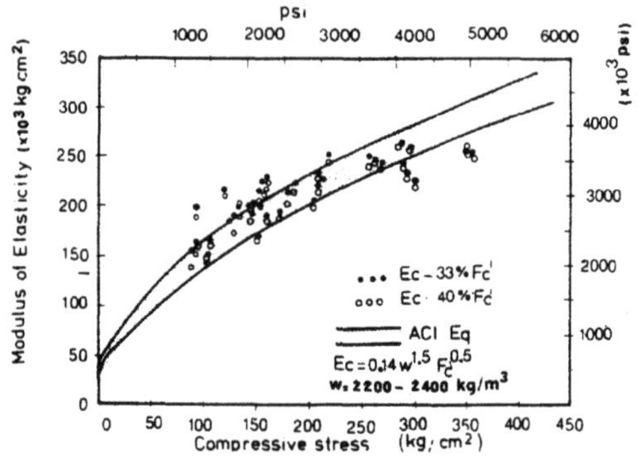

Fig. 7 Modulus of Elasticity of RHA Concrete

3.6 Efficiency of RHA concrete

Haque (1985) defined efficiency as "the ratio of the strength of Fly-ash concrete to an equivalent plain concrete, which has the same total cementitious content, age, and curing history." Generally, the efficiency of concrete decreases as the percentage of fly ash increases. The same behaviour can be seen with the use of the rice husk ash in Figure 8. Among the series, the low grade LG-13N (80:20) gives the highest effeciency value of 93.06% at 56-day age compression and 89.5% at 28-day age split tension. However, the other grade series possessed a relatively low efficiency values especially at 29.5% (60:40) and 38.7% (50:50) ash content. Potential optimum proportions are at 13.6% by weight of cement or 80:20 by volume. Based from this result, the potential optimum proportion is at 80:20 (cement:RHA) by volume.

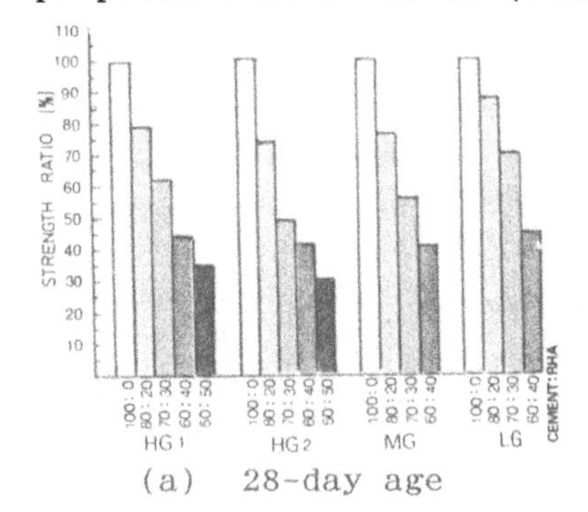

(a) 28-day age

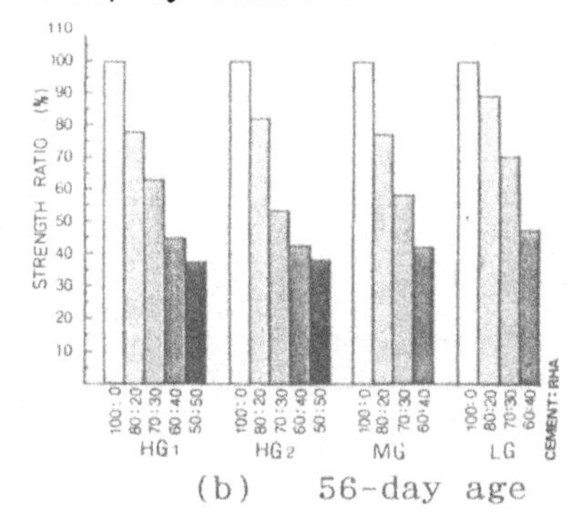

(b) 56-day age

Fig.8 Compressive Strength Efficiency of RHA concrete

3.7 Economic Efficiency of RHA concrete

A present economy study was made to examine the economic efficiency of the use of RHA as cement extender. If savings or losses are purely based on material costs only, and the cost of RHA is considered negligible since it is abundant in the country side only as agricultural waste, then the total material costs will only depend on the quantity of cement. In this study, the strength/cost ratio was used as an evaluating factor of the cost performance of RHA concrete as compared with the plain portland cement concrete.

For the grade series HG1, HG2, and MG, economic efficiencies decreses abruptly as the percentage of ash increases, see Fig.9. This implies that the reduction in material costs due to the use of RHA was counterbalanced by an uneconomical reduction in the compressive strength. On the other hand, the low grade series LG at 13% ash has an economic efficiency of 105 to 110%, and this value is still expected to increase as time goes on. In terms of economic efficiency, the 13% by weight replacement of ash is the most effective proportioning especially for low grade concrete. From earlier observations, the highest compressive strength efficiency can also be found at this proportion.

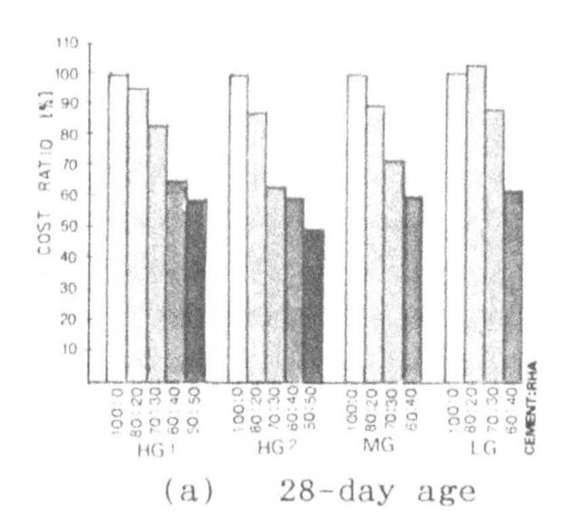

(a) 28-day age

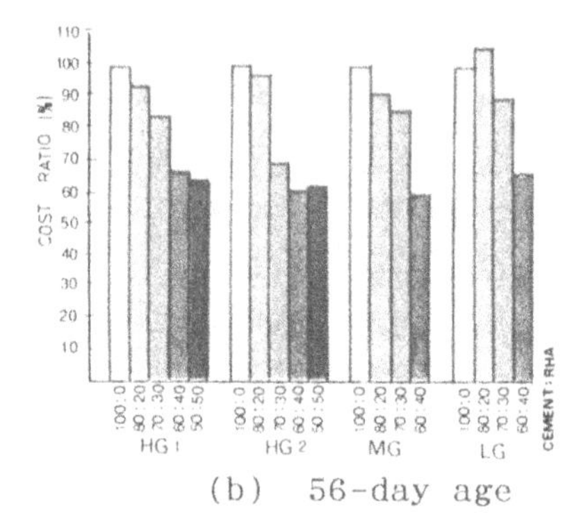

(b) 56-day age

Fig. 9 Economic efficiency of RHA concrete

4 Conclusion

This study indicates that the utilization of rice husk ash can solve to significant problems, viz. elimination of solid wastes, and the provision of a valuable construction material. The study presents some significant effects of this kind of ash to concrete as a cement extender.

(a) The amount of water required to maintain a constant consistency increases as the percentage of RHA as cement extender increases.

(b) Harshness and reduction of air content was observed as the percentage of ash increases.

(c) Both the initial and final setting time of concrete retards as the percentage of ash increases.

(d) Use of finer ash like RHA-B and RHA-C resulted to an increase in strength by 8 to 12% compared with RHA-N concrete.

(e) The use of RHA as cement extender can be most effective for low grade concrete or low strength materials, e.g. less than 2000 psi, or 250 kg/m^3 cement content.

(f) The modulus of elasticity can be conservatively and conveniently estimated by the use of standard ACI Equation. The equations of the tensile properties of RHA concrete, are shown as follows:

$$\text{Split tension} : F_t = 1.592\,F_c'^{\,0.522}$$

$$\text{Mod rupture} : F_r = 0.652\,F_c'^{\,0.684}$$

5 References

American Concrete Institute. (1977) Recommended Practice for Selecting Proportions for Normal and Heavyweight Concrete, ACI Standard 211.11-77, ACI, Michigan, USA.

American Society for Testing and Materials. (1988), Concrete and Mineral Aggregates, 1988 Annual Book of ASTM Standards, Sec.4, Vol. 4.02, 4.03, ASTM, USA.

ACI Committee 318, (1983) Building Code Requirements for Reinforced Concrete ACI Standard 318-83, ACI, USA, 111pp.

Chen, W.F. (1982) Plasticity in Reinforced Concrete, USA.

Chopra, Y.S. (1978), Cementitious Binder from Rice husk: An Overview, Cement research Institute, India.

Dass, Arjun (1978), Pozzolanic Behavior of Rice husk ash, Developing Countries Viewpoint, UDC 691.34; AIT Thailand, pp. 301-311.

Haque, Day, Langan. (1988), Air-entrained Concrete strength with and without flyash, ACI Materials Journal, Vol.85, No.4; ACI, USA, pp. 241-247.

Jorillo, P.Jr.A. (1989), Fresh and Mechanical Properties of RHA Concrete, M.Sc.Thesis, University of the Philippines, Diliman Quezon City, Philippines

Loo and Weregama. (1978), Mix Design of Rice husk ash concrete, Developing Countries Viewpoint, UDC 691.328; AIT, Thailand, pp. 361-366.

Mehta, P.K. (1975), Rice Hull Ash...High quality, Acid resisting, ACI Journal, Vol72, ACI, USA, pp. 235-236.

Popovics,S. (1979), Concrete Making Materials, Hemisphere Publication Corp., USA.

Price, W.H. (1975), Pozzolan - A Review, ACI Journal , Vol.72; ACI, USA, pp. 225-232

Acknowledgment

The authors sincerely thank the staff of CE-Division of the Integrated Research and Training Center of the Technological University of the Philippines for their help in this project. Special thanks to Prof Jose Ma. De Castro of the University of the Philippines and Japanese International Cooperation Agency (JICA) for their invaluable support.

35 OPTIMIZATION OF RICE HUSK ASH PRODUCTION

M.A. CINCOTTO, V. AGOPYAN and V.M. JOHN
Institute of Technological Research and Polytechnic School of University of São Paulo, Brazil

Abstract
Brazil is a main rice producer and large amount of rice husk is available close to the urban centers. A fluidized bed boiler (FBB), where the residence temperature and time can be controlled has been used in order to maximize the energy output as well as the ash pozzolanic activity. For non ground ashes the mechanical properties and pozzolanic activity of FBB ashes are far better than those from ordinary burnt ashes.
Keywords: Rice Husk Ash, Cements, Alternative Cements, By-products, Wastes, Pozzolans, Fluidized Bed Boiler.

Resumé
Le Brésil est un grand producteur de riz et un volume considérable de déchet est disponible dans des grandes villes où il est écorcé. Le reciclage de ce déchet a été donc étudié, comme combustible, dans un four en lit fluidisé; le bût a été celui de determiner les conditions optimales relies à rendement thermique de la combvustion et activitée pouzzolanique de la cendre élevés. Les caractéristiques de cette cendre et d'une autre obtenue en four vertical d'essai de résistance au feu ont été comparés. Les essais mécaniques sur mortiers confectionés avec des mélanges ciment Portland - cendre ont montré que la cendre obtenue en lit fluidisé a une qualité supérieure.
Móts Clés: Cendre d'écorce de riz, Pouzzolane, Ciments, Ciments Alternative, Déchet, Four en Lit Fluidisé.

1 Introduction

Rice husk ash (RHA) is pointed out as the most promising Brazilian agricultural residue to be mixed with ordinary Portland cement. Therefore in a comprehensive study of the use of by-products in the Building Industry carried out at IPT the development of the alternative low alkaly cements

Table 1. Rice husk availability in Brazil (1986)

States	Rice with husk (10^3 tonnes)	Rice Husk (10^3 tonnes)
Rio Grande do Sul	3208	607
Maranhão	1604	303
Minas Gerais	813	154
Goiás	1726	326
São Paulo	914	173
Other States	2310	437
Brazil	10575	2000

for natural fibre composite based on RHA (John et al., 1990) was also included.

The main uses of the rice husk are: mixture with the soil (agricultural use), production of plywood and burning as a fuel. In Brazil, most of the rice processing industries give the husk free of charge because their equipments are electrical ones. On the otherhand, in some industries rice husk is burnt in ordinary boilers for the drying processes which results in about 18% of ashes.

IPT approach for the study of rice husk ash cement is to recicle the husk as a fuel obtained under controlled temperature in order to obtain the best pozzolanic activity of the ash. Published studies present new furnace developments (Salas, et al, 1986; Mehta and Pitt, 1977; Ojha et al., 1977), howewer at IPT persued the use of conventional furnace and avoid the grinding of the resulting ash was pursued.

2 Rice husk availability in Brazil

Although rice production is widespread in the country, only seven states produce the equivalent to 80% of the total amount. They are: Rio Grande do Sul, Maranhão, Goiás, Minas Gerais, Mato Grosso, São Paulo and Santa Catarina. Most of the rice is processed at the production region by small processing industries. Only 17% of te rice is processed at different regions from the production ones. Therefore the husk is available mainly at the rice production regions.

Table 1 presents the states of Brazil where husk is more available. In this table the amount of the husk available was estimated at 18.9% of the rice production. It is possible to conclude that the husk is available in large amounts in the South and Southeast regions of the country which are the most populated ones. About 1/3 of the husk is available in six medium size cities, most of which near the large urban centers.

3 Properties of the husk

The main properties of the ashes produced at IPT are compared with the data available.

Salas et al. (1986) mention that the value of calorific power of the husk, presented by various authors varies from 13810 to 15070kJ.kg^{-1}. In their work they have obtained 18590kJ.kg^{-1} which is substantially greater than any other value mentioned by existing literature. The rice husk in our study has the calorific power of 13045kJ.kg^{-1} (higher calorific value, dry basis). The last value is similar to one presented by Mehta and Pitt (1977). However if we consider the husk available in the rice processing plants which has humidity of 19.6%, the calorific power of the humid material is reduced to 10488kJ.kg^{-1}. It is necessary to point out that this property varies according to different varieties of paddy (Maheshwari and Ojha, 1977).

Bulk density of the husk is 101 kg.m^{-3}, its apparent specific gravity is 900kg.m^{-3} and the true specific gravity is 1400kg.m^{-3}.

4 Burning method

Rice husk can be burnt as a fuel in ordinary furnaces however not all the energy can be easily obtained by this simple process nor is the resulting ash usually suitable for use as a pozzolanic material. Beagle (1977) informs that 50% of energy can be wasted. Therefore a fluidized bed boiler (FBB), where the residence temperature and time can be controlled has been used, so the energy output is maximized as well as the ash produced has adequate pozzolanic activity.

With the help of a theoretical model (de Souza Santos, 1989) and a computer, it was possible to simulate an operation using a Babcok & Wilcox boiler where the mineral coal was replaced by husk.

The results of this simulation demonstrated the feasibility of the use of husk as a fuel. This material was tested in a reactor of 500 KW which has an internal diameter of 0.5m, a bed of 1m and total height of 3.5m to prove that the ash produced has pozzolanic activity and also to confirm that it is possible to operate the boiler in a steady-state way. The feeding of the husk into the boiler is done pneumatically and the fly ash is collected by the cyclone.

The first test produced ashes with a high rate of carbon as the loss on ignition varied from 12% to 17% of the weight. For this reason, it was decided to recycle part of the ashes by increasing the residence time to 1.33h. In this case it was possible to reduce the loss on ignition to less than 2% by weight. The results of pozzolanic activity of the ashes were as high as 700mgCaO/g of ash (Chapelle Method) but their activity can be reduced to 400mg CaO/g of ash if the maximum temperature of the free board of the boiler is higher than $900^{o}C$.

In the last attempt it was decided to reduce the temperature of the boiler which reduced the carbon concentration in the bed by both reducing the rate of introduction of the materials into the boiler and the fluidizing air flow. With these procedures it was possible to operate the boiler under better conditions. The feeding rate of the husk was $45kg.h^{-1}$, and the ash production $3.6kgh^{-1}$. In this case only 8% of the weight of the husk was collected by the cyclones as fly ash. The remainder is bottom ash which is mixed with the sand of the boiler.

5. Properties of the ash

The chemical composition of the ash is: 94.7% SiO_2; 0.09% Al_2O_3; 1.46% Fe_2O_3; 0.99% CaO; 0.95% MgO; 0.04% Na_2O; 1. 75% K_2O and 0.21% SO_3.

The ordinary ash (burnt in static furnace, starting with a fuel and smouldering during 24 hours. at temperatures bellow $800^{o}C$) tested by the Chapelle Method has a lime consumption of 606 $mgCaO.g^{-1}$ of ash which is lower than the range of results of the ashes prepared in FBB (623 - 699 $mgCaO.g^{-1}$ of ash) without grinding. For standard fineness of 325 mesh the results are 700 $mgCaO.g^{-1}$ and 730 $mgCaO.g^{-1}$ for ordinary and FBB ash respectively. From the results of the Chapelle Test Method, one concludes that all of the samples have good pozzolanic activity (higher than $500mgCaO.g^{-1}$ of ash).

Figures 1 and 2 present the micrographs by scanning electron microscopy of the ordinary ash and of an ash obtained in FBB as described in this section. The microstructures of the ashes are different. The ash from FBB has a very porous structure like the husk itself while the ash from an ordinary boiler has a fraction of compacted and granulated structure, this probably happended due to long residence period.

The ordinary ash has a fineness (Blaine method) of $145m^2.kg^{-1}$ and a specific gravity of $1880kg.m^{-3}$. The FBB ash has a fineness of $419m^2.kg^{-1}$ and specific gravity of $2110kg.m^{-3}$. Mehta (1986) shows results of specific area by nitrogen absorption between 50 and $60m^2.g^{-1}$, however the ash prepared by this author probably had a high carbon content as he informed that the resulting concrete is black (Mehta and Pitt, 1977). The high carbon content may be the reason for this high specific area. In our case with low carbon content (less than 2%), the specific areas are $18.5m^2g^{-1}$ and 6.0 m^2g^{-1} for ordinary burnt ash and FBB ash, respectively.

The FBB ash collected in the cyclone has a litle quantity of sand from the bed. An average sample has 43.5% of particles small than 0.044mm and only 4.3% above 0.084mm, which are mainly sand as observed in an optical

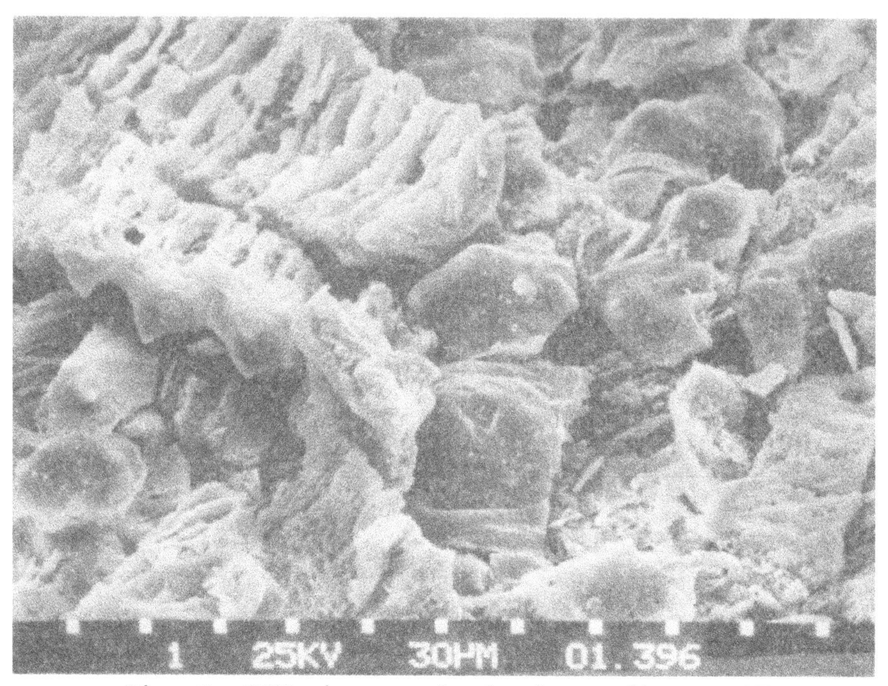

Fig. 1. SEM micrograph showing ordinary RHA.

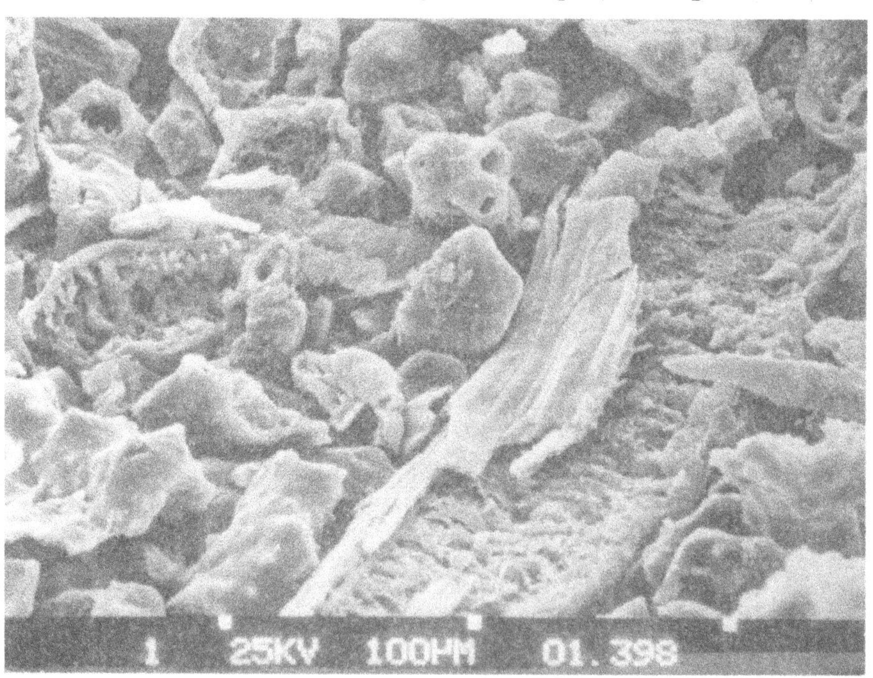

Fig. 2. SEM micrograph showing FBB rice husk ask.

mycroscope.

Figure 3 presents the results of X-ray diffraction analysis under different burning conditions: a band centered at 21.7 degrees, which is caracteristic of amorphous silica for all ashes; however for FBB burnt at temperature higher than 800°C but not exceeding 900°C it was observed cristobalite and alpha-quartz which is from the sand (curve b); moreover for high residence time tridimite peaks are also observed (curve c). Curve 'a' refers to an ordinary burnt ash.

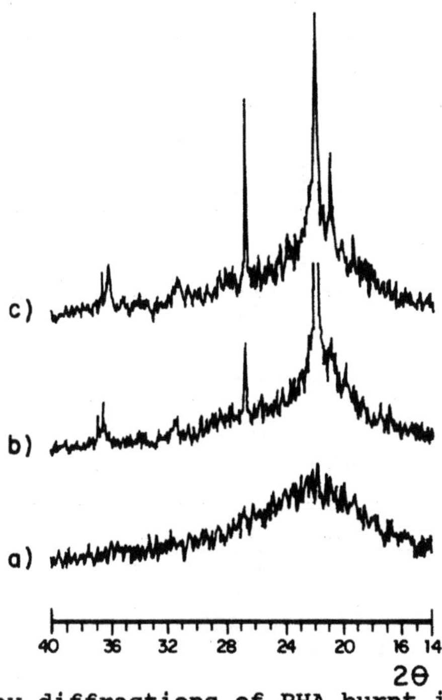

Fig. 3. X-ray diffractions of RHA burnt in ordinary furnace (a), in FBB during 1.33 hours (b) and in FBB of residence time more than 1.33 hours.

6 Mechanical properties

For non ground ashes, mixtures of up to 50% of FBB ash have higher compressive and bending strengths than those ones of 30% of ordinary ash as can be seen in figure 4. This happens because of higher pozzolanic activity and lower water content and air content of the FBB ash mixtures, for the same workability, as can be seen in figures 5 and 6, respectively. All of the tests were performed with binder:sand ratio of 1:2.5 and workability of the mortar of 250mm (flow table). Therefore one can say that the FBB ashes can reduce OPC contents in the mortars with the same mechanical performance.

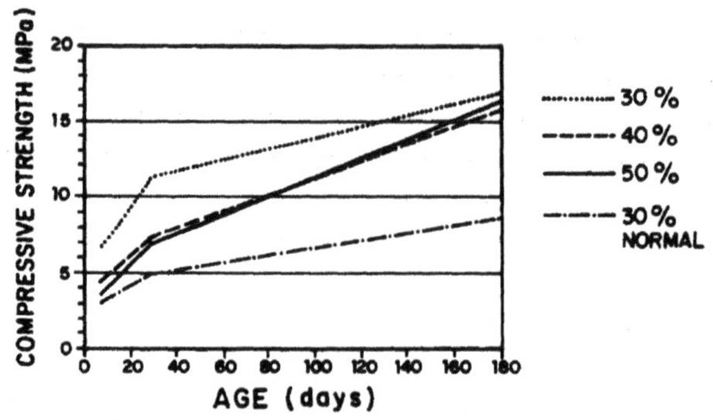

Fig. 4. Compressive strength of mortars produced with differents RHA contents production methods.

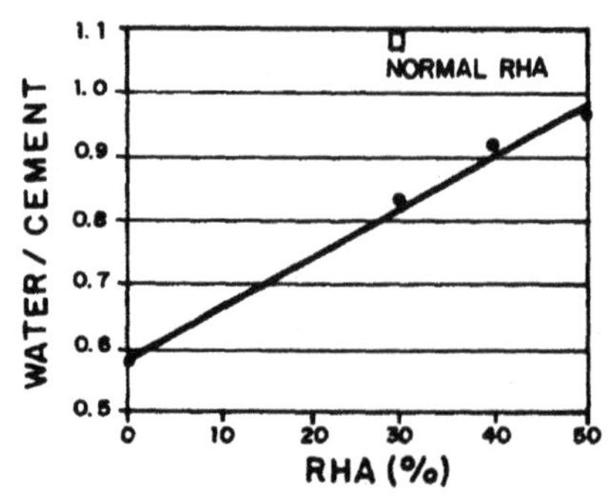

Fig. 5. Water/cement ratio and RHA content for constant workability.

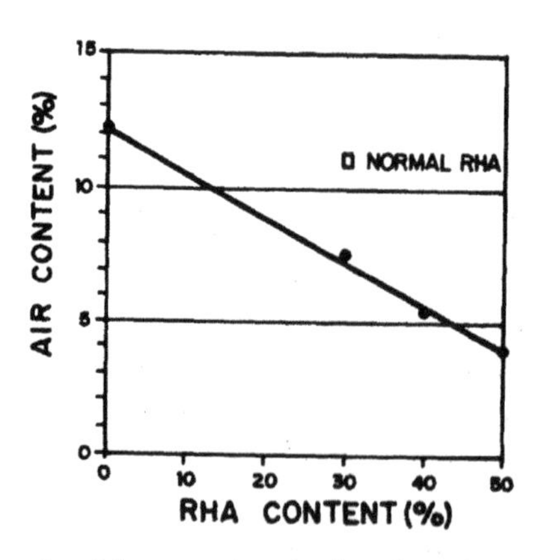

Fig. 6. Air content in fresh mortars.

7 Final comments

The authors do not consider the research as finished as some details have not yet been yet clarified. For instance, the temperature and residence time of the burning of the husk in the fluidized boiler bed can be better optimized. In addition experiments with the bottom ash mixed with sand must be made because the amount of this ash is as much as the fly ash used in this research.

The results obtained are very encouraging, therefore further studies are recommended. The binder developed in this research can be regarded as a mansory cement following the BSI and ASTM specifications, however this must be better analysed.

8 References

Beagle, E. C. (1977) Basic and Applied Research Needs for Optimizing Utilization of Rice Husk, in **Rice by-products utilization Int. Conf.**, Instituto de Agroquímica y Tecnologia de Alimentos, Valencia, v. 1, pp. 1-44.

de Souza Santos, M.L. (1989) Comprehensive Modeling and Simulation of FBB and Gaseifiers. **Fuel**, 68, 1507-1521.

John, V. M. Agopyan, V. and Derolle, A. (1990) Durability of BFS Based Cement Mortar Reinforced with Coir Fibres, in **2nd Int. Symp. on Vegetable Plants and their Fibres as Building Materials**, RILEM, Salvador (in printing).

Maheshwari, R.C. and Ojha, T. P. (1977) Fuel Characteristics of Rice Husks, in **Rice By-products Utilization Int. Conf.**, Instituto de Agroquímica y Tecnologia de Alimentos, Valencia, v. 1, pp. 67-76.

Mehta, P. K. (1986) **Concrete: Structure, Properties, and Materials.** Prentice-Hall Inc., Englewood Cliffs.

Mehta, P.K. and Pitt, N. (1977) A New Process of Rice Husk Utilization, in **Rice by-products utilization Int. Conf.**, Instituto de Agroquímica y Tecnologia de Alimentos, Valencia, v. 1, pp. 45-58.

Ojha, T.P. Maheswari, R. C. and Shukla, B.D. (1977) Optimizing Technologies of Rice Husk Utilization as Source of Fuel and Industrial Board, in **Rice by-products utilization Int. Conf.**, Instituto de Agroquímica y Tecnologia de Alimentos, Valencia, v. 1, pp. 77-87.

Salas, J. et al. (1986) Use of Rice Husk Ash: an Addition in Mortar. **Materiales de Construccion**, v.36, n.203, 21-39.

Acknowledgements

This paper presents a part of vegetable fibre reinforcement studies made by the authors at IPT. The authors would like to thanks the Secretariat of Science, Technology and Economical Development of the State of São Paulo and the IDRC- International Development Research Centre (Canada) for the financial support. They also would like to aknowledge the help of the Thermal Group of the Mechanical and Electricity of IPT, led by Dr. M. L. de Souza Santos.

36 FIQUE LIQUOR AS RAW CONSTRUCTION MATERIAL

R. De GUTIERREZ
University of Valle, Cali, Colombia

Abstract
An admixture for use in Cement Portland pastes, mortars and concretes is obtained from the juice remained in the process of pulling out the fiber from the pulpy leaves of the fique plant (Furcraea Macrophylla). The fique juice represents 84% of the whole plant. The developed liquor enlarges the plasticity of the fresh cementicious materials and decreases the water/cement ratio up to 10% unchanging the original workability. In addition, it is possible to increase 50% the compressive strength with respect to the standard material (without admixture) cured at 28 days. The behavior of the fique admixture is similar to that of the commercial superplasticizers.
Keywords: Admixture, Superplasticizer, Fique Juice.

1 Introduction

Fique is a hard fiber extracted from the cabuya plant. It is cropped in fifteen departments of Colombia being the most importants: Antioquia, Santander, Cauca, Nariño and Boyacá. More than one million inhabitants live from the Cabuya.

In the Cauca department, Fique is being grown in 8000 hectares by indigenous people. The predominant Fique variety is the FURCRAEA MACROPHYLLA. This receives the name of Eagle Nail because its sharp printed green leaves have rounded pins at its edges, (Fig. 1).

In the fiber pulling out process from the Fique leaves remains 96% by weight of waste composed of a liquid substance and a fibrous pulp, (Fig. 1). Up to now the long fiber has been utilized as raw material for bags, ropes, and craftsman articles. However, the manmade polymers compete advantageously to displace the natural fibers utilization. Because of this situation, it is neccesary to search for new use and applications of the Fique fiber. Two of them are the manufacture of fine papers and fiber reinforced building materials.

Fig. 1. FIQUE a) Plant. b) Fique Fiber (Cabuya).
c) Waste.

An Universidad del Valle research program (Supported by grants provided by a Colombian governmental agency to promote cientific research, COLCIENCIAS and the Agriculture Ministry) is working on an integral use of the plant Fique components. One of the main objectives of the project is to develop methods in order to increase the durability and the dimensional stability of the fiber to produce fiber reinforced building materials elements. Table 1 establishes a comparison between the properties of the Fique fiber and other types of fibers. This program also pretends to develop admixtures for cement mortars and concretes from the juicy part of the Fique pulp. Other

TABLE 1. PROPERTIES OF SOME FIBERS

FIBER	PROPERTIES			
NAME - SPECIE (ORIGIN)	DENSITY (Kg/m3)	TENSILE STRENGHT (MPa)	D. (μm)	ELONG. AT BREAK (%)
FIQUE (Cabuya) - Furcraea Macrophylla (Colombia)	1009	293	48-338	2.6
JUTE - Corchorus (S. Asia)	1500	217	-	1.6
BAGASSE - (Tropics)	1240	20	-	-
BAMBOO - Dendrocalamus Strictus	1500	350 500	-	-
COTTON - Gossypium Family (U.S., Russia, S. America, Africa, Asia)	1445	280 840	10-20	5-10
SISAL - Agave Sisalan (E. Africa, S. America, Mexico)	1340	278 839	7-48	2.9-3.7
HENEQUEN-Agave Fourcroydes (Cuba-Mexico)	-	205	-	5.5
COIR - Cocas Nucifera (Tropics)	1150	72 530	100-450	15-40
FLAX - Linum Usital I. (Europe, Russia Asia)	1500	385	11-50	1.8
CELULOSE -	1200	400	-	-
ASBEST - (Mineral)	2550	200 3500	0.02-30	2-3
POLYPROPILENE - (Manmade)	910	400	100-4000	8-18

researchers are interested to produce alcohols, fertilizers, concentrates, steroids, insecticides, and many other products.

2 Methodology

The main chemical, physical and mechanical characteristics of the cement used as standard are presented in Table 2. Portland Cement pastes and mortars were prepared using different types of admixtures. These juices were produced from the juice by a selective organic solvent process of extraction. It was evaluated the effect of those additions on the normal consistency, setting time, and compressive strengths when curing at 7 and 28 days.

TABLE 2. PHYSICAL PROPERTIES AND CHEMICAL ANALYSIS OF CEMENT

CHEMICAL COMPOSITION, (% w/w):

SiO_2	Fe_2O_3	Al_2O_3	CaO	MgO	SO_3	CaO (free)	L.o.I.
21.07	4.69	5.41	63.04	2.02	1.51	1.63	1.52

PHYSICAL PROPERTIES:

Normal consistency (%)	23.66
Specific Gravity (Kg/m^3)	3120
Setting time (initial) (min)	95
Fineness:	
- 74 μm (retained) (%)	4.00
- Surface Area, Blaine (m^2/Kg)	368

3 Results and Discussion

The original Fique Plant juice, without previous treatment, was added to a cement paste replacing 10% of the water to obtain normal consistency with a water/cement ratio of 0.208. In the cement paste without admixture was measured a ratio of 0.237 to get the same consistency. This indeedly shows the plasticizing effect on the mix of the juice incorporation. A better result was obtained by removing the vegetable pulpy wastes, principally chloroplasts, from the original juice.

The setting time of the original and modify pastes were almost the same but, in the first minutes, the viscosity of the added pastes was greater than the Portland cement pastes without admixture. This effect is similar to that observed in commercial plasticizers.

Portland Cement added mortars (composed of one part by weight of cement and 2.75 parts of Ottawa sand were casted in 1 x 2 inch cylinder molds) showed to have similar characteristics to those observed in the pastes. As can be seen in Table 3, the incorporation of 2%, with respect to the cement of the original juice (samples 1 and 2), reported greater plasticity than the standard mortar (sample 0). When the water/cement ratio was disminished to 0.434 to get the same workability, the strength lost was minor. The same Table 3 show the strength results in mortars including different types of admixtures produced by filtering, cold or hot extraction in organic phase (CC14, CHCL3, CH2C12, AcEt, C6H6, ethanol), evaporation and dilution. The samples 3, 4, 6 and 16 report the best results with strength increases up to 55% with respect to the standard mortar at 7 curing days and up to 70% at 28 curing days. These samples correspond to the liquors extracted from the Fique plant by solvents as CH2CL2, CHC13 and ethanol. These admixtures are similar to saponins because of the brown color, sweetness and foaming in water.

TABLE 3. MORTAR STRENGTHS

SAMPLE	WATER/CEMENT RATIO (w/c)	FLUIDITY (%)	COMPRESSIVE STRENGTH (MPa) CURING AGE (days)	
			7	28
0	0.485	81	18.0	20.8
1	0.485	100	0.4	3.1
2	0.485	108	2.0	9.5
1	0.434	95	7.1	9.1
2	0.434	100	7.1	10.6
3	0.434	84	20.9	34.5
4	0.434	100	19.8	29.6
5	0.434	92	23.1	25.9
6	0.434	84	27.9	35.6
7	0.434	82	14.4	18.5
8	0.434	100	0.3	12.2
9	0.434	82	15.2	22.1
10	0.434	78	16.6	22.8
11	0.434	80	16.7	19.4
12	0.434	78	15.9	16.8
13	0.434	68	14.2	16.7
14	0.434	73	12.6	17.6
15	0.434	75	13.6	17.0
16	0.434	85	24.0	34.4

Table 4 presents the strength results of one half inch cylinder cement pastes prepared by mixing water to which

TABLE 4. CEMENT PASTE STRENGHTS, 28 days (MPa)

SAMPLE	ADDITION (%)	RATIO water/cement (w/c, %)						
		25.6	25	24	23	22	21	20
STANDARD	0	188	229	233	237	248 (216)	-	-
ALCOHOL EXT.	1.0	268	-	290 (202)	239	282	-	-
HOT ALCOHOL EXT.	0.4	-	-	-	298 (246)	285	-	-
CH2CL2, EXT.	1.7	-	-	-	263 (220)	259	-	-
HOT CH2CL2 EXT.	1.0	-	-	-	283	284 (271)	-	-
MELMENT	0.5	-	-	-	-	261	280	232
MELMENT	1.0	-	-	-	-	-	265	257

has been added an admixture obtained from Fique juice by filtering, extracting in presence of ethanol or CH2C12 followed by a concentration process at a maximun of 70° C. Each result corresponds to a mean value of at least thirty datas with a maximun variance of 6.3%. The compressive strengths at 28 curing days are reported for mortars with different w/c ratios but the same fluidity. Comparatively, it is reported the values obtained in pastes with addition of 0.5 and 1.5% of the Melment with respect to the pure cement. The outlined datas are the optima where the workability agrees with the greatest strengths. Seven days curing strengths appear between parenthesis. The pure cement paste reported a normal consistency of 25.6%. It was found, in all cases, that the strengths generated by incorporation of the Fique admixture were greater than the standard and Melment added pastes.

By now, the research work tends: to prove the effect of the Fique admixtures on concretes, to determine its main active constituents and to design the making process.

4 Conclusions

On the basis of this study, the following conclusions can be drawn.

a. The extracted Fique plant admixture increases the workability, if the water cement ratio is kept up unchanged.

b. In order to obtain the same workability it is possible to reduce up to 10% of the mixing water by incorporation of the Fique admixture.

c. The addition of 2% of the Fique admixture increases the strenght in order of 25% for cement pastes and 50% for cement mortars. 4. In accordance with the Fique admixture behavior and its effect by incorporation in cement mixes, the new product obtained from the pulpy Fique plant can be considered a plasticizer water reduction admixture.

d. In accordance with the Fique admixture behavior and its effect on the incorporation in cement mixes, the new product obtained from the pulpy Fique plant can be considered a plasticizer water reductor admixture.

5 References

Acuña, O. y Romo, G. (1983) Posibilidades técnicas de elaboración de pulpa y papel a partir de las cabuyas blanca y negra. Politécnica, VIII, 4, Ecuador.

Domínguez, A. (1976) **Métodos de Investigación Fitoquímica.** Capítulo XI.

Hannant, D.J. (1978) **Fibre Cements and Fibres Concretes.** John Willey & Sons, New York.

Hewlet, P.C. (1984) Superplasticised Concrete: Part 1 and 2. Concrete.

Moreno, M.C. (1988) **Plan Indicativo del Fique.** Ministerio de Agricultura, Bogotá.

Muñoz, L.A. (1988) **Producción de Cortisona a partir del Jugo de Fique.** Tesis, Univ. Pontificia Bolivariana, Medellín.

Venuat, M. (1972) **Aditivos y tratamientos de morteros y hormigones.** Editores Técnicos Asociados, España.

________. (1986) Proyecto de una planta piloto para la producción de 2.5 ton/turno de pulpa de celulosa a partir de fique. Ministerio de Agricultura, Dri-Pan y Techmipetrol, Bogotá.

________. (1988) **Vegetable-Fibre Cement Board.** Prepared by the Institute for Material and Environmental Research and Consulting (Intron), United Nations.

37 THE PERFORMANCE OF A PULP AND PAPER INDUSTRY BY-PRODUCT AS A WATER-REDUCER IN CONCRETE AND A CORROSION INHIBITOR TO STEEL REINFORCEMENT

H.A. El-SAYED
Building Research Institute, Dokki, Cairo, Egypt

Abstract
Black liquor (orlye)-by product wasted in the pulp and paper industry in Egypt-mainly composed of soda lignin-has been evaluated from the standpoints of : a) functioning as a water-reducing admixture in concrete. Its effects on the reinforcement passivity and its corrosion resistance towards surrounding aggressive media as well as it impact on the engineering properties of concrete, namely, its compressive strength and bond strength with the embedded reinforcement have been evaluated , and b) corrosion inhibition afforded when black liquor is incorporated in a coating to the steel reinforcement in concrete admixed with or, immersed in aggressive media. Also, the effect of such steel coating on the bond strength between the steel and the surrounding concrete has been investigated.
Keywords: Black liquor, Corrosion, Inhibition, reinforcement,Compresive, bonding

1 Introduction

A water-reducer can be defined as " an admixture that reduces the amount of mixing water of concrete for a given workability" . It improves the properties of hardened concrete and, in particular, increases strength and durability. A reduction in the amount of water required to make concrete is commonly obtained by using admixtures based on lignosulfonates, inorganic materials and complexes of organic hydroxy carboxylic acids.

In the present study, black liquor (or lye)-a commercial by-product wasted in the pulp and paper industry in Egypt, mainly composed of soda lignin-has been utilized as a water-reducing concrete admixture. Its effects on the passivity of steel reinforcement and corrosion resistance against surrounding aggressive media as well as its impact on the engineering properties of concrete, namely, its compressive strength and bond strength with the encased steel reinforcement have been investigated.

Where exposure conditions are severe, screening the steel from the concrete and thus from the aggressive medium could be the proper method to protect reinforcement from corrosion and,hence, consequent damage of concrete. In the present investigion, black liquor has been tried as an admixture to a cement slurry to develop a new protective coating to the steel. The corrosion prevention efficiency of this coating has been evaluated using anodic polarization of the steel in concrete when corrosive ions are mixed with the concrete and when aggressive ions diffuse from outer media through the concrete. The steel/concrete bonding characteristics as affected by the coating have been also determined.

2 Experimental

The ingredients of the concrete mix used in the present study were ordinary Portland cement, sand and gravel. The chemical composition and Blaine specific surface area of the used cement is given in Table 1. The maximum size of the coarse aggregate was 10 mm. This size has been found relevant for use with the different sizes of moulds utilized. The mix proportions were 1:2:4, by weight, for cement:sand: gravel. The water/cement ratio was 0.6 by weight. Tap water was used for concrete mixing. The concrete samples containing black liquor were prepared in the same way. Laboratory measurements indicated that, each 1000 ml of black liquor of density 1 gm/cm^3 contains 16.2 gm of solid material. An admixture concentration of 0.2%-which is normally used in practice for water-reducers has been used. The equivalent solution was added to the mixing water before adding it to the dry concrete mix. In the presence of black liquor, the W/C ratio has been decreased from that of the control sample i.e. 0.60 to 0.55 to give the same workability as measured by the slump test (B.S.1881, 1970).

The reinforcing steel used was mild steel bars, 6 mm in diameter. Table 2 gives the chemical analysis for C,S,P,Mn and Si as well as the mechanical properties of such steel. The rods were mechanically polished to remove the firmly adherent mill-scales on the steel surface and then coated with wax at two places in such a manner that a definite predetermined area was always exposed between the waxed areas. In case of steel coating, the sequence of the coating was as follows:
a-De-rusting of the steel surface: The steel rods were dipped into 5% H_2SO_4 acid to which 0.1% thiourea inhibitor has been added for about 20 minutes. The rods were removed from solution as soon as the oxides next to the metal surface were removed satisfactorily and bright surfaces were obtained.
b-Acid neutralization: Acid pickling was immediately followed:by cleaning the rods with a wet waste cloth carrying sodium carbonate so as to neutralize any acid that may be lurking in crevices.
c-Phosphating : It is known that when steel is treated under appropriate conditions with phosphoric asid, alone or containing certain metal phosphates in solution, an adherent phosphate film is produced on the metal surface. This type of coating does not suffice by itself to prevent the rusting of steel for a long time, but, it forms an admirable basis for painting schemes. The advantage of such phosphate coat lies in the acid attack of the steel and the production of a network of porous iron phosphate crystals firmly bonded to the steel surface. This will ensure a good adherence of subsequent coating and

Table 1: Chemical composition and Blaine specific surface area of ordinary Portland cement.

Component	Loss on Ignition	SiO_2	Fe_2O_3	Al_2O_3	CaO	MgO	SO_3	Free Lime	Na_2O	K_2O	Cl^-	Total	Blaine $(cm^2 g^{-1})$
Percentage	1.17	20.10	3.26	5.19	62.06	2.62	2.37	0.30	2.48	0.75	0.0	100.6	3659

Table 2: Chemical analysis and mechanical properties of reinforcing steel

Chemical analysis, %					Mechanical properties		
C	S	P	Mn	Si	Yield stress $(Kg.mm^{-2})$	Ultimate stress $(Kg.mm^{-2})$	Elongation (%)
0.12	0.022	0.063	0.43	Trace	38.8	49.3	23.0

decrease the tendency for corrosion to undercut the coat film at scratches or other defects in the coat at which corrosion could initiate. Phosphating (sometimes called phosphatizing or pakerizing or bonderizing) was applied by brushing the clean surface of steel by a cold 3% zinc orthophosphate with orthophosphoric acid ($Zn_3(PO_4)_2 \cdot 4H_2O + H_3PO_4$ till acidity) then leaving the phosphating solution on the steel surface for 45-60 minutes. The thickness of the phosphate film thus produced was about 3 μm

d- Steel coating : cement slurry was prepared by stirring 1000 gm of Portland cement, passed 240 mesh sieve, with 500 ml of distilled water for two hours. The pH of the cement slurry should be about 12.5. 0.2% black liquor, by weight of cement in the slurry, was added to the water before mixing. A cement slurry coating was applied by brushing a first coat to the steel in the same day of performing the steps of derusting, neutralizing and phosphating. After 24 hours of air drying, brushing with a sealing solution (80% $K_2Cr_2O_7$ solution) was carried out then applying the second coat of cement slurry. Then, after 24 hours of air drying , the coat was brushed with the sealing solution and after 4 hours of air drying again the sealing solution was applied. The entire procedure needed about 3 days.

The corrosion behavior of steel-in-concrete electrodes was studied by means of the galvanostatic anodic polarization technique. The steel-in-concrete was made as an anode in a circuit using an auxiliary platinum electrode as cathode. A constant current density of 10 $\mu A\ cm^{-2}$ was applied to the steel electrode and the corresponding potential was recorded as a function of time.

For carrying out the compressive strength measurements, concrete cubes 10x10x10 cm were prepared. The cubes were cured for 24 hrs, demoulded and continuously cured in a humidity chamber till the time of testing at 7, 28, 90, 180 and 365 days. A set of 3 cubes were used for each compressive strength determination.

For carrying out the bond strength measurements, the concrete mix was cast into standard 10x20 cm cylindrical moulds in which steel rods were centered. The concrete samples were demoulded 24 hrs after casting and then cured in the humidity chamber up to 6 months. The bond strength was then determined by carrying out the pull-out tests for the steel embedded in the concrete cylinders and recording the respective loads at which initial and then ultimate slips occur. A set of 3 tests were carried out for each measurement.

3 Results and discussion

3.1 Utilization of black liquor as a concrete admixture

3.1.1 Anodic polarization behavior of reinforcing steel embedded in concrete admixed with black liquor.

Previous investigations (Gouda and Halaka, 1970; El-Sayed and Sherbini, 1984; El-Sayed 1985) proved that anodic polarization of steel in concrete at low current densities is the most reliable accelerated corrosion test for determining whether a given medium is corrosive or inhibitive. Hence, the anodic polarization behavior of steel in concrete admixed with 0.2% black liquor, relative to cement weight, at a constant applied current density of 10 $\mu A\ cm^{-2}$ has been determined. The curve obtained is given in Fig. 1 which presents also the curve pertaining to the control sample (admixture-free).

It can be seen that, both curves exhibit the same general trend where the steel potential rises steadily towards noble values till reaching the oxygen evolution potential. In presence of black liquor the steel reinforcement has become passive after only 5 minutes where the steady oxygen evolution potential has reached a value of +1180mV compared to 6 minutes to reach passivation at a steady state potential of + 1120 mV for the control sample.

3.1.2 Role of black liquor upon maintaining the passivity of steel reinforcement in concrete subjected to different surrounding aggressive media.

The effect of admixing the concrete with black liquor upon suppressing the corrosion of steel reinforcement in concrete during its service in aggressive environments has been evaluated . Thus , a series of experiments has been carried out in which steel-in-concrete electrodes were prepared. The concrete was mixed with 0.2% black liquor. The electrodes were immersed for 28 days in media simulating some of the most severe conditions that could confront the concrete in service, namely, 3, 4, 5, and 6% NaCl solutions and 5, 6 and 7% Na_2SO_4 solutions as well as sea water. Then , anodic polarization measurements were carried out. The results obtained are summarized in Table 3 giving the aggressive medium and its concentration and the time required for attaining steel passivation. Such recorded time has been taken as a measure for the degree of steel passivation. Thus, if the time is $\leqslant$ 7, 7-15 and $\geqslant$ 15 minutes, the degree of passivation has been regarded as " high" , "moderate" and "low" , respectively.

Table 3 shows that steel reinforcement could affectively sustain the deleterious effects of aggressive ions in the environment surrounding the concrete when the later has been mixed with 0.2%

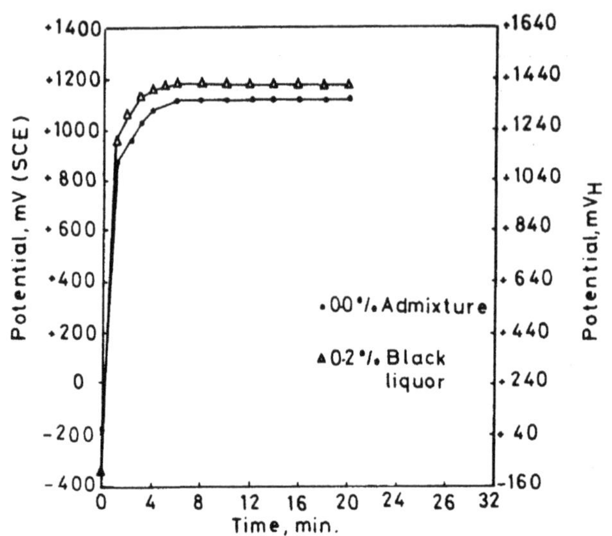

Fig.1. Anodic polarization of steel embedded in Portland cement concrete admixed with 0.2% black liquor relative to the control sample at a current density of 10 μA cm^{-2} tested in sat. $Ca(OH)_2$ solution.

black liquor. It can be seen that, while steel passivity has been destroyed in the control concrete upon soaking in 4%NaCl solution, it has been maintained in concrete incorporating 0.2% black liquor upon soaking in NaCl solutions up to a concentration of 6% . The degree of steel passivity is "moderate" on soaking in 4% NaCl and "low" on soaking in 5% and 6% NaCl. Also, in presence of black liquor, steel reinforcement in concrete manifests a " high " degree of passivation upon soaking in Na_2SO_4 solution up to a concentration of 7%. In sea water, the degree of passivation is " moderate".

Table 3. Effect of black liquor upon maintaining the passivity of steel reinforcement in concrete immersed in different concentrations of sodium chloride, sodium sulfate and in sea water.

Admixture conc.	Aggressive medium and its conc.	Time required for passivation, min.	Degree of steel passivation
Control sample (Admix.-free)	NaCl		
	3 %	12	Moderate
	4 %	..	Borderline case
	Na_2SO_4		
	5 %	6	High
	6 %	7	High
	7 %	7	High
	Sea water	16	Low
0.2%Black liquor	NaCl		
	3 %	7	High
	4 %	13	Moderate
	5 %	18	Low
	6 %	19	Low
	Na_2SO_4		
	5 %	7	High
	6 %	7	High
	7 %	7	High
	Sea water	9	Moderate

3.1.3 Effect of black liquor on some engineering properties of hardened concrete

This part of the study has been carried out to investigate the effect of black liquor-as a water reducing admixture- on the compressive strength of concreteas well as its bond strength with the embedded reinforcement.

3.1.3.1 Compressive strength

The compressive strength values obtained for concrete admixed with 0.2% black liquor compared to the values obtained for the control sample as a function of curing time from 7 days up to 1 year have been determined. The results are given in Table 4 and are diagramatically represented in Fig 2.

Table 4. Compressive strength of concrete admixed with 0.2% black liquor compared to control concrete as a function of curing time, Kg,cm^{-2}.

Curing age, days / Admixture	7	28	90	180	365
Control sample	138.8	167.3	171.4	204.0	235.0
0.2% black liquor	181.3	189.3	255.6	296.0	324.0

It can be seen that, an appreciable gain of compressive strength resulted due to admixing concrete with 0.2% black liquor reaching 30.6% over the compressive strength of the control sample after curing for 7 days. The compressive strength progressively increased up to 1 year where the increase over the control sample reached 37.9%.

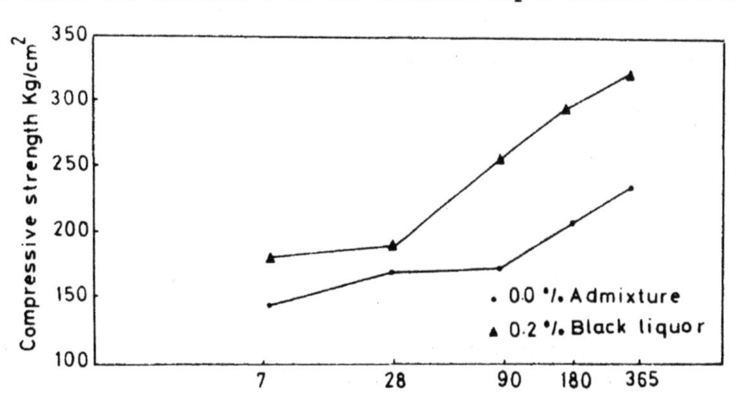

Fig.2. Compressive strength of concrete admixed with 0.2% black liquor compared to control sample as a function of curing time.

3.1.3.2 Bond strength

Table 5 presents the results of bond strength measurements for reinforcing steel embedded in concrete admixed with 0.2% black liquor compared to admixture-free concrete cured for 6 months.

Table 5. Bond strength (Kg cm^{-2}) for reinforcing steel in concrete incorporating 0.2% black liquor compared to control concrete cured for 6 months

Admixture	Initial strength	Ultimate strength
Control sample	21.5	22.5
0.2% black liquor	28.4	29.6

It is evident that, the initial as well as the ultimate bond strength values increase upon black liquor inclusion relative to those obtained for the control concrete.

Such enhancement of the engineering properties of concrete upon admixing with black liquor is mainly attributed to the reduction in the amount of the initial mixing water, thus, decreasing the total porosity of the hardened concrete and consequently increasing the compressive strength as well as the bond strength with the embedded reinforcement.

It is worth mentioning that, scant literature has been found concerning the performance of black liquor as an admixture in concrete inspite of the extensive studies devoted to pure calcium lignosulfonate and its effect as a concrete admixture. Rixom (1975) reported that, by-products from the sulfite process for the production of pulp i.e. lignosulfonates could be useful in increasing the Workability of concrete or for allowing higher strengths by the utilization of lower water/cement ratios. Also, in an unpublished work, Gouda, et al, indicated that admixing concrete with black liquor did not adversely affect the mechanical properties of concrete.

3.2 Utilization of black liquor as a steel coating

3.2.1 Anodic polarization behavior of coated reinforcing steel in concrete mixed with or immersed in different aggressive media

The anodic polarization behavior of reinforcing steel coated with cement slurry incorporating 0.2% black liquor compared to that coated with plain cement slurry or uncoated steel embedded in concrete mixed with 1% NaCl+ 1% $Na_2 SO_4$ or immersed in sea water has been determined. It is to be mentioned that , such salt content i.e. 1%NaCl+ 1% Na_2SO_4 has been selected as a test medium since it has been found frequent in concrete specimens obtained from several deteriorated buildings at different locations in Egypt (El-Sayed, et al , 1986; El-Sayed,1986).

Figure 3 presents the anodic polarization curves of reinforcing steel coated with cement slurry admixed with 0.2% black liquor compared to that coated with plain cement slurry and uncoated steel embedded in concrete admixed with 1% NaCl+ 1% Na_2SO_4. It is worth noting that in this medium the uncoated steel potential rises very slowly towards the noble direction and could attain the oxygen evolution potential but after considerably long time reaching 64 minutes . On the other hand, the coated steels exhibited considerably shorter times to reach passivation. These are 13 and 10 minutes for steel coated with plain cement slurry and that incorporating 0.2% black liquor, respectively.

Figure 4 depicts the anodic polarization behavior of reinforcing steel coated with cement slurry incorporating 0.2% black liquor compared to steel coated with plain cement slurry and uncoated steel embedded in concrete immersed in sea water. It can be seen that, all the polarization curves exhibit the same general trend where the steel potential rises steadily towards noble values till reaching the oxygen evolution potential. The steady oxygen evolution potential has attained after 16 minutes for uncoated steel. This time interval has reduced upon steel coating to reach 12 and 9 minutes for steel coated with plain cement slurry and that mixed with 0.2% black liquor, respectively.

These results reveal that, a steel coating of cement slurry containing 0.2% black liquor is highly efficient in suppressing the corrosive effect of even inimical double salt such as 1% NaCl+1%Na_2SO_4 incorporated in concrete or stringent conditions that could confront the reinforced concrete in service such as subjecting to marine environment.

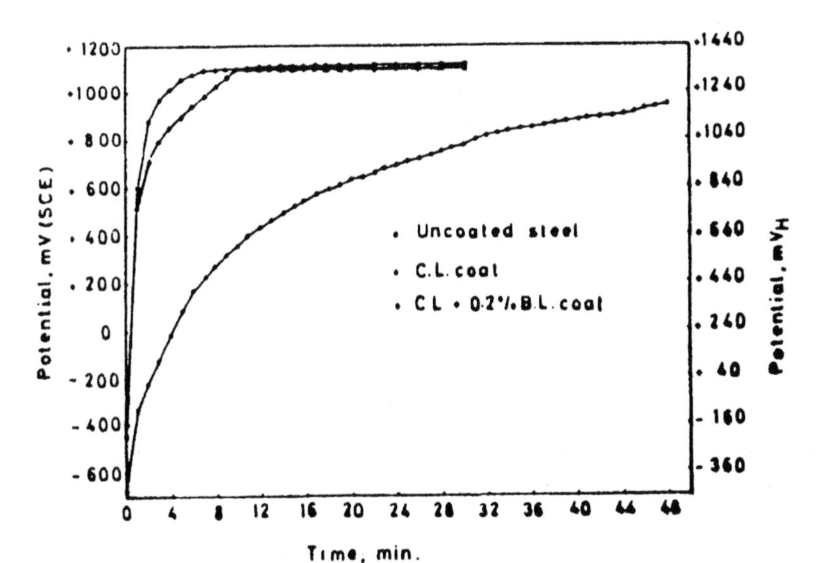

Fig.3. Anodic polarization of steel coated with plain cement slurry(C.L.) and C.L.+0.2% black liquor relative to uncoated steel embedded in concrete admixed with 1% NaCl+1% Na_2SO_4, at a current density of 10 μA cm^{-2}.

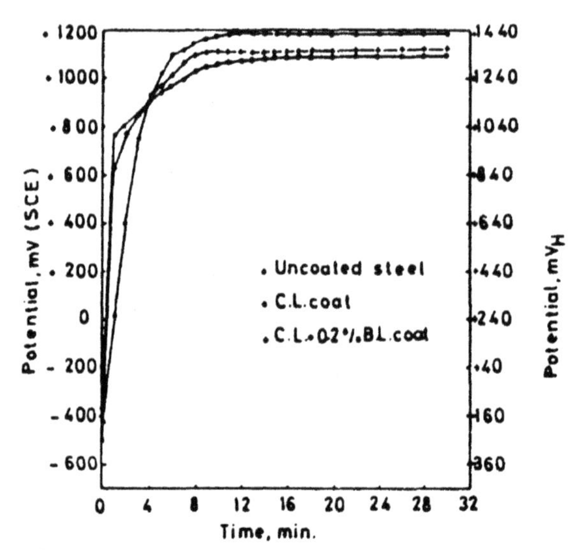

Fig.4. Anodic polarization of steel coated with plain cement slurry (C.L.) and C.L.+0.2% black liquor relative to uncoated steel embedded in concrete immersed in sea water, at a current density of 10 μA cm^{-2}.

3.2.2 Bond strength measurements for coated reinforcing steel in concrete mixed with or immersed in different aggressive media

Previous investigations (Lewis, 1962; Clifton, 1976) reported that treatments of steel surfaces might involve a loss of the steel/concrete bond strength, either before or after some slight corrosion has occurred. The bond between concrete and coated steel should not be significantly less than that between concrete and uncoated steel.

This part of the study has been carried out to assess the effects of the applied steel coating on the concrete/steel bond strength relative to the bond between concrete and uncoated steel. Bond strength measurements have been carried out after subjecting the steel -in-concrete for 6 months to the same media utilized in anodic polarization measurements i.e. concrete mixed with 1%NaCl+1%Na_2SO_4 and concrete immersed in sea water. The results are presented in Table 6.

These results demonstrate that coating the steel with 0.2% black liquor affords the highest initial and ultimate bond strength values towards the surrounding concrete whether incorporating aggressive salts or immersed in a hostile environment such as sea water.

Table 6. Bond strength (Kg cm^{-2}) for reinforcing steel in concrete mixed with and cured in 1% NaCl+1%Na_2SO_4 and for steel in concrete immersed in sea water for 6 months.

Medium	Coating	Initial strength	Ultimate Strength
1%NaCl+1%Na_2SO_4	No coat	29.6	30.8
	Cement slurry (C.L.)	31.2	32.6
	C.L.+0.2%black liquor	33.1	33.0
Sea water	No coat	27.6	28.8
	Cement slurry (C.L.)	30.5	31.4
	C.L+0.2% black liquor	31.4	32.3

4 Conclusions

A- As a concrete admixture, black liquor behaves as a reasonable anodic corrosion inhibitor and looks very promising as a useful material for allowing high compressive strength of concrete and improving its bond with steel reinforcement.

b- As a steel coating cement slurry incorporating black liquor presents a versatile steel treatment highly efficient in protecting steel reinforcement against hostile surrounding environment as well as improving its bond strength with the surrounding concrete.

5 References

B.S.1881, Part 2 (1970) Methods of testing concrete.

Clifton, J.R. (1976) Protection of reinforcing bars with organic coatings. Mater. Perf., 15,14-17.

El Sayed, H.A.and Sherbini, G.M.(1984) Investigation of the factors inducing early deterioration of a reinforced concrete construction. Surf. Technol., 23,291-300.

El-Sayed, H.A.(1985) Case study of an early failure of a concrete construction due to reinforcement corrosion. Durab. Buildg. Mater., 3,23-34.

El-Sayed, H.A., Didamony, H.and Ali,A.H, (1986) Case histories of premature failures of concrete structures due to reinforcement corrosion under different environmental conditions. Corros.Prev. and Contr. J., 33(4), 88-92 .

El-Sayed , H.A.(1986) Corrosion of steel in concrete and its prevention. Corros. Prev. and Contr. J., 33(4), 92 - 99.

Gouda, V.K.and Halaka, W.J. (1970) Corrosion and corrosion inhibition of reinforcing steel. Br. Corros. J., 5,204-208.

Lewis, D.A.(1962) Some aspects of the corrosion of steel in concrete, 1 st Inter. Congr. on Metall. Corros., Butter worths , London, PP. 547-555.

Rixom, M.R. (1975) The future for concrete admixtures. Chem. and Indust., 4,162 -167.

38 STUDY FOR BRAZILIAN RICE HUSK ASH CEMENT

J.S.A. FARIAS
Federal University of Rio Grande do Sul, Porto Alegre, Brazil

F.A.P. RECENA
Foundation for Science and Technology,
RGS, Porto Alegre, Brazil

Abstract
This paper describes the procedures for preparing Rice Husk Ash Cement for application on low cost civil construction. The preliminary chemical and physical analysis with raw materials, such as Rice Husk Ash (RHA), Lime and Ordinary Portland Cement (OPC), stand for the first procedure. After that, a variety of RHA Cement mixes performed during the laboratorial work are presented and afterwards submitted to characterization and adherence test.
Keywords: Rice Husk Ash, Lime, Agglomerant, Ordinary Portland Cement.

1 Introduction

In Brazil the RHA have not been used in all its full potential. At the most, the rice husk has functioned as heat generator through its own combustion during parabolysed rice production and drying processes. Still, the residual ashes are not re-used resulting in negative impacts on the environment.

With the objective of stablishing an effective recycle for RHA and, thus, turn it into an important raw material, this research was aimed to developing a hydraulic cement through the blending RHA, lime and OPC.

2 Materials characterization

2.1 Rice Husk Ash

It was obtained in rice processing industry, where the husk was utilized to produce heat energy.

The ash was grinded in a ring mill for 3 minutes to increase its specific surface and to improve the reaction with lime.

Chemical analysis of RHA: SiO_2 = 91.72%; MgO= 0.52%; Na_2O= 0.11%; SO_2 = 0.02%; MnO= 0.20%; TiO_2 = 0.05%; CaO= 0.5% Al_2O_3 =0.60%; K_2O= 1.30%; Fe_2O= 0.40%; P_2O_5 = 0.50%; Cl=1.80%

Physical analysis of RHA: the X-Ray Diffraction was utilized to verify the morphological structure, showing the amorphous structure of RHA.

The study of RHA pozzolanic activity demonstrated its suitability to application as pozzolanic material. It presented 87.04% of Pozzolanic Activity Index.

Specific gravity: 2.14g/cm³.

2.2 Lime

Chemical analysis of lime: SiO_2= 4.0%; MgO= 28.6%; CO_2= 2.1%; SO_2 < 0.15%; CaO= 47.7%; CaO(disposal)= 33.97%; Al_2O_3+ Fe_2O_3= 1.0%; loss in ignition= 20.3%; portion not soluble=5.4%.

Physical analysis of lime:

Fineness: residue on sieve nº200 (0.0075mm)= 14.9%.

Specific gravity: 2.44g/cm³.

Standard consistence of paste: 20mm of penetration with 192g of water.

Stability: the specimen surface did not present bubbles.

Plasticity: P= 146.

Standard consistence in mortar: the mortar reached 215mm of consistence with 450g of water (water-agglomerant relation= 0.90%).

Water retention: R= 81.0%.

Sand incorporation: see Table 1.

Table 1. Analysis of sand incorporation

Sample (g)	Std sand nº30,50&90	Proportion (weigth)	Water (g)	Consist. (mm)	Height not exuded (mm)
250	750	1:3.00	220	206	0
214	750	1:3.50	210	214	10
200	750	1:3.75	200	211	37

2.3 Ordinary Portland Cement

Chemical analysis of OPC: SiO_2= 19.0%; SO_2= 3.16%; MgO= 3.40%; Na_2O 0.20%; MnO_2= 0.05%; Al_2O_3= 4.80%; Fe_2O_3= 3.32%; TiO_2= 0.25%; CaO= 62.0%; P_2O_5= 0.76%; loss in ignition= 3.70%; portion not soluble= 0.60%.

Physical analysis of OPC:

Fineness: residue on sieve ABNT 0.0075mm (R)= 5.46%.

Specific gravity: 3.11 g/cm³.

Settle time: initial set= 4 hours.
final set= 7 hours and 49 minutes.

Expansibility determination: at the cold analysis, it was not constated any increase in volume. The results of hot analysis are on Table 2.

Table 2. Results of hot analysis

Specimen	Initial	3 hs	5 hs	7 hs
1	0.0	0.0	0.0	0.0
2	0.4	0.4	0.4	0.0
3	0.2	0.2	0.2	0.0

Compressive strength (σ): Table 3.

Table 3. Analysis of compressive strength

Age of specimen (days)	σ (MPa)
3	22.8
7	30.2
28	39.8

3 Adequate rate of lime and RHA

At this stage, OPC share on the mixture was fixed at 10%. The resulting compression strength differences were analysed by varying lime-RHA rate and keeping all of others variables constant.

The different rates of lime-RHA are shown on Table 4.

Table 4. Proportions realized on mixes

Mix	OPC (%)	Lime (%)	RHA (%)	Total (%)
1	10	80	10	100
2	10	65	25	100
3	10	50	40	100
4	10	35	55	100
5	10	20	70	100
6	10	10	80	100
7	10	5	85	100

Cylindrical specimens with 50 mm diameter and 100 mm height were casted at the proportion of 1:3 (mix:standard sand). Sufficient water was added to each mortar at the water-mix relation of 0.80. The consistences were measured by Flow Table Test and the results are presented on Table 5.

The Specimens were submitted to drying cure, since at this second phase the objective was to compare the mixes.

The compressive strength (σ) and standard deviation (s) were determinated at 7 and 28 days after moulding. The presented results on Table 5 are average of six specimens.

Table 5. Consistence and compressive strength of mortars

Mortar	Consistence (mm)	7 days		28 days	
		σ (MPa)	s (MPa)	σ (MPa)	s (MPa)
M1	227	0.83	0.02	1.07	0.16
M2	248	1.53	0.06	2.12	0.15
M3	257	2.21	0.19	3.08	0.20
M4	251	2.45	0.12	4.52	0.63
M5	262	3.07	0.14	5.83	0.51
M6	267	3.18	0.07	3.93	0.17
M7	269	2.53	0.19	2.51	0.04

The mortar of RHA cement is very workable and although of the high consistence, they presented cohesion on Flow Table Test.

The Specimens made with the mortar 5 had more compressive strength increases between the 7 and 28 days. Thus, the research extended in this mix by studying the water-mix relation and cure kind.

4 Comparing compressive strengths by modifying the water-mix relation and types of cure setting

At this phase, mix 5 was moulded utilizing water-mix relation of 0.48 and 0.80 in a wetting setting. (Table 6)

After two days of wetting storage, the specimes were demoulded and storaged under water at 23 ± 3°C.

Table 6. Compressive strength of RHA Cement mortar

Water/Mix (%)	3 days		7 days		28 days	
	σ (MPa)	s (MPa)	σ (MPa)	s (MPa)	σ (MPa)	s (MPa)
48	1.30	0.04	4.43	0.25	13.60	0.40
80	ND*	ND*	0.67	0.02	4.59	0.22

This analysis showed the importance of water-mix relation and cure kind.

*Measurement was impossible due the high water-mix relation.

5 Influence of ash grinding time on the compressive strength

The RHA was grinded on a ring mill for different periods of time, measuring the fineness by Blaine Air Permeability Apparatus. (Table 7)

Table 7. Specific surface at different grinding runs

Grinding Time (minutes)	Specific Surface (m^2/Kg)
3	1,286.75
5	1,333.97
9	1,321.00
10	1,586.62
15	1,704.08
20	1,816.66
30	1,974.97
40	1,949.06
50	1,965.24
60	1,871.69
70	1,881.57

This results shows that ash grinding runs of 30 minutes responded for greater specific surface than that of 15 minutes, which resulted on a 14% smaller surface.

Afterwards, specimens were moulded with water-mix relation of 0.48 and 0.80, using, for each pair of different water-mix relations, ash grinded at 15 and 30 minutes runs. The results are on Table 8.

Table 8. Grinding runs and compressive strength

Grinding runs (minutes)	Water Mix (%)	7 days		28 days	
		σ (MPa)	s (MPa)	σ (MPa)	s (MPa)
15	80	0.67	0.04	6.67	0.43
15	48	7.09	0.08	19.22	0.33
30	80	0.24	0.05	7.15	0.54
30	48	5.05	0.24	18.18	0.53

This results demonstrated the importance of water-agglomerant relation. Thus others grinding runs were investigated in water-agglomerant relation of 0.48.(Table 9)

Table 9. Comparing compressive strength by modifying grinding runs

Grinding runs (minutes)	Water Mix (%)	7 days		28 days	
		σ (MPa)	s (MPa)	σ (MPa)	s (MPa)
9	48	7.43	0.20	17.70	0.85
22	48	7.22	0.23	18.60	0.76

The analisis demonstrated that the ash with 9 minutes of grinding run is suitable for the agglomerant, since it had more strength than the one set by the cement masonry standards.

Thus the next step was to investigate the reduction of OPC percentage in the agglomerant mix in order to accomplish a lower cost product.

6 Reduction of Ordinary Portland Cement percentage in the mix

For this purpose specimens were moulded with a proportion of 1:3 (mix:standard sand) and with water-mix relation of 0.48 at different rates of OPC, keeping the lime-RHA relation at 1:3.5. The specimens were submitted to wetting setting.

The new mixtures percentage and the obtained results are shown on Table 10 and 11 respectively.

This step demonstrated that only 2% of OPC is due to accelerate the setting time in the agglomerant. Thus the adequate proportion is 2:21.78:76.22 (OPC:lime:RHA).

Table 10. Variation of OPC percentage in the mix 5

Mix	OPC (%)	Lime (%)	RHA (%)	Total (%)
C1	2	21.78	76.22	100
C2	4	21.33	74.67	100
C3	6	20.89	73.11	100
C4	8	20.44	71.56	100
C5	10	20.00	70.00	100

Table 11. Consistence and compressive strength by varying OPC in mix 5

Mortar	Consistence (mm)	7 days σ (MPa)	7 days s (MPa)	28 days σ (MPa)	28 days s (MPa)
C1	132	4.80	0.36	11.95	0.29
C2	130	4.56	0.31	11.70	0.19
C3	134	3.91	0.23	11.45	0.19
C4	132	4.37	0.26	12.98	0.53
C5	133	7.43	0.20	17.70	0.85

7 Final characterization of hydraulic RHA cement

The characterization was performed following the brazilian standards (ABNT- Associação Brasileira de Normas Técnicas) to cement and lime.

Fineness: residue in ABNT sieve 0.0075 mm (R)= 7.10%.

Specific gravity: 2.270 g/cm^3.

Specific surface: 1,356 m^2/Kg.

Expansibility determination: at the cold analysis, it was not constate any increase in volume.

Compressive strength: results are on Table 12.

Table 12. Results of compressive strength analysis

Age of specimen (days)	σ (MPa)
7	4.80
28	11.95

Paste normal consistence: (Table 13)

Table 13. Results of paste normal consistence analysis

Water (g)	Penetration (mm)
180	9
185	14
189	17

The analysis of RHA cement paste showed that it reaches

normal consistence with less water that the lime.

Stability: the specimen surface did not present burbles.

Plasticity: P= 112.94 (Table 14).

Table 14. Results of plasticity analysis

F Read	t Time (min.)	$P=(F^2+(10t)^2)^{1/2}$ Plasticity
100	11.58	116.23
100	10.92	109.66
Average	11.25	112.95

Mortar standard consistence:the mortar reached 206 mm of consistence on Flow Table Test with 0.316 kg of water and water-RHA cement relation of 0.72.

The analysis of standard consistence of mortar showed that RHA cement needs less water than lime mortar.

Water retention: R= 86.42%.

Sand incorporation:the results are on Table 15.

Table 15. Results of sand incorporation analysis

Sample (g)	Standard sand-nº30 50 & 100 (g)	Proportion (weight)	Water (g)	Consistence (mm)	Height not exuded (mm)
250	1000.0	1:4.00	230.2	212.0	4.50
250	875.0	1:3.50	195.6	206.0	4.10
250	812.5	1:3.25	190.0	207.5	2.40

8 Measurement of masonry flexural bond strength

This step was made following the ASTM C1072/86

For these tests two kinds of solid bricks were used, being one with good quality (T1) and the other of a worse quality than the first (T2). Four prisms of 4 bricks were made with mortar at 1:3 (RHA cement:sand), measuring the wrench strength of each joint. (Tables 16 and 17)

Table 16. Results in masonry made with bricks T1 (MPa)

Specimen	Joint 1	Joint 2	Joint 3	Average
Prism 1	0.5150	0.3399	0.4782	0.4444
Prism 2	0.3085	0.4364	– –	0.3725
Prism 3	0.3805	0.5521	0.4839	0.4722
Prism 4	0.4928	0.5121	0.4661	0.4903
Average	0.4242	0.4601	0.4761	0.4514

Table 17. Results in masonry made with bricks T2 (MPa)

Specimen	Joint 1	Joint 2	Joint 3	Average
Prism 1	– –	0.2747	0.2769	0.2758
Prism 2	0.3027	0.3361	0.3477	0.3288
Prism 3	0.3065	0.3244	0.3248	0.3186
Prism 4	0.3219	0.3555	0.3183	0.3319
Average	0.3104	0.3227	0.3169	0.3172

The rupture occured at the brick in the prisms made with brick T2 and in the joint when the brick T1 was used.

These tests demonstrated the RHA cement has suitable adherence to masonry units.

9 Conclusions

1. RHA burned in rudimental kiln type can be used in manufacturing of RHA cement since the carbon content is low.
2. The proportion of 2:21.78:76.22 (OPC:lime:RHA) is suitable to use in masonry.
3. The RHA cement fulfils all the requirements for lime standards and for masonry cement.
4. The increase of RHA fineness improves the workability, the water retention and the sand incorporation in the mortar.

10 References

Associação Brasileira de Normas Técnicas. (1986) **Cal hidratada para argamassa,** ABNT, NBR 7175, Rio de Janeiro.

---. (1986) **Cal hidratada para argamassa: determinação da retenção de água,** ABNT, NBR 9290, Rio de Janeiro.

---. (1985) **Cal hidratada para argamassa: determinação da capacidade de incorporação de areia no plastômero de Voss,** ABNT, NBR 9207, Rio de Janeiro.

---. (1985) **Cal hidratada para argamassa: determinação da estabilidade,** ABNT, NBR 9205, Rio de Janeiro.

---. (1986) **Cal hidratada para argamassa: determinação da finura,** ABNT, NBR 9289, Rio de Janeiro.

---. (1985) **Cal hidratada para argamassa: determinação da plasticidade,** ABNT, NBR 9206, Rio de Janeiro.

---. (1979) **Cimentos: métodos de determinação de atividade pozolânica em pozolanas,** ABNT, MB 960, Rio de Janeiro.

---. (1982) **Ensaio de cimento portland,** ABNT, NBR 7215, Rio de Janeiro.

---. (1986) **Método de ensaio para determinação da massa específica de ciemntos,** ABNT, MB 346, Rio de Janeiro.

---. (1977) **Pozolanas: determinação do índice de atividade pozolânica com cimento portland,** MB 1153, Rio de Janeiro.

Beagle, E.C. (1978) Rice Husk - Conversion to Energy. **FAO Agricultural Services Bulletim,** Food Organization of the United Nations, Rome, Italy, pp. 128-143.

Cook, D.J.; Pama, R.P. & Paul, B.K. (1987) Rice Husk Ash--lime-cement mixes for use in masonry units. **Building and Environment,** 12, pp. 281-288.

Metha, P.K. (1977) Properties of blended cements made from rice husk ash. **Journal of the American Concrete Institute,** 74(9), pp. 440-442.

Smith, R.G. (1984) Rice hush ash cement small scale production for low cost housing, in **International Conference on Low Cost Housing for Developing Countries,** Roorkee, India, pp. 687-695.

Smith, R.G. & Kamwanja, G.A. (1986) The use of rice husk for making a cementitious material, in **Use of Vegetables plants and fibres as building materials,** Joint Symposium RILEM/CIB/NCCL, Baghdad, Iraq,pp. E85-E94.

INDEX

This index has been compiled using the keywords provided by the authors of the individual papers, with some editing and addition to ensure consistency. The numbers refer to the first page of the relevant paper.

Accelerated testing 87
Acid extraction 29
Admixtures 343, 349
Adobe blocks 139
Alkalis 29, 120
Ammomium carbonate 239
Anodic polarization 349
Asbestos replacements 39, 77

Bamboo 1, 9, 50, 60, 295, 305
Barley fibre 199
Bending 193, 248
Black liquor 350
Blastfurnace slag 87, 239
Boards 193
Bonding 350
Bricks 314
Building
  blocks 139
  conservation 21
  panels 29
  systems 255

Calcium chloride 193
Carbonation 87, 173
Cellulose 29
Cement-bonded particle board 239
Cement composites 1
Cement substitutes 321, 334
Cipla 139
Clove oil 120
Coatings 350
Coconut fibres 77, 87, 139, 161
Coir fibres 87, 98, 150
Colophony 120
Compaction 39
Compressive strength 50, 193, 360
Concrete 321
Connections 295
Consistence 360
Constitutive law 108
Corrugated sheets 77
Costs 199, 305

Cracking 77, 108, 130, 173
Craft techniques 21
Curaguilla 199

Deflection 39, 77
Degradation mechanisms 120, 277
Dune sand 224
Durability 1, 9, 29, 87, 98, 120, 239, 277, 350

Earth construction 182
Elastic constants 248, 266
Elasto-plastic behaviour 130
Electrochemical effects 350
Electron microscopy 87, 334
Embankments 214
Engineering performance 1

Fibre concrete 204
Fibre properties 334
Finite elements 214
Fique liquor 334
Fire-resistant materials 60
Flexural testing 39, 69, 77, 87, 120, 130, 161
Floor slabs 173
Fluidised bed boiler 334
Fly ash 60
Forestry thinnings 295
Formwork 173

Glass fibres 50, 161
Glulam 266
Grass fibres 29, 224
Grinding 360
Gypsum 21, 87, 161

Hay 314
Hemicellulose 29,
Housing 1, 277, 314

Impact strength 130, 150, 161
Impregnation 120
Inhibitors 350
Insulation 286

Joints 295

Laminated timber 266
Lightweight panels 255
Lignin 29
Lime 87, 360
Load-deflection properties 161, 173
Loading 255, 286
Low-cost building systems 286

Magney fibre 199
Maintenance 21, 314
Management 204
Manufacturing 199, 204, 239
Marsh 214
Masonry 360
Matrix 1, 130
Mechanical properties 193, 286, 321
Megass 69
Microstructure 1

Mineral fibres 21
Mix proportions 321, 360
Mixing procedure 69, 224
Modulus of elasticity 1, 248, 371
Modulus of rupture 266
Moment-curvature relationships 108, 173
Mortar 69, 130, 150, 360
Natural fibres 1,

Palm fruit fibres 9, 29
Papyrus-cement 9, 93
Panels 150, 161, 182, 255, 286
Permeability 182
Piche 139
Pipes 305
Polyacrylonitrile 98
Polyvinyl alcohol 98
Potassium carbonate 239
Pozzolanas 9, 321, 334
Preservative treatments 305
Pulp 39

Reed fibres 9, 224, 314
Reed reinforcement 214
Reinforced concrete 108, 173
Reinforced earth 182
Reinforced sand 224
Reinforcement 9, 39, 69, 350
Rendering 182
Review 9
Rice husk 9, 60
Rice husk ash 9, 87, 321, 334, 360
Roof construction 314
Roofing materials 77, 199, 204
Roundwood 295

Scale effects 1, 266
Scanning electron microscopy 87, 334
Seawater 350
Setting time 239, 321
Shear modulus 266
Sheet materials 204
Sisal 21, 98, 108, 120, 130, 139, 161, 173

Slope stabilization 21
Soil properties 214
Soil stabilization 182, 214
Soil testing 139
Space structures 295
Stability 255
Standards 248
Strength development 39, 321, 334, 343, 350, 360
Specific surface 360
Structural analysis 108, 255
Structural elements 108
Structural properties 305
Structures 255
Sugar cane fibres 9, 29, 69
Sun-dried bricks 314
Superplasticizers 343
Swelling 193

Tannin 120
Temperature effects 239
Tensile strength 98, 150, 50
Termites 277
Thermal insulation 21
Thermal properties 60, 286
Thin sheets 77
Tiles 204
Timber 255, 305
Toughness 39
Traditional buildings 314
Trass 239
Trusses 295

Vacuum process 161
Vegetable plants 9, 314
Vibration 199

Walls 21
Waterabsorption 77, 120, 139, 193
Water/cement ratio 321
Water extraction 29
Waterglass 239
Water purity 305
Water reducing admixture 350
Water repellent treatments 139
Water resistance 239
Water retention 77
Water-soluble ceramics 60
Water storage 305
Wheat fibre 199
Wire lacing tool 295

Wood 39, 277
products 266
properties 248
waste 286
Workability 343

X-ray diffraction 334, 360
Xylophages 277

www.ingramcontent.com/pod-product-compliance
Ingram Content Group UK Ltd.
Pitfield, Milton Keynes, MK11 3LW, UK
UKHW020402010325
455677UK00021B/580